# Cycling for Sustainable Cities

**Urban and Industrial Environments**

Series editor: Robert Gottlieb, Henry R. Luce Professor of Urban and Environmental Policy, Occidental College

For a complete list of books published in this series, please see the back of the book.

# Cycling for Sustainable Cities

Edited by Ralph Buehler and John Pucher

The MIT Press
Cambridge, Massachusetts
London, England

This book was set in Stone Serif and Stone Sans by Westchester Publishing Services. Printed and bound in the United States of America.

Library of Congress Cataloging-in-Publication Data

Names: Buehler, Ralph, editor. | Pucher, John R., editor.
Title: Cycling for sustainable cities / edited by Ralph Buehler and John Pucher.
Description: Cambridge, Massachusetts : The MIT Press, [2021] |
    Series: Urban and industrial environments | Includes bibliographical references
    and index.
Identifiers: LCCN 2020012187 | ISBN 9780262542029 (paperback)
Subjects: LCSH: Bicycle commuting—Social aspects. | Bicycle commuting—
    Health aspects. | Urban transportation—Environmental aspects. |
    Sustainable urban development.
Classification: LCC HE5736 .C9284 2021 | DDC 388.3/47—dc23
LC record available at https://lccn.loc.gov/202001218787

10  9  8  7  6  5  4  3  2  1

Ralph Buehler dedicates this book to Nora, Niels, Tillman, Liesel, Schorsch, Steffi, and Tizian for all their support.

John Pucher dedicates this book to Chris Kerzman, his aunt, who inspired his four decades of car-free living, and Alan Alshuler, his dissertation advisor at MIT, who enabled his lifelong career in urban transport research.

# Contents

Editors    ix

Contributors    xi

Preface    xv

Acknowledgments    xvii

1  Introduction: Cycling to Sustainability    1
   John Pucher and Ralph Buehler

2  International Overview of Cycling    11
   Ralph Buehler and John Pucher

3  Cycling and Health    35
   Jan Garrard, Chris Rissel, Adrian Bauman, and Billie Giles-Corti

4  Cycling Safety    57
   Rune Elvik

5  Bicycling Infrastructure for All    81
   Peter G. Furth

6  Bicycle Parking    103
   Ralph Buehler, Eva Heinen, and Kazuki Nakamura

7  Programs and Policies Promoting Cycling    119
   Eva Heinen and Susan Handy

8  Evaluation of Cycling Policies and Projects    137
   Bert van Wee

9  E-bikes in Europe and North America    157
   Christopher R. Cherry and Elliot Fishman

10  Bikesharing's Ongoing Evolution and Expansion    173
    Elliot Fishman and Susan Shaheen

11  Women and Cycling: Addressing the Gender Gap   197
    Jan Garrard

12  Children and Cycling   219
    Noreen McDonald, Eleftheria Kontou, and Susan Handy

13  Older Adults and Cycling   237
    Jan Garrard, Jennifer Conroy, Meghan Winters, John Pucher,
    and Chris Rissel

14  Social Justice and Cycling   257
    Karel Martens, Aaron Golub, and Andrea Hamre

15  Cycling in China and India   281
    John Pucher, Geetam Tiwari, Zhong-Ren Peng, Rong Cao,
    and Yuan Gao

16  Cycling in Latin America   301
    Carlosfelipe Pardo and Daniel A. Rodriguez,
    with Lina Marcela Quiñones

17  Cycling in New York, London, and Paris   321
    John Pucher, John Parkin, and Emmanuel de Lanversin

18  Cycling in Copenhagen and Amsterdam   347
    Till Koglin, Marco te Brömmelstroet, and Bert van Wee

19  Implementation of Pro-bike Policies in Portland and Seville   371
    Roger Geller and Ricardo Marqués

20  Cycling Advocacy in Europe, North America, and Australia   401
    John Pucher, Bernhard Ensink, Tim Blumenthal, Bill Nesper,
    Ken McLeod, Andy Clarke, Jean-François Pronovost,
    Dave Snyder, Robin Stallings, Fiona Campbell, and Peter Bourke

21  Cycling to a More Sustainable Transport Future   425
    Ralph Buehler and John Pucher

Index   441

# Editors

**Ralph Buehler** is professor of Urban Affairs and Planning at Virginia Tech Research Center, Arlington, Virginia. Most of his research has an international comparative perspective, contrasting transport policies, transport systems, and travel behavior in western Europe and North America. Specifically, his research interests include the influence of transport policy, land use, and sociodemographics on travel behavior; active travel and public health; and public transport demand, supply, regional coordination, and financial efficiency. He is author or coauthor of over 65 refereed articles in academic journals, the book *City Cycling* (MIT Press), and reports to federal and local governments, NGOs, and industry organizations in the United States and abroad. Between 2012 and 2018, he served as chair of the Committee for Bicycle Transportation of the Transportation Research Board (TRB) of the National Academies.

**John Pucher** is professor emeritus at the Bloustein School of Planning and Public Policy at Rutgers University, New Brunswick, New Jersey. He conducts research on a wide range of topics in urban transport. For almost four decades, he has examined differences in travel behavior, transport systems, and transport policies in Europe, Canada, the United States, and Australia. In recent years, his research has focused on walking and bicycling, and what North American and Australian cities can learn from European cities to improve the safety and convenience of these sustainable means of urban travel. His particular passion is to make cycling possible for everyone, including all ages and abilities. His research emphasizes walking and cycling for daily travel to increase mobility, improve transport equity, and enhance overall public health through physical activity. His research has been published in over 100 peer-reviewed journal articles, 20 book chapters, and three books.

# Contributors

**Adrian Bauman**, professor, Sydney School of Public Health, University of Sydney, Australia

**Tim Blumenthal**, executive director, PeopleForBikes, Boulder, Colorado

**Peter Bourke**, executive director, We Bike Australia, Canberra, Australia

**Marco te Brömmelstroet**, associate professor, Amsterdam Institute for Social Science Research, University of Amsterdam, the Netherlands

**Fiona Campbell**, manager of cycling strategy, City of Sydney, Australia

**Rong Cao**, PhD student, Department of Urban and Regional Planning, University of Florida, Gainesville, Florida

**Christopher R. Cherry**, professor, Civil and Environmental Engineering, University of Tennessee, Knoxville, Tennessee

**Andy Clarke**, director of strategy, Toole Design; former executive director, League of American Bicyclists; former executive director, Association of Pedestrian and Bicycle Professionals; former secretary general, European Cyclists' Federation.

**Jennifer Conroy**, post-graduate student, Gerontology Program, Simon Fraser University, Vancouver, British Columbia, Canada

**Rune Elvik**, researcher, Norwegian Institute of Transport Economics, Oslo, Norway

**Bernhard Ensink**, secretary general, European Cyclists' Federation, Brussels, Belgium

**Elliot Fishman**, director, transport innovation, Institute for Sensible Transport, Melbourne, Australia

**Peter G. Furth**, professor, Department of Civil and Environmental Engineering, Northeastern University, Boston, Massachusetts

**Yuan Gao**, senior lecturer, School of Public Administration, Zhengzhou University, China

**Jan Garrard**, senior lecturer, School of Health and Social Development, Deakin University, Melbourne, Australia

**Roger Geller**, bicycle coordinator, Bureau of Transportation, City of Portland, Oregon

**Billie Giles-Corti**, distinguished professor, Centre for Urban Research, RMIT University, Australia

**Aaron Golub**, associate professor, Toulan School of Urban Studies and Planning, Portland State University, Portland, Oregon

**Andrea Hamre**, research associate, Western Transportation Institute, Montana State University, Bozeman, Montana

**Susan Handy**, professor, Department of Environmental Science and Policy, University of California, Davis, California

**Eva Heinen**, academic research fellow, Institute for Transport Studies, University of Leeds, Leeds, United Kingdom

**Till Koglin**, associate professor, Department of Technology and Society, Lund University, Lund, Sweden

**Eleftheria Kontou**, assistant professor, Civil and Environmental Engineering, University of Illinois, Urbana-Champaign, Illinois

**Emmanuel de Lanversin**, director, Directorate for Housing, Urban Development and Landscapes, French Ministry of Transport, Paris, France

**Ricardo Marqués**, associate professor, Faculty of Physics, University of Seville, Seville, Spain

**Karel Martens**, associate professor, Faculty of Architecture and Town Planning, Technion, Israel Institute of Technology, Haifa, Israel, and Institute for Management Research, Radboud University, Nijmegen, the Netherlands

**Noreen McDonald**, professor, Department of City and Regional Planning, University of North Carolina, Chapel Hill, North Carolina

**Ken McLeod**, policy director, League of American Bicyclists, Washington, DC

**Kazuki Nakamura**, associate professor, Department of Civil Engineering, Meijo University, Nagoya, Japan

**Bill Nesper**, executive director, League of American Bicyclists, Washington, DC

**Carlosfelipe Pardo**, senior manager, City Pilots, New Urban Mobility Alliance, Washington, DC.

**John Parkin**, professor, Centre for Transport and Society, University of the West of England, Bristol, United Kingdom

**Zhong-Ren Peng,** professor, International Center for Adaptation Planning and Design, College of Design, Construction and Planning, University of Florida, Gainesville, Florida

**Jean-François Pronovost,** executive director, Vélo Québec, Montréal, Québec

**Lina Marcela Quiñones,** adviser, Despacio, Bogota, Colombia

**Chris Rissel,** professor, Sydney School of Public Health, University of Sydney, Australia

**Daniel A. Rodriguez,** professor, College of Environmental Design and Institute for Transportation Studies, University of California, Berkeley, California

**Susan Shaheen,** adjunct professor, Civil and Environmental Engineering and Transportation Sustainability Research Center, University of California, Berkeley, California

**Dave Snyder,** executive director, California Bicycle Coalition, Sacramento, California

**Robin Stallings,** executive director, BikeTexas, Austin, Texas

**Geetam Tiwari,** professor, Department of Civil Engineering and Transportation Research and Injury Prevention Programme, Indian Institute of Technology, Delhi, India

**Bert van Wee,** professor, Faculty of Technology, Policy and Management, Delft University of Technology, Delft, the Netherlands

**Meghan Winters,** associate professor, Faculty of Health Sciences, Simon Fraser University, Burnaby, British Columbia, Canada

# Preface

This book is the follow-up to the 2012 book *City Cycling*, also published by MIT Press and edited by the same co-editors. As suggested by the new title, however, it is a very different book. *Cycling for Sustainable Cities* offers chapters about 11 new topics not examined in the 2012 book. Even the nine chapter topics that have been retained from *City Cycling* have been completely rewritten and updated, in some cases by different authors with different perspectives than those in the 2012 book.

In addition to a wider variety of chapter topics covering more of the world (such as the new chapters on Latin America, China, and India), the diversity of chapter authors has also increased. They include roughly as many women as men, ranging in age from 25 to 75. The chapter authors come from 16 countries on five continents, thus providing an international perspective throughout the book. They represent countries with low but growing cycling levels (for example, Australia, Canada, and the United States) as well as northern European countries with traditionally high levels of cycling (for example, the Netherlands and Germany). The authors also provide a multidisciplinary perspective that draws on a wide range of academic backgrounds, including geography, urban planning, environmental science, transport planning, civil engineering, and public health. The diversity of the chapter authors mirrors their variety of analytical approaches to their chapter topics, as well as their advocacy for cycling facilities and programs that encourage cycling among the widest possible range of social, demographic, income, and ethnic groups.

All the chapters report the state of scientific knowledge and best practices in each of their topic areas. Recent decades have witnessed dramatic growth in interest in cycling by researchers, transport professionals, city planners, the media, and the public at large. In the decade since the publication of *City Cycling* in 2012, published scientific research about cycling has more than doubled. More important, cities throughout Europe, North America, South America, and Australia have made massive investments in expanded, improved, and better-connected cycling facilities, resulting in impressive growth in cycling levels in many cities. In short, much has happened since 2012, and this book provides the latest information available on a wide range of issues.

As suggested by the title *Cycling for Sustainable Cities*, a key theme throughout this new book is the important role of cycling in making cities more sustainable. We define sustainability broadly, similar to the World Bank's widely accepted concept of social, economic, and environmental sustainability. Every chapter in the book deals with at least one—and often many—of the different aspects of sustainability. These include the environment; energy use; economic efficiency; mobility and accessibility; equity and social justice; cycling for different demographic, economic, and ethnic groups; traffic safety; individual and societal health; livable cities; and urban development patterns that enable less car dependence.

Throughout the book, the focus is on practical lessons that can be applied and adapted to the specific situations in different kinds of cities and for different groups. Thus, the book has been written to be as clear and understandable as possible, avoiding technical jargon and complicated statistical analysis to the extent possible. In addition to examining past and recent trends, every chapter also examines future prospects in each topic area; for example, future developments in e-bikes and bikesharing, which have been changing rapidly in recent years. Indeed, the second half of the concluding chapter explicitly examines future developments in a range of different areas.

Finally, we emphasize that this book was written before the onset of the COVID-19 pandemic in late December 2019 and early January 2020, which continues in summer 2020, as the book is prepared for press. All 21 chapters rely on information and data collected by mid-December 2019, which thus were not affected by the devastating pandemic. As co-editors, we have asked chapter authors not to attempt to assess the current impact of the COVID-19 pandemic on cycling, nor to speculate on the likely future impact. Data for 2020 are unavailable and, at any rate, would be unreliable and unrepresentative due to the extraordinary nature of the pandemic and the large geographic variation in its intensity.

# Acknowledgments

We are indebted to our editors at MIT Press for having initiated the idea for this book and then guiding us along to its completion. In particular, we thank Robert Gottlieb, editor of the MIT Press's Urban and Industrial Environments series, and Beth Clevenger, our acquisitions editor for the MIT Press. We also thank Jennifer Burton, Kate Elwell, Heather Goss, Hal Henglein, Gregory Hyman, Marcy Ross, Molly Seamans, and Anthony Zannino.

This book is the result of a team effort requiring intensive cooperation and coordination between the authors of each chapter as well as across the various chapters. We would, above all, like to thank each of the chapter authors for the time, effort, and thought they invested in their chapters.

We also thank the three anonymous referees for their suggestions for improving the draft book manuscript. Both before and after the official peer review, outside experts on the topic of each chapter made important contributions by providing chapter authors with extensive information and datasets, reviewing successive drafts of the chapters, and offering various other kinds of assistance that were essential to the content and quality of the chapters.

John Pucher would like to thank his long-time colleagues, co-authors, and friends Alan Altshuler (MIT and Harvard University), Martin Wachs (UCLA and UC Berkeley), and David Banister (Oxford University) for their career-long dedication to equity and social justice in transport, which inspired Pucher's own work on transport equity as well as the theme of equity and social justice that permeates every chapter of this book.

The co-editors would also like to thank their colleague and friend Ria Hilshort, the bike planner for the City of Amsterdam, who provided the photos of cyclists in Amsterdam used for the book cover. We also thank Ria for her kind and generous help over almost two decades of our research into cycling trends, infrastructure, and policies in Amsterdam and the

Netherlands as a whole, including information used in several chapters of this book. We thank Lucas Harms, executive director of the Dutch Cycling Embassy, and Marianne Weinreich, executive director of the Cycling Embassy of Denmark, for the extensive information they provided on Dutch and Danish cycling, which was used in chapters throughout the book.

The authors of each chapter have chosen to thank the persons listed here for especially important contributions to their chapters. The list is organized by chapter and then alphabetically within each chapter.

Chapter 2: Lewis Dijkstra, Veronique Feypell, Lucas Harms, Holger Haubold, Ria Hilshorst, Thomas Krag, Manfred Neun, Marc Panneton, Gabe Rousseau, and Marianne Weinreich

Chapter 4: Rebecca Sanders and Kay Teschke

Chapter 5: Brian Gould, Marc Jolicoeur, Kornel Mucsi, Bill Schultheiss, and Kyle Wagenschutz

Chapter 6: Quentin Champauzac, Matthew Dyrhdal, Mike Goodno, Bernhard Gutzmer, and Roland Jannermann

Chapter 8: Maria Börjesson, Irma Keserü, Cathy Macharis, Kees van Ommeren, and Lake Sagaris

Chapter 9: John MacArthur

Chapter 10: Paul DeMaio and Russell Meddin

Chapter 13: Jelle Van Clauwenberg

Chapter 15: Madhav Badami, Chris Cherry, Jingjing Chang, and Shengyong Yao

Chapter 16: Chester Harvey

Chapter 17: Rich Conroy, Nathalie Daclon, Julia Kite, Charles Komanoff, Alexander Longdon, Sebastian Marrec, Ken Podziba, Beatrice Ras, Michael Replogle, and Ted Wright

Chapter 18: Lucas Harms, Ria Hilshorst, Thomas Krag, Marjolein de Lange, Thomas Alexander Sick Nielsen, and Marianne Weinreich

Chapter 19: Manuel Calvo-Salazar and Vicente Hernández-Herrador

Chapter 20: Niles Barnes, Lucas Harms, Holger Haubold, Ria Hilshorst, Marc Jolicoeur, Fabian Küster, Dennis Markatos-Soriano, Kevin Mayne, Erin O'Melinn, Marc Panneton, Judi Varga-Toth, and Marianne Weinreich

In addition to those individuals specifically listed, we thank the countless individuals and organizations who contributed information, data, photos, and advice throughout the three years it took to produce this book.

# 1  Introduction: Cycling to Sustainability

John Pucher and Ralph Buehler

In our 2012 book *City Cycling*, we argued that cycling is probably the most sustainable urban transport mode, feasible not only for short trips but also for medium-distance trips too long to make by walking (Pucher and Buehler 2012). Cycling causes virtually no environmental damage, promotes health through physical activity, takes up little space, and is economical, both in direct user costs and in public infrastructure costs. In short, cycling is environmentally, socially, and economically sustainable. Those benefits have generated widespread support for cycling among proponents of sustainable transportation, energy, and urban development. Most academic studies as well as government policy statements and plans explicitly cite the multifaceted sustainability of cycling as an important reason for investing in cycling infrastructure and programs.

In the years since publication of the 2012 book, there have been important developments in the world of cycling, which are examined in this new book. Perhaps the most important development over the past two decades has been the continued growth in cycling in major cities throughout the world, facilitated by a massive expansion and improvement in separate cycling facilities, especially protected bike lanes and paths and, most recently, express bikeways (cycle superhighways). As shown in figure 1.1, the bike shares of trips increased in 23 major cities of western Europe, North America, South America, and Australia. The most dramatic growth has been in cities where cycling had not previously been a regular means of daily travel. In most of the North American cities shown, for example, cycling's mode share tripled or quadrupled between 1990 and 2016. In South America, cycling doubled in Santiago (Chile) and increased sixfold in Buenos Aires (Argentina) and elevenfold in Bogota (Colombia). In Europe, the cities with the fastest growth in cycling were Seville (sixfold), Paris (fourfold), and

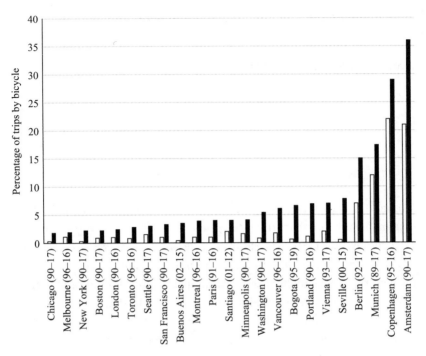

**Figure 1.1**
Increasing bicycle mode shares of trips in large cities of western Europe, North America, South America, and Australia, 1990–2017. *Note*: The timespan for the data varies by city, as shown in parentheses after each city's name. *Sources*: Based on travel surveys conducted for each city.

London and Vienna (roughly tripling)—none of which had historical cycling cultures. Equally impressive are the increases in Amsterdam and Copenhagen, where large increases were in addition to already high cycling levels.

Accompanying and fostering the growth in cycling have been a proliferation of professional and academic conferences on cycling; increasingly influential, widespread, and interconnected local, state, national, and international cycling organizations; internet sites devoted to sharing knowledge about best practices; and various other information-sharing forums, including social media. The growing worldwide network of researchers, planners, advocates, and cyclists has inspired, enabled, and actively promoted cycling.

Information exchange about the best ways to increase cycling has been especially important. The first and most extensive efforts to promote more and safer cycling were in western Europe, especially in the Netherlands,

Denmark, and Germany, starting in the 1970s and continuing today. The successful cycling infrastructure, programs, and policies in Dutch, Danish, and German cities were adopted in many other European cities during the 1980s and 1990s, as well as in North American, South American, and Australian cities, especially since 2000. The geographic spread of best practices has facilitated the impressive growth in cycling.

The growing importance of cycling as a mainstream means of travel is reflected in the dramatic increase in published research on cycling over recent decades. According to the Web of Science, a standard index of scientific publications, the average annual number of peer-reviewed cycling articles (in the field of transport) rose from only 35 per year during the period 2001–2005 to 492 per year during the period 2014–2018, a fourteenfold increase.

## Purpose of the Book

As in *City Cycling*, the dual purposes of this new book are to portray trends in cycling and to identify which measures are most effective for increasing cycling levels, improving cycling safety, and making cycling possible for all segments of society. Our focus is on promoting daily cycling in cities for utilitarian purposes such as visiting friends and making trips to work, school, shopping, church, and doctors' offices. As documented in many of the chapters, success depends on a coordinated package of mutually supportive infrastructure, programs, and policies. Although cycling infrastructure is essential, it must be supported by a range of complementary measures.

This book is not simply an updated version of *City Cycling* but an almost completely new book. Even the eight chapters on topics retained from the 2012 book have been completely rewritten, reflecting the latest data and the massive amount of new research on cycling over the past decade. Several of those chapters have new authors with perspectives different from those of the earlier book, and they examine new issues within their topic areas. The most dramatic difference in this new book, however, is the inclusion of 11 completely new chapter topics.

To expand the geographic scope of the book, we have included a new chapter about cycling in China and India and another chapter about cycling in Latin American cities. There is also a new chapter comparing the cycling trends and policies of Copenhagen and Amsterdam, widely viewed as world leaders in cycling. To expand the thematic scope of the book, we have included eight other new chapters, examining topics such as electric-assist bicycles (pedelecs or e-bikes), bike parking, cycling promotional programs, economic evaluation of cycling projects, cycling advocacy, and the politics and process of implementing the measures needed to increase cycling. There is also a chapter on the crucial issue of equity and social justice, a theme that is also considered in other chapters of the book. Related to that theme, we have added a new chapter on cycling for older adults, complementing the thoroughly revised and updated chapters focusing on children and women.

## Structure of the Book

Chapter 2 provides a broad overview of cycling trends, safety, and policies in countries throughout the world. Of all the chapters in this book, the overview chapter is most similar in structure and content to its corresponding chapter (also chapter 2) in the 2012 book, but all information has been completely updated. We have also expanded the geographic scope of the chapter by including Japan in most of the country comparisons and by including South American, Japanese, Chinese, and Indian cities in the city comparisons. The most detailed analyses are for the Netherlands, Denmark, Germany, the United Kingdom, Japan, and the United States, the countries for which the available national statistics are most comparable.

Chapters 3 and 4 deal with two important impacts of cycling on public health. Chapter 3 provides an updated review of the rapidly mounting scientific evidence of the physical, mental, and social benefits of cycling. Even though they rely on a wide variety of locations, timing, study designs, and statistical techniques, virtually all available studies find that cycling is healthy for the individual. For the population as a whole, the health benefits of cycling far outweigh the health costs of traffic injuries. In addition, cycling for daily travel can help reduce car use and thus generate societal health benefits by lessening the noise, air pollution, greenhouse gases, and overall traffic danger caused by motor vehicles. Reduced health costs from increased cycling not only exceed the costs of cycling infrastructure but also often provide a higher ratio of economic benefits to costs than other kinds of transport investments.

An important goal of transport policy is to reduce the number and severity of traffic injuries. Chapter 4 shows that much can be done to improve cycling safety, which would not only reduce injuries but also encourage more cycling. In many cities throughout the world, large increases in cycling have usually been accompanied by reduced risks of cycling, as measured by fatalities and serious injuries per bike trip or per kilometer cycled. That suggests a "safety in numbers" effect, as more cycling is correlated with safer cycling. As chapter 4 shows, however, improved cycling safety does not automatically result from increased cycling. Greater cycling safety requires a range of measures, such as improved cycling infrastructure—especially facilities physically separated from motor vehicle traffic—as well as simple protective measures that cyclists can take, such as wearing helmets and reflective clothing and using lights and bells on their bicycles.

Chapters 5, 6, and 7 analyze government policies to encourage cycling and make it safer. Chapter 5 examines a wide range of infrastructure provisions for cyclists: roads with no special provisions for cyclists at all; traffic-calmed residential streets; roads with painted bike lanes; roads with physically separated, protected bike lanes or cycle tracks; and paved off-road bike paths. Chapter 5 compares the bikeway design standards of various countries. It also assesses the extent to which the separation of cycling facilities from motor vehicle traffic improves cycling safety while also raising cycling levels. Although for decades Europe has led the way in such protected cycling facilities, hundreds of kilometers of protected bike lanes and paths have been built over the past 10–15 years in the Americas and Australia. The updated chapter 5 enables an evaluation of the impacts of those new facilities.

Chapter 6 examines the crucial issue of bike parking. Good bike parking is as important to cycling as car parking is to driving a car. Sufficient, sheltered, and secure bike parking at trip destinations is an essential aspect of cycling infrastructure and is crucial to encouraging cycling. Chapter 6 summarizes the scientific literature on bicycle parking and examines the best bike parking practices in 11 bicycle-friendly cities on three continents, providing lessons for improving bike parking facilities in other cities.

Chapter 7 examines the important role that noninfrastructural policies and programs have to play in the integrated package of measures needed to increase cycling and make it safer. Such programs usually complement infrastructure investments, ensuring that such investments pay off to their fullest. In some situations, however, programs may be implemented as the main intervention and not simply to enhance infrastructure. Chapter 7 examines a wide variety of programs and policies that can increase cycling, including promotions, incentives, laws, and built environment policies. Government policies that discourage driving are an essential complement to those that encourage bicycling.

Chapter 8 explores alternative ways to evaluate the economic costs and benefits of the various types of infrastructure and programs examined in chapters 5, 6, and 7. Such economic evaluation techniques enable comparison among alternative bicycle projects and with other public expenditures in general. Because funding for cycling projects is limited, it is important to channel that funding to the projects with the greatest benefit for the money spent. Chapter 8 examines the strengths and weaknesses of three evaluation tools: cost-effectiveness analyses (CEA), multicriteria analysis (MCA), and cost-benefit analysis (CBA).

Chapters 9 and 10 examine two of the most rapidly changing aspects of cycling: electric-assist bikes (e-bikes) and bikesharing. Various types of e-bikes have been evolving since the early 1990s, first in China and then spreading to other parts of the world, especially to Europe but recently to North America and Australia as well. Chapter 9 examines the range of electric-assist vehicles, differentiating between pedal bikes with battery-powered motor assistance and other kinds of vehicles that are often called e-bikes but rarely involve pedaling at all, such as e-motorcycles and e-scooters. Chapter 9 considers both the advantages and disadvantages of e-bikes, concluding that they will probably increase cycling in Europe and North America.

Bikesharing has experienced truly dramatic growth in the past decade, particularly since 2015, coinciding with the emergence of dockless bike-sharing systems. Chapter 10 reviews the four generations of bikesharing and the technological innovations that have continued to widen its appeal. It then explores some of the alternative business models underpinning third- and fourth-generation bikesharing systems. Finally, it assesses the range of benefits of bikesharing and examines how cities can get the most out of bikesharing by designing a user-oriented system.

Chapters 11, 12, and 13 address the special needs and preferences of women, children, and older adults. Because recent surveys indicate a low or declining level of cycling by women, children, and older adults in most countries, it is essential to adapt cycling infrastructure and programs to the special needs of these groups. Chapter 11 examines variations in women's cycling rates among countries, cities, and city neighborhoods. Where women cycle as much as men, the overall bike share of trips is high and cycling safety is excellent. Chapter 11 considers the many factors that affect women's cycling, finding that women are often deterred from cycling by concerns about personal safety and traffic risks. Providing safe, connected cycling infrastructure—separated from motor vehicle traffic—is the most important measure to encourage more cycling by women. Safe cycling infrastructure must, however, be accompanied by complementary strategies to adapt cycling policies to the special concerns of women.

It is especially important to promote cycling among children, because habits learned while young tend to persist throughout life. As documented in chapter 12, cycling can provide children with a range of physical, mental, and social benefits, greatly enhancing their mobility and independence. Unfortunately, cycling rates for children have fallen in most countries in recent decades. The downtrend can be explained by increased car ownership among parents, longer distances between home and school, increased traffic danger, and parental fears about the personal safety of their children. In addition, children have been shifting to other forms of recreation, such as organized sports, internet games and messaging, and watching television. Chapter 12 examines ways to reverse the downtrend in child cycling based on successful policies in some countries and cities. Improving cycling safety, especially for the trip to school, is a key approach, but other complementary programs are also required.

Older adults are the fastest-growing age cohort in almost all economically developed countries. As chapter 13 documents, cycling can provide valuable physical activity, mobility, independence, and social interaction for people as they age. Older age need not be a barrier to cycling, which remains physically possible for many people well beyond the age of 65, provided conditions are safe and convenient, as demonstrated by several countries where cycling rates among older adults are as high as for younger adults. The most important measure to enable cycling by older adults is the provision of cycling infrastructure that is safe, well designed, well maintained, well connected, and, to the extent possible, separated from motor vehicle traffic. Recent developments in bicycle technology—such as e-bikes, tricycles, and recumbent bikes—also facilitate more cycling by older adults.

Chapter 14 is devoted to equity and social justice, which should be key considerations in all areas of transport policy, including cycling. On the one hand, cycling has the potential to greatly increase the mobility and accessibility of lower-income and minority groups, helping them to escape the isolation of spatial segregation within cities, but especially to reach jobs, shopping, services, and family and friends. Cycling could also help increase the mobility and independence of children and adolescents, providing them more convenient and more frequent access to friends and recreational opportunities. Yet cycling policies and infrastructure provision rarely reflect the specific needs of disadvantaged and marginalized groups. Chapter 14 documents the many dimensions of social injustice in cycling policies. It recommends the development and implementation of laws and regulations requiring that the needs of everyone, including and especially the disadvantaged, be explicitly considered in all aspects of urban transport policy, planning, funding, and decision-making.

Chapters 15 and 16 examine cycling in areas of the world that were only briefly mentioned in the 2012 book: Asia and Latin America. Chapter 15 is devoted to China and India, which together account for over one-third of the world's population. For many decades, the bicycle has been an important means of travel in both China and India, accounting for much higher shares of travel than in almost all other countries. In China, the bicycle share of urban trips fell from over two-thirds in the 1980s to about one-third by 2015, but with much variation among cities. The dramatic growth of bikesharing and e-bikes in China has led to renewed interest in cycling and to a turn-around in cycling's decline in some cities since 2015. Cycling levels in Indian cities have remained at around one-fifth of trips in recent decades, with much variation among cities. Chapter 15 provides an overview of cycling trends, cyclist demographics, traffic safety, and government policies in the two countries, concluding with recommendations for increasing cycling and making it safer.

Chapter 16 examines cycling trends and policies in Latin America. Most of the improvements in cycling conditions, and accompanying growth in cycling levels, have been in a few major cities, such as Bogota, Cali, Buenos Aires, Santiago de Chile, Mexico City, Rosario, and Montevideo. In most Latin American cities, however, the bicycle share of trips is only 1% or 2%, with cycling remaining a marginal mode of transport. As docu-mented in chapter 16, however, several large cities have vastly expanded and improved their cycling facilities over the past decade, with plans for further infrastructure expansion in the coming years. Inspiring examples of recreational "open street" programs (starting with Bogota's Ciclovía in

1974) have spread throughout Latin America and beyond to the rest of the world, popularizing cycling for recreation and physical activity. As cycling conditions improve, cycling will become a safer, more convenient way for everyone to get around, especially for daily utilitarian trips.

Chapter 17 describes the recent evolution of cycling in New York, London, and Paris. Over the past few decades, all three cities have implemented policies aimed at making cycling a feasible, convenient, and safe way to travel. Each city's mix of policies has been unique, but all three cities have vastly expanded and improved cycling infrastructure as the foundation for increasing cycling. To varying degrees, the cities have also increased bike parking, implemented bikesharing, expanded cycling training programs, reduced speed limits for cars, traffic-calmed residential neighborhoods, expanded car-free zones, restricted car parking, and, in the case of London, imposed congestion pricing for the city core. The result has been a large increase in cycling levels and a rise in cycling's share of total trips: doubling in London, tripling in New York, and increasing tenfold in Paris compared to the early 1990s. As noted in chapter 17, future growth in cycling seems certain given the continued investments in cycling planned for the coming years.

Amsterdam and Copenhagen have been at the leading edge of innovations in cycling infrastructure, programs, and policies, as documented in chapter 18. Both cities already had a long history of cycling as a major travel mode, but in recent decades they have made massive investments in their cycling infrastructure while setting the world's standard for infrastructure design and a range of policies that encourage cycling and discourage car use. As shown in figure 1.1, the result has been large increases in the bicycle share of trips. Chapter 18 examines the reasons underlying their success and, in particular, the policies that have enabled cycling to flourish, providing lessons for other cities seeking to increase cycling. The comparative analysis identifies differences between the two cities in their land use and urban design, overall transport policies, and specific mixes of policies and programs to promote cycling.

Chapter 19 tells the stories of how Portland and Seville became the most bike-oriented cities in the United States and Spain, respectively. There are differences as well as similarities in how the two cities dramatically increased cycling while making it safer. Confirming the experience of other cities examined in this book, both Portland and Seville demonstrate the crucial need to provide a safe, convenient, and comfortable network of integrated cycling facilities as a foundation for all other policies and programs to increase cycling. Their mixes of cycling facility types differ, however, as do other complementary policies and programs. Moreover, Portland's transformation into a cycling city took more than two decades, while in Seville it took only four years. The two cities provide important lessons on how to get the necessary pro-cycling policies implemented through the active support and cooperation of local political leaders, dedicated and well-trained professional staff, and cycling advocacy organizations.

Chapter 20 examines the crucial role of national and state cycling advocacy in the growth of cycling in cities throughout the world. It first examines Europe, where the European Cyclists' Federation (ECF) is the leading advocacy group. ECF helps to coordinate and encourage cycling advocacy at the national level in addition to its international advocacy with the European Union and, more recently, other continents as well. Chapter 20 uses national organizations in the Netherlands, Denmark, Germany, and the United Kingdom as examples to show the range of advocacy efforts undertaken at the national level in Europe. Moving on to North America, the chapter first examines the two main national organizations in the United States and then presents case studies of state advocacy in California and Texas. In Canada, the focus is on Vélo Québec, which is one of the most important and innovative cycling advocacy organizations on the continent and has been a model that others have followed. Finally, Chapter 20 examines national and state cycling advocacy groups in Australia.

Chapter 21 concludes the book with an overview and synthesis of what works best for raising cycling levels and improving cycling safety under a variety of situations. It draws on evidence from throughout the book to suggest ways to increase the equity of cycling policies by enabling cycling for as many people as possible: women as well as men, all income and ethnic groups, and of all ages and abilities. The wide range of countries and cities examined in this book offers an extensive base of experience for planners and government officials to draw on when developing effective policies to promote cycling.

**References**

Pucher, John, and Ralph Buehler, eds. 2012. *City Cycling*. Cambridge, MA: MIT Press.

# 2 International Overview of Cycling

Ralph Buehler and John Pucher

There are large differences in cycling levels around the world, even among countries with similarly high levels of economic development and car ownership. At the low end, the bike share of trips is only 1.0% in the United States, 1.4% in Australia and Canada, and 1.5% in the United Kingdom (see figure 2.1). At the upper end, the bike share is 28% in the Netherlands, 14% in Denmark, 13% in Japan, and 11% in Germany.

For most countries shown in figure 2.1, the bike share refers to daily trips for all trip purposes, as derived from national travel surveys. Australia and Canada do not have national travel surveys, however, and their censuses report only on trips to work. Census data on work trips probably underestimate overall levels of cycling, as is seen most clearly by comparing the 2016 and 2017 surveys for the United States. The US Census Bureau's 2016 *American Community Survey* (*ACS*), which only includes work trips, reports a bike share two-thirds as high (0.6% vs. 1.0%) as the 2016–2017 *National Household Travel Survey* (*NHTS*), which includes all trip purposes. There are also methodological differences in travel surveys among countries, which limit their comparability. Nevertheless, it is clear that cycling rates in most northern European countries and in Japan are much higher than in North America and Australia.

The national differences presented in figure 2.1 hide the variation in cycling levels among cities within each country. Figure 2.2 shows bike mode shares for selected cities in Japan, Australia, North America, and South America. Figure 2.3 shows bike mode shares for selected cities in Europe. The range of bike mode shares among cities is especially large in the United States, the United Kingdom, France, and Italy, countries with low bike mode shares overall. Similarly, there is also much variation among

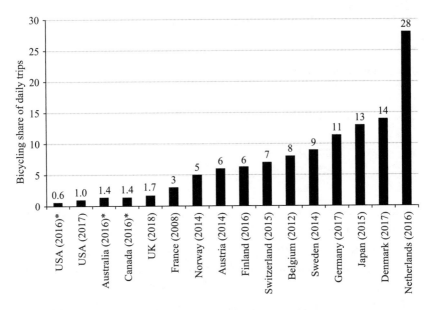

**Figure 2.1**
Bicycling share of daily trips in Europe, North America, Australia, and Japan for 2008–2017. *Note*: The latest available travel surveys were used for each country, with the survey year noted in parentheses after each country name. The modal shares shown in the figure reflect travel for all trip purposes except for those countries marked with an asterisk (*), which report only journeys to work, derived from their censuses. Differences in data collection methods, timing, and variable definitions across countries and over time limit the comparability of the modal shares shown in the figure. Data for Japan report the average modal share for 70 large cities. *Sources*: Australian Bureau of Statistics (2018); Austrian Ministry of Transport (2017); Buehler et al. (2017); German Ministry of Transport (2018); Danish Ministry of Transport (2017); Department for Transport (2017); Japanese Ministry of Transport (2016); Netherlands Ministry of Transport (2017); Statistics Canada (2018); USDOC (2018); USDOT (2018).

cities in Latin America. There is much less variation among cities in the Netherlands and Denmark, which have high levels of cycling overall.

With few exceptions, the most bike-oriented cities in the United States, Canada, and Australia have lower bicycle shares of trips than the average cycling levels in the Netherlands, Japan, Denmark, and Germany. For example, Portland (Oregon, United States), Vancouver (British Columbia, Canada), and Melbourne (Victoria, Australia) have bike mode shares much lower than in most Dutch, Japanese, Danish, and German cities.

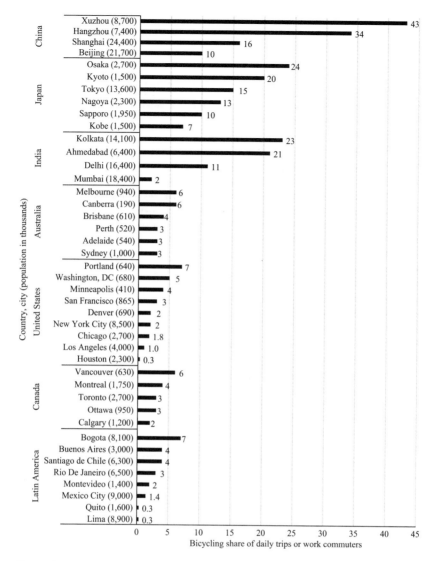

**Figure 2.2**

Bicycling share of daily trips or work commuters in selected cities in Japan, India, China, Australia, and the Americas for 2016. *Note*: Except for Australia, the population and bicycle share numbers refer only to the political boundaries of the central government jurisdiction, not the metro area. For Australia, both population and bicycle share numbers are based not on political boundaries but rather on the area within 10 km of the central point of the metro area. This was necessary because of the extreme variation in the percentage of central government jurisdictions within metropolitan areas in Australia. For all countries, differences in data collection methods, timing, geographic boundaries, and variable definition across cities and over time limit comparability. *Sources*: Data for Australia, India, and Canada are derived from their 2011 and 2016 censuses because they do not have national travel surveys. Data for the United States are based on the *American Community Survey* (USDOC 2018). Data for Japan and South America are based on travel surveys in each individual city. The census data refer to the bicycling share of daily work commuters. Travel survey data refer to the bicycling percentage of total daily trips (for all purposes).

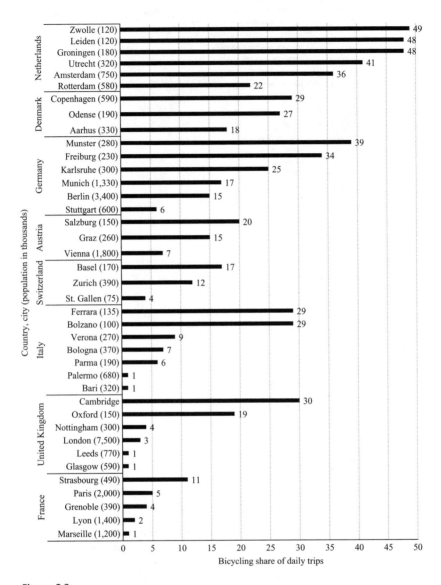

**Figure 2.3**

Bicycling share of daily trips in selected cities in Europe for 2010–2016. *Note*: Unlike in figure 2.1, all bicycling mode shares shown here refer to the percentage of daily trips made by bike. Nevertheless, differences in data collection methods, timing, geographic boundaries, and variable definitions across cities and over time limit comparability of the bicycling mode shares shown. *Sources*: Data collected directly from city travel surveys, national statistical offices, and the European Commission.

## Trip Distance and Purpose

The higher share of trips by bicycle in Dutch, Danish, and German cities is only partly explained by shorter trip distances than in American, Canadian, and Australian cities because of more mixed-use development, less suburban sprawl, and higher population densities in Europe (Buehler et al. 2017; Heinen, van Wee, and Maat 2010; Krizek, Forsyth, and Baum 2009). In the Netherlands, Denmark, and Germany, 42%–45% of all trips (by all modes) are shorter than 3.0 km, compared to 35%–36% in the United States and the United Kingdom (German Ministry of Transport 2018; Danish Ministry of Transport 2017; Department for Transport 2017; Statistics Netherlands 2017; USDOT 2018).

Even within the same trip distance categories, however, there are large differences among countries in bike mode share. Americans and Britons cycle for only 2% of trips shorter than 3 km, compared to 39% in the Netherlands, 23% in Denmark, and 16% in Germany. For trip distances from 3.0 to 8.0 km, the bike share of trips is only 1% in the United States and less than 2% in the United Kingdom, compared to 31% in the Netherlands, 22% in Denmark, and 11% in Germany. In short, the Dutch, Danes, and Germans cycle for much higher percentages of trips than Americans and Britons over all distance categories.

Trip purpose also varies by country. Cycling is mainly for recreation in the United States, accounting for 55% of all bike trips in 2017, compared to 33% in Germany, 32% in the United Kingdom, 24% in the Netherlands, and 19% in Denmark (German Ministry of Transport 2018; Danish Ministry of Transport 2017; Department for Transport 2017; Statistics Netherlands 2017; USDOT 2018). Most cycling in northern Europe is for practical, utilitarian purposes such as getting to work or school, shopping, running errands, and visiting friends.

## Who Cycles?

Cycling is common among all demographic groups in Germany, the Netherlands, Denmark, and Japan. For example, as shown in figure 2.4, women account for 49% of bike trips in Germany, 52% in Denmark, 54% in the Netherlands, and 55% in Japan. In contrast, women account for only 26% of bike trips in the United Kingdom and Australia, 30% in the United States,

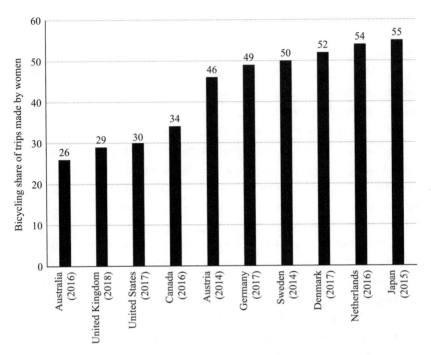

**Figure 2.4**
Bicycling share of trips made by women in the United Kingdom, Australia, the United States, Canada, Austria, Germany, Denmark, the Netherlands, and Japan for 2015–2017 (as a percentage of all bike trips). *Sources*: Australian Bureau of Statistics (2018); Austrian Ministry of Transport (2017); German Ministry of Transport (2018); Danish Ministry of Transport (2017); Department for Transport (2017); Japanese Ministry of Transport (2016); Netherlands Ministry of Transport (2017); Statistics Canada (2018); USDOC (2018); USDOT (2018).

and 34% in Canada. (For a detailed analysis of cycling by women, see Garrard, chapter 11, this volume.)

Similarly, cycling is common in all age categories in Germany, the Netherlands, Denmark, and Japan. As shown in figure 2.5, children and adolescents have the highest cycling levels in four of the five countries, with the United Kingdom being the only exception. (For a detailed analysis of cycling by children, see McDonald, Kontou, and Handy, chapter 12, this volume.) Cycling levels remain high, however, for adults of all ages in Germany, Denmark, the Netherlands, and Japan. Indeed, cycling rates increase slightly for the 65–69 age group and fall only slightly for the 70 and older

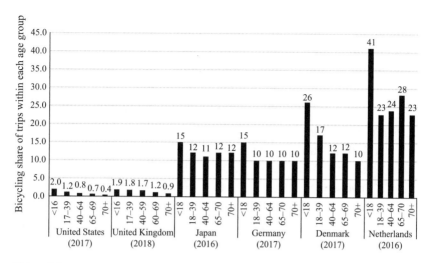

**Figure 2.5**
Bicycling share of trips within each age group in the United States, the United Kingdom, Japan, Germany, Denmark, and the Netherlands for 2015–2017 (as a percentage of trips by all modes). *Sources*: German Ministry of Transport (2018); Danish Ministry of Transport (2017); Department for Transport (2017); Japanese Ministry of Transport (2016); Netherlands Ministry of Transport (2017); USDOC (2018); USDOT (2018).

age group, which makes 10% of their trips by bike in Denmark and Germany, 12% in Japan, and a truly remarkable 23% in the Netherlands. In stark contrast, the 70 and older age group makes only 0.4% of their trips by bike in the United States and 0.9% in the United Kingdom. The examples of Denmark, Germany, Japan, and the Netherlands clearly show that older age need not be a barrier to cycling. As discussed in detail by Garrard et al. (chapter 13, this volume), cycling is physically possible well beyond the age of 65, provided that conditions are safe and convenient.

**Impacts of Car Ownership**

Increased car ownership and use discourages cycling by offering direct competition as a mode of travel and worsening traffic dangers for cyclists on roads. The very high motorization rates (cars and light trucks per 1,000 residents) in the United States (757), Canada (618), and Australia (551) probably contribute to their very low cycling rates (Buehler, Pucher, and Bauman 2020; OECD 2018; EC 2017). Japan (478), the Netherlands (477), and Denmark (419) have much lower motorization rates and much higher levels of cycling.

Nevertheless, even those lower motorization rates are among the highest in the world; most Dutch, Danish, and Japanese households have at least one car. These countries are also among the most affluent countries in the world, so their high cycling levels do not result from an inability to afford a car. Germany is an especially good example of this. Its motorization rate (548) is almost as high as Australia's (551), but Germany has a bicycling mode share eight times that of Australia (11% vs. 1.4%) and almost as high as Denmark's (14%). Conversely, the United Kingdom has the same motorization rate as the Netherlands (477) but a bicycling mode share that is one-twentieth as high (1.5% vs. 28%). In short, high incomes and high motorization rates are not necessarily incompatible with high rates of cycling.

Higher taxes on car ownership and use in Europe help explain lower levels of automobile ownership and use compared to the United States, Canada, and Australia. In Europe, taxes account for roughly 65% of the retail price of gasoline, compared to 50% in Japan and much lower tax shares in Australia (38%), Canada (32%), and the United States (21%) (IEA 2018). As a result, in 2017 retail prices per liter of gasoline (in US dollars) were roughly twice as high in the Netherlands ($1.92), Germany ($1.76), the United Kingdom ($1.67), Denmark ($1.51), and Japan ($1.36) as in Canada ($1.01), Australia ($0.93), and the United States ($0.71) (IEA 2018).

Europeans also pay higher taxes and fees on new car purchases. In Denmark, the tax rate on new car purchases (including fees and import charges) ranges from 85% to 150%, depending on the value, size, efficiency, and pollution emissions of the vehicles (ACEA 2018). In the Netherlands, tax rates and fees range from about 20% to 60%, also depending on the type of vehicle and emissions. Tax rates on car purchases are lower in Germany (19%) and the United Kingdom (20%), but as in Denmark and the Netherlands, there are significant registration fees—in addition to taxes—that range widely by vehicle type. The lowest taxes and fees are in North America and Australia. They vary greatly from state to state and by type of vehicle but average about 6% in the United States, 14% in Canada, and 15% in Australia (American Automobile Association 2017; Australian Automobile Association 2017; Canadian Automobile Association 2017).

In addition to higher taxes on car ownership and use, better and more convenient public transport systems in Europe—integrated with comprehensive bikeway and walkway networks—reduce the need to own and drive an automobile (Buehler et al. 2017; Newman and Kenworthy 2015; Pucher and Buehler 2008). The superior public transport systems in northern Europe provide an essential complement to cycling. Extensive and high-quality bike parking facilities at metro and suburban rail stations encourage the use of bikes to cover the short distances to access public transport, which is then used to cover longer distances. Except in small towns, public transport is essential for anyone not owning a car and indeed makes car-free living possible, even for those who make most of their trips by bike.

## Policy Shifts to Promote Cycling

In the 1950s and 1960s, increasing motorization levels, sprawling suburban development, and government policies in most western European countries favored car use and contributed to a sharp decline in cycling. For example, the number of daily bike trips in Berlin fell by 75% from 1950 to 1975 (City of Berlin 2003). Other German, Dutch, and Danish cities reported declines in the bicycling share of trips from roughly 50% to 85% in 1950 to only 14% to 35% of trips in 1975 (Dutch Bicycle Council 2010). During that period, many European cities focused on expanding their roadway and car parking supply while largely ignoring the needs of cyclists (Hass-Klau 2015; Newman and Kenworthy 2015; Schiller and Kenworthy 2018).

Increasing car use in cities led to environmental pollution, roadway congestion, and a sharp rise in traffic injuries and fatalities. Those harmful impacts of car use provoked a dramatic reversal of the transportation policies of most German, Dutch, and Danish cities. Instead of further adapting themselves to the car, most cities chose to restrict car use and increase its cost while promoting public transport, walking, and cycling (Buehler et al. 2017; ECMT 2004; Hass-Klau 2015; Pucher and Buehler 2008).

Greatly expanded and improved cycling infrastructure contributed to a rebound in cycling. Between 1975 and 1995, cycling levels rose by about 25% in the same sample of German, Dutch, and Danish cities that had witnessed a drastic decline in cycling prior to 1975 (Dutch Bicycle Council 2010). In Berlin, the number of daily bike trips increased by 368% between 1975 and 2012 (City of Berlin 2018). National data also show a considerable increase in cycling since the policy shift of the 1970s. From 1978 to 2017, average daily kilometers cycled per inhabitant increased from 0.6 km

to 1.2 km in Germany, from 1.3 km to 1.6 km in Denmark, and from 1.7 km
to 2.5 km in the Netherlands. Over the same period, daily cycling levels
declined in the United Kingdom from 0.3 km to 0.2 km and were roughly
constant in the United States (0.1 km) (German Ministry of Transport 2017;
Danish Ministry of Transport 2017; Department for Transport 2017; Statistics Netherlands 2017; USDOT 2018).

## Cycling Safety

Many studies document that traffic danger is a deterrent to cycling, especially for women, the elderly, and children (UN Habitat and European
Commission 2016; LAB 2018; OECD 2018). Thus, safer cycling conditions
are an important reason for more cycling among all groups in Germany,
Denmark, and the Netherlands.

Figure 2.6 shows declining cyclist fatality rates per 100,000 people from
1990 to 2018 in the United States, Canada, Australia, the United Kingdom,

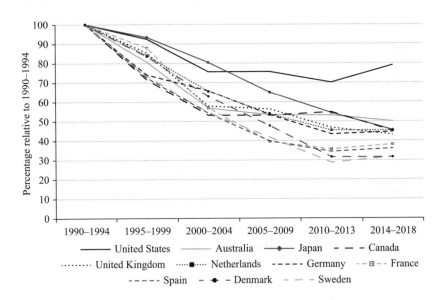

**Figure 2.6**
Trend in cyclist fatality rates per 100,000 people for 1990–2018, five-year annual
averages relative to the 1990–1994 average (=100%). *Note*: To control for annual fluctuations, five-year averages were used for cyclist fatalities per capita. *Source*: Calculated by the authors based on data provided by ITF/OECD (2020).

Japan, the Netherlands, Sweden, Denmark, Germany, France, and Spain. Rates are shown as averages for each of the six periods to reduce year-to-year statistical fluctuations, with each period's rate calculated relative to the rate in the first period, 1990–1994. By far the least progress has been made in the United States, whose cyclist fatality rate per capita fell by only 21%, compared to 55%–68 % in western Europe, 55% in Canada, 54% in Japan, and 50% in Australia. These per capita rates adjust for population changes but do not adjust for changes in cycling levels per capita over time. Nevertheless, the large percentage reductions over the 28-year period suggest long-term improvement in cycling safety. However, over the most recent period (2010–2018), cyclist fatalities per capita rose in five countries: by 13% in the United States, 10% in Sweden, 7% in France, 5% in Spain, and 3% in Germany. By comparison, the per capita fatality rate fell by 18% in Canada, 16% in Japan, 8% in the United Kingdom, and 6% in Australia. The rate remained virtually unchanged in the Netherlands (–0.4%) and in Denmark (–0.1%). The contrast between the United States (+13%) and Canada (–16%) is especially striking given the many similarities between the two countries in land use, car ownership, and travel behavior.

Controlling for exposure levels, the rates shown in figure 2.7 are calculated as annual cyclist fatalities relative to kilometers of cycling per year in each country. The Netherlands, Denmark, and Germany have the safest cycling among the six countries. For the period 2014–2016 (three-year average), cyclist fatalities per 100 million km were 0.8 in the Netherlands, 0.9 in Denmark, and 1.1 in Germany. Those fatality rates are about half as high as in Japan (1.7) and the United Kingdom (1.9). Worst of all is the United States, with a fatality rate per 100 million km of 5.6, which is about six times higher than in the Netherlands, Denmark, and Germany and about three times as high as in Japan and the United Kingdom.

Not only is cycling in the United States the most dangerous of these six countries, but it has also been getting more dangerous, with the fatality rate per kilometer rising by 19% between the two most recent periods (three-year averages for 2008–2010 and 2014–2016), while fatality rates per kilometer have fallen in the five other countries. The alarming rise in cycling fatalities per kilometer is consistent with the 13% increase in per capita cycling fatalities for the United States from 2010 to 2018, as shown in figure 2.6. Schneider, Vargo, and Sanaizadeh (2017) also report a sharp increase in cyclist fatality rates in the United States in recent years—whether measured

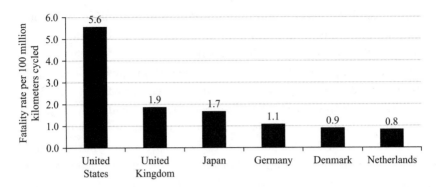

**Figure 2.7**
Cyclist fatality rates per 100 million km cycled for 2014–2016. *Note*: To control for annual fluctuations, three-year averages were used for cyclist fatalities. Trips and kilometers for cycling exposure levels were derived from the latest national travel survey data for each country. *Sources*: German Ministry of Transport (2017); Danish Ministry of Transport (2017); Department for Transport (2017); Japanese Ministry of Transport (2016); Netherlands Ministry of Transport (2017); USDOC (2018); USDOT (2018).

per trip, per kilometer, or per hour cycled. The high and increasing danger of cycling in the United States is probably one of the reasons its mode share of cycling has remained at about 1%for many years.

The aggregate national data for the United States hide the impressive progress of some American cities that have vastly expanded and improved their cycling infrastructure, providing more separation from motor vehicle traffic on roadways (Buehler, Pucher, and Bauman 2020; Pucher and Buehler 2016). For example, New York City expanded its bikeway network by 381% from 2000 to 2017, resulting in a 207% increase in the number of bike trips and a 72% reduction in the rate of cyclist fatalities and serious injuries (overnight hospitalization) per 100,000 bike trips (see Pucher, Parkin, and de Lanversin, chapter 17, this volume). Similarly, Chicago, Los Angeles, Philadelphia, Washington, Seattle, San Francisco, Minneapolis, and Portland all expanded and improved their bike infrastructure, increased the number of bike trips, and reduced the number of cyclist fatalities, serious injuries, and crashes per 100,000 trips (Pucher and Buehler 2016). Thus, cycling safety is improving greatly in some American cities while declining for the country as a whole.

Many cities and countries in Europe, North America, and Australia now have Vision Zero in their overall transport policies, particularly for cycling and walking safety. First developed and adopted in Sweden, Vision Zero argues that traffic fatalities and serious injuries are not inevitable and that they can be reduced to low levels by implementing the right policies, especially improved infrastructure and technology (Pucher and Buehler 2016). It is unlikely that any country or city can actually achieve zero traffic fatalities and serious injuries. Nevertheless, Vision Zero officially acknowledges the crucial importance of traffic safety and motivates a comprehensive approach to improving safety in all aspects of transport policy and infrastructure design. Several studies emphasize the special importance of Vision Zero for cycling safety (Ahangari et al. 2016; Ahangari et al. 2017; Andersen et al. 2018; Cushing et al. 2016). Elvik (chapter 4, this volume) analyzes in detail the important issue of cycling safety, including the problem of serious cyclist injuries, which are many times more common than cyclist fatalities.

## National Cycling Strategies

Most policies that increase cycling and make it safer are implemented at the local level. National governments, however, influence cycling through national cycling policies, dedicated funding, traffic regulations, roadway and bikeway design standards, and dissemination of cycling expertise. Motivations for national governments to promote cycling vary but often include environmental and public health benefits, reduced traffic congestion and noise, improved traffic safety, and tourism (LAB 2018; ECF 2018; EC 2015; TRB 2016).

National cycling policies and plans vary greatly in content, level of detail, legal status, and financial commitment. Australia and Canada are two major countries with neither a national cycling policy nor federal funding for cycling. Some Australian states and Canadian provinces, however, have adopted general policies encouraging more sustainable and active travel (including cycling) to promote health, safety, and the environment. The Quebec Ministry of Transport, for example, works closely with local governments throughout the province to develop, fund, and implement cycling projects.

In most of the other countries we examined, national cycling policies establish the general goal of increasing cycling levels and making cycling safer. For example, Germany's national cycling plan calls for significant increases in cycling without quantifying specific goals (German Ministry of Transport 2015). In some countries, national governments set specific goals for cycling. For example, the UK National Cycling Investment Strategy of 2017 set the goal of doubling cycling levels by 2025 relative to 2013 (Department for Transport 2015). Similarly, the Dutch national cycling plan set the goal of increasing kilometers of cycling by 20% between 2017 and 2027 (Netherlands Ministry of Transport 2016). In the United States, the federal government set a goal to reduce pedestrian and cyclist fatalities and severe injuries by 80% over 15 years and to increase the combined share of short trips made by foot (< 1 mi) and bicycle (<5 mi) to 30% by 2025—a 50% increase compared to 2009 (USDOT 2016).

National cycling policies generally provide a vision for cycling that can guide lower levels of government in their own efforts to increase cycling and make it safer (ECF 2018). In Denmark and the Netherlands, for example, national cycling policies are intended to help local jurisdictions develop their own bicycle plans (ECF 2018; Netherlands Ministry of Transport 2016). The Netherlands established an intergovernmental collaborative group called Tour de Force to increase funding and coordinate efforts to promote cycling (Netherlands Ministry of Transport 2016). In most countries, the national cycling policies recommend improved data collection and benchmarking efforts to increase knowledge about the state of cycling over time. Some national governments also coordinate the dissemination of information about best practices or cutting-edge planning tools (e.g., the Danish "ideas catalog").

### National Funding for Cycling

Some countries provide central government funding to finance the implementation of the measures proposed in the national cycling plan (ECF 2018). In Germany, for example, annual federal funding dedicated for bicycling rose from €50 million in 1990 to €125 million in 2016 (ADFC 2017; German Ministry of Transport 2015). In the United Kingdom, the national Cycling and Walking Investment Strategy provides £1 billion in national funding over five years (2015–2020), which local governments can use to facilitate cycling and walking (Department for Transport 2015). In most countries, however, central government funding is not dedicated specifically for cycling but rather for urban transport in general.

States and local governments usually have some flexibility in their use of national transport funds. Cycling projects, for example, are eligible for federal Transportation Alternatives (TA) earmarked funds in the United States and for federal Urban Transport funds in Germany (German Parliament 2016; USDOT 2017). In the United States, almost all federal funding for urban transport is technically eligible to be spent on cycling. States and metropolitan planning organizations decide how much of the federal funding is spent on cycling projects (USDOT 2017; LAB 2018). The use of federal funding for bicycling and walking projects increased from only $23 million per year in 1992 to over $1.02 billion per year in 2019—a forty-four fold increase over 27 years (USDOT 2019). As in many other countries, it is not possible to separate funding for bicycling and walking for the United States, as they are combined into one category for funding purposes. Nevertheless, the significant increase in funding for both modes combined suggests a large increase in federal spending on cycling.

As noted previously, the federal government of Canada has no role in urban transport, including cycling. Canadian provinces, however, play a major role. Quebec and Ontario, for example, provide dedicated funding to match local government funding for cycling infrastructure improvements needed to achieve provincial goals such as increasing bicycling's modal share and decreasing cyclist fatalities and serious injuries.

In some countries, national governments fund cycling infrastructure along national highways, often as part of roadway improvement programs. Moreover, some national governments contribute toward funding programs to improve cycling training and safety, such as the Safe Routes to School activities in the United States, the Bikeability Program in the United Kingdom, and the dedicated fund for cycling safety in Denmark (Danish Ministry of Transport 2011; Department for Transport 2015; ECF 2018; USDOT 2018). In some countries, part of the national funding is competitive, such as the Cykelpuljen program in Denmark, the Sustainable Transport Fund in the United Kingdom, and the former Urban Transport Showcase Fund in Canada (Danish Ministry of Transport 2011; Department for Transport 2018; Transport Canada 2010).

Overall, in the countries we examined, governments at all levels have generally increased their support for cycling over the past few decades, with the expectation that a modal shift from driving to cycling would help combat societal problems such as obesity, traffic congestion, energy use, and air pollution. Garrard et al. (chapter 3, this volume) discuss in detail the range of important individual and societal health benefits of cycling that justify government support for policies to promote cycling.

## Traffic Regulation and Training for Drivers and Cyclists

Australian states, Canadian provinces, and American states are responsible for driver training and traffic regulations for motorists as well as cyclists (LAB 2018; Australian Government 2004; Transport Canada 2018). Following federal recommendations, all Australian states passed laws requiring cyclists of all ages to wear helmets. In Canada, 5 of 13 provinces and territories require all cyclists to wear helmets, and three provinces require cyclists under the age of 18 to wear helmets. Despite federal government endorsement of helmet use in the United States, no state requires bike helmets for adults, and only half the states require helmets for children (IIHS 2018). Denmark, Germany, the Netherlands, and the United Kingdom do not require bike helmets for children or adults but strongly encourage bike helmets for children (EC 2018).

All European countries have national traffic signage and regulations. In the Netherlands, Denmark, and Germany, national regulations prioritize the traffic safety of cyclists and pedestrians, with special protection for children and seniors, who are especially vulnerable (Buehler et al. 2017; ECF 2017; ECF 2018; Pucher and Buehler 2008). Nationally standardized driver training courses and strict motorist licensing tests in Europe emphasize the importance of protecting vulnerable road users. Driver training is expensive in western Europe, where obtaining a driver's license typically costs between €1,000 and €2,000 (EDSA 2018). Driver training in the United States, Canada, and Australia is much less expensive and less rigorous than in Europe and does not stress the legal obligation of motorists to avoid endangering pedestrians and cyclists (Australian Government 2011; Transport Canada 2018; USDOT 2017).

Most children in Germany, Denmark, and the Netherlands participate in cycling training and testing in school (German Ministry of Transport 2015; Danish Ministry of Transport 2011; Netherlands Ministry of Transport 2018). In both Denmark and the Netherlands, safe cycling courses for schoolchildren are required by the national government. Safety courses are financed by the national government in Denmark but by municipalities in the Netherlands. About 80% of Dutch schools participate voluntarily in a national cycle testing program for children—focusing on practical on-road cycling skills beyond classroom safety lessons (Netherlands Ministry of Transport 2018).

In Germany, all states have adopted bike training as an integral part of the school curriculum in the third or fourth grade (German Ministry of Transport

2015). Courses vary from state to state but usually include classroom instruction about cycling safety and traffic regulations, police-administered training sessions on special off-street cycling-training facilities, and in-traffic cycling training with police officers on local streets. Only a few American, Canadian, and Australian schools provide cycling training for children, and participation by students is voluntary (Pucher, Buehler, and Seinen 2011; Pucher, Garrard, and Greaves 2011). Nongovernmental organizations such as CAN-Bike in Canada and the League of American Bicyclists in the United States provide cyclist training for all ages and skill levels, but such courses are not offered in most cities (CAN-Bike 2018; LAB 2018). Moreover, they charge a fee and reach only a small percentage of the population.

### Speed Limits and Design Standards for Roadways and Bikeways

In all countries, municipalities determine policies such as speed limits and traffic calming of residential streets. Starting in the 1970s, German and Dutch municipalities progressively traffic-calmed most neighborhood streets (Hass-Klau 2015; Pucher and Buehler 2008). Speed limits are typically 30 kilometers per hour (km/h) or less, and sometimes as low as 15 km/h or even 7 km/h (walking speed) (Buehler et al. 2017). In North America and Australia, speed limits in residential neighborhoods are much higher, and traffic calming is usually limited to a few streets, not area-wide as in European cities. Some North American cities, such as Vancouver (British Columbia) and Seattle (Washington), have implemented traffic-calming measures for decades, but none as extensively as in Europe (Ewing 2009; Hass-Klau 2015). Quebec recently introduced the concept of shared streets in the province's Road Safety Code (20 km/h), giving pedestrians priority over motorists and facilitating cycling as well because of the lower speed limits for cars on such streets. The new law also permits and defines bike boulevards, where motor vehicle speeds are limited to 30 km/h, thus improving cycling safety and reducing traffic stress. Bike boulevards generally increase average cycling speed by offering more direct connections for cyclists on key routes. They also allow cyclists the full use of traffic lanes, thus enabling them to ride side-by-side (Quebec Ministry of Transport 2018).

National and state governments are responsible for speed limits and road design on national and state highways. Even in urban areas, such roads are subject to specific national or state standards. Municipalities must seek the approval of higher levels of government to make any physical changes to these national or state roads. National governments or national nongovernmental organizations often publish design guidelines for roadway and cycling facilities. In the United States, for example, the federal government publishes the *Manual on Uniform Traffic Control Devices*, which must be followed by all states (USDOT 2009). The American Association of State Highway and Transportation Officials (AASHTO) and the National Association of City Transportation Officials (NACTO) publish detailed design guidelines for bicycle facilities in the United States (AASHTO 2012; NACTO 2014). Similar nongovernmental organizations in the Netherlands and Germany regularly update and disseminate guidelines for best practices in bikeway and roadway design in those two countries (CROW 2017; Roadway and Transport Research Center 2010). Canadian provinces and Australian states have their own road and bikeway standards, which are often based on national guidelines (Austroads 2010; TAC 1998).

## Conclusions

As documented in the first half of this chapter, there is great variation among countries in levels of cycling, the demographics of cyclists, bike trip distance and purpose, and cycling safety. In general, countries with the highest levels of cycling have the best cycling safety rates, the highest proportion of women and older cyclists, the most utilitarian cycling, and the longest average bike trip distances.

Correspondingly, government policies toward cycling vary greatly among countries. Because of limited space, this chapter focuses on variation among countries in national cycling policies and funding, although variation in cycling policies among cities within countries is at least as important. Our brief overview of government policies finds that national and state governments in many countries provide local governments with funding as well as policy guidelines, technical assistance, and coordination of cycling planning and promotion efforts.

In all countries, however, local governments have the ultimate responsibility for adopting and implementing specific cycling infrastructure and programs. The most innovative infrastructure, programs, and policies—including land use and car restraint policies—have often been developed and implemented at the local level. The crucial role of local governments is examined in detail for Latin American cities by Pardo and Rodriguez (chapter 16, this volume), for New York, London, and Paris by Pucher, Parkin, and de Lanversin (chapter 17, this volume), for Copenhagen and Amsterdam by Koglin, te Brömmelstroet, and van Wee (chapter 18, this volume), and for Portland and Seville by Geller and Marqués (chapter 19, this volume).

## References

Ahangari, Hamed, Carol Atkinson-Palombo, and Norman W. Garrick. 2016. Progress towards Zero, an International Comparison: Improvements in Traffic Fatality from 1990 to 2010 for Different Age Groups in the USA and 15 of its Peers. *Journal of Safety Research* 57: 61–70. https://doi.org/10.1016/j.jsr.2016.03.006.

Ahangari, Hamed, Carol Atkinson-Palombo, and Norman W. Garrick. 2017. Automobile-Dependency as a Barrier to Vision Zero, Evidence from the States in the USA. *Accident Analysis & Prevention* 107: 77–85. https://doi.org/10.1016/j.aap.2017.07.012.

Allegemeine Deutscher Fahrradclub (ADFC). 2017. *Cycling in Germany. Now!* Berlin: Allegemeine Deutscher Fahrradclub (German National Bicyclist Association).

American Association of State Highway and Transportation Officials (AASHTO). 2012. *Guide for the Development of Bicycle Facilities*. Washington, DC: American Association of State Highway and Transportation Officials.

American Automobile Association (AAA). 2017. *Your Driving Costs 2017*. Washington, DC: American Automobile Association.

Andersen, L. B., A. Riiser, H. Rutter, S. Goenka, S. Nordengen, and A. K. Solbraa. 2018. Trends in Cycling and Cycle Related Injuries and a Calculation of Prevented Morbidity and Mortality. *Journal of Transport and Health* 9: 217–225.

Association des Constructeurs Européens d'Automobiles (ACEA). 2018. *ACEA Tax Guide*. Brussels: Association des Constructeurs Européens d'Automobiles (European Automobile Manufacturers Association).

Association of Australian and New Zealand Road Transport and Traffic Authorities (Austroads). 2010. *Traffic Control Devices*. Sydney: Association of Australian and New Zealand Road Transport and Traffic Authorities.

Australian Automobile Association. 2017. *Transport Affordability Index*. Canberra: Australian Automobile Association.

Australian Bureau of Statistics. 2018. *2016 Census of Population and Housing, Journey to Work Files*. Canberra: Australian Bureau of Statistics.

Australian Government. 2004. *Monograph 17: Cycle Safety*. Sydney: Australian Transport Safety Bureau.

Australian Government. 2011. *Apply for a Driver's License.* Sydney: Australian Government.

Austrian Ministry of Transport. 2017. *Austrian National Travel Survey.* Vienna: Austrian Ministry of Transport.

Buehler, Ralph, John Pucher, and Adrian Bauman. 2020. Physical Activity from Walking and Cycling for Daily Travel in the United States, 2001–2017: Demographic, Socioeconomic, and Geographic Variation. *Journal of Transport and Health* 16 (March): 1–14. https://doi.org/10.1016/j.jth.2019.100811.

Buehler, Ralph, John Pucher, Regine Gerike, and Thomas Götschi. 2017. Reducing Car Dependence in the Heart of Europe: Lessons from Germany, Austria, and Switzerland. *Transport Reviews* 37(1): 4–28.

Canadian Automobile Association (CAA). 2017. *Driving Costs Calculator.* Ottawa: Canadian Automobile Association.

CAN-Bike. 2018. *CAN-Bike Program.* Ottawa: CAN-Bike.

City of Berlin. 2003. *Cycling in Berlin 2003.* Berlin: City of Berlin, Department of Urban Development.

City of Berlin. 2018. *Cycling in Berlin.* Berlin: City of Berlin, Department of Urban Development.

CROW. 2017. *ASVV Recommendations for Traffic Provisions in Built-Up Areas.* Ede, Netherlands: CROW.

Cushing, Matthew, Jonathan Hooshmand, Bryan Pomares, and Gillian Hotz. 2016. Vision Zero in the United States versus Sweden: Infrastructure Improvement for Cycling Safety. *American Journal of Public Health* 106(12): 2178–2180.

Danish Ministry of Transport. 2011. *Bicycling in Denmark.* Copenhagen: Danish Ministry of Transport.

Danish Ministry of Transport. 2017. *Danish National Travel Surveys.* Copenhagen: Danish Ministry of Transport.

Department for Transport. 2015. *Cycling and Walking Investment Strategy.* London: Department for Transport.

Department for Transport. 2017. *National Travel Survey.* London: Department for Transport.

Department for Transport. 2018. *Local Sustainable Transport Fund.* London: Department for Transport.

Dutch Bicycle Council. 2010. *Bicycle Policies of the European Principals: Continuous and Integral.* Amsterdam: Dutch Bicycle Council.

European Commission (EC). 2015. *Clean Transport. Urban Transport. Cycling.* Brussels: European Commission.

European Commission (EC). 2017. *Energy and Transport in Figures.* Brussels: European Commission, Directorate General for Energy and Transport, Eurostat.

European Commission (EC). 2018. *Pro and Cons Regarding Bicycle Helmet Legislation.* Brussels: European Commission.

European Conference of Ministers of Transport (ECMT). 2004. *National Policies to Promote Cycling.* Paris: European Conference of Ministers of Transport.

European Cyclists' Federation (ECF). 2017. *EU Cycling Strategy.* Brussels: European Cyclists' Federation.

European Cyclists' Federation (ECF). 2018. *National Cycling Policies Overview.* Brussels: European Cyclists' Federation.

European Driving Schools Association (EDSA). 2018. *Driver's License Costs in Europe.* Munich: European Driving Schools Association.

Ewing, Reid. 2009. *U.S. Traffic Calming Manual.* Washington, DC: Routledge.

German Ministry of Transport. 2015. *German National Bicycling Master Plan.* Berlin: German Ministry of Transport.

German Ministry of Transport. 2018. *National Household Travel Survey: Mobility in Germany 2017/2018.* Berlin: German Ministry of Transport.

German Parliament. 2016. *Federal Funding for Local Transport Projects.* Berlin: German Parliament.

Hass-Klau, Carmen. 2015. *The Pedestrian and City Traffic.* London: Routledge.

Heinen, Eva, Bert van Wee, and Kees Maat. 2010. Bicycle Use for Commuting: A Literature Review. *Transport Reviews* 30(1): 105–132.

Insurance Institute for Highway Safety (IIHS). 2018. *Bicycle Helmet Use.* Arlington, VA: Insurance Institute for Highway Safety.

International Energy Agency (IEA). 2018. *Energy Prices and Taxes.* New York: International Energy Agency.

International Transport Forum (ITF) and Organization for Economic Cooperation and Development (OECD). 2020. *Traffic Safety Statistics.* Paris: International Transport Forum and Organization for Economic Cooperation and Development.

Japanese Ministry of Transport. 2016. *Travel in 70 Large Cities.* Tokyo: Japanese Ministry of Transport.

Krizek, Kevin, Ann Forsyth, and Laura Baum. 2009. *Walking and Cycling International Literature Review.* Melbourne: Victoria Department of Transport.

League of American Bicyclists (LAB). 2018. *2018 Benchmarking Report on Bicycling and Walking in the United States*. Washington, DC: League of American Bicyclists.

National Association of City Transportation Officials (NACTO). 2014. *Urban Bikeway Design Guide*. Washington, DC: National Association of City Transportation Officials.

Netherlands Ministry of Transport. 2016. *Bicycle Agenda: Tour de Force*. The Hague: Netherlands Ministry of Transport.

Netherlands Ministry of Transport. 2017. *Dutch National Travel Survey*. Rotterdam: Netherlands Ministry of Transport, Public Works, and Water Management.

Netherlands Ministry of Transport. 2018. *Cycling Facts*. Rotterdam: Netherlands Ministry of Transport, Public Works, and Water Management.

Newman, Peter, and Jeffrey Kenworthy. 2015. *The End of Automobile Dependence: How Cities Are Moving beyond Car-Based Planning*. Washington, DC: Island Press.

Organization for Economic Cooperation and Development (OECD). 2018. *OECD Statistics*. Paris: Organization for Economic Cooperation and Development.

Pucher, John, and Ralph Buehler. 2008. Making Cycling Irresistible: Lessons from the Netherlands, Denmark, and Germany. *Transport Reviews* 28(4): 495–528.

Pucher, John, and Ralph Buehler. 2016. Safer Cycling through Improved Infrastructure. *American Journal of Public Health* 106(12): 2089–2090.

Pucher, John, Ralph Buehler, and Mark Seinen. 2011. Bicycling Renaissance in North America? An Update and Re-assessment of Cycling Trends and Policies. *Transportation Research Part A: Policy and Practice* 45(6): 451–475.

Pucher, John, Jan Garrard, and Stephen Greaves. 2011. Cycling Down Under: A Comparative Analysis of Bicycling Trends and Policies in Sydney and Melbourne. *Journal of Transport Geography* 19(2): 332–345.

Quebec Ministry of Transport. 2018. *Road Safety Code*. Montreal: Quebec Ministry of Transport.

Roadway and Transport Research Center. 2010. *Guidelines for City Streets*. Cologne: Roadway and Transport Research Center.

Schiller, Preston, and Jeffrey Kenworthy. 2018. *Introduction to Sustainable Transportation: Policy, Planning, and Implementation*. London: Routledge.

Schneider, Robert, Jason Vargo, and Aida Sanatizadeh. 2017. Comparison of US Metropolitan Region Pedestrian and Bicyclist Fatality Rates. *Accident Analysis and Prevention* 106: 82–98.

Statistics Canada. 2018. *Canadian Census 2016*. Ottawa: Statistics Canada.

Statistics Netherlands. 2017. *Transportation Statistics*. Amsterdam: Statistics Netherlands.

Transport Canada. 2010. *Leading by Example: Urban Transportation Showcase*. Ottawa: Transport Canada.

Transport Canada. 2018. *Driving in Canada*. Ottawa: Transport Canada.

Transportation Association of Canada (TAC). 1998. *Manual of Uniform Traffic Control Devices for Canada*. Ottawa: Transportation Association of Canada.

Transportation Research Board (TRB). 2016. *NCHRP Report 803: Pedestrian and Bicycle Transportation along Existing Roads—ActiveTrans Priority Tool Guidebook*. Washington, DC: Transportation Research Board.

UN Habitat and European Commission. 2016. *State of European Cities Report*. Brussels: European Commission.

US Department of Commerce (USDOC). 2018. *American Community Survey 2016*. Washington, DC: US Department of Commerce, US Census Bureau.

US Department of Transportation (USDOT). 2009. *Manual on Uniform Traffic Control Devices*. Washington, DC: US Department of Transportation, Federal Highway Administration.

US Department of Transportation (USDOT). 2016. *Strategic Agenda for Pedestrian and Bicycle Transportation*. Washington, DC: US Department of Transportation, Federal Highway Administration.

US Department of Transportation (USDOT). 2017. *The Transportation Alternatives Set-Aside Program*. Washington, DC: US Department of Transportation, Federal Highway Administration.

US Department of Transportation (USDOT). 2018. *National Household Travel Survey 2016/2017*. Washington, DC: US Department of Transportation, Federal Highway Administration.

US Department of Transportation (USDOT). 2019. *Federal-Aid Highway Program Funding for Pedestrian and Bicycle Facilities and Programs*. Washington, DC: US Department of Transportation, Federal Highway Administration.

# 3 Cycling and Health

Jan Garrard, Chris Rissel, Adrian Bauman, and Billie Giles-Corti

Cycling has many health benefits, particularly as a form of moderate to vigorous physical activity. Cycling for transport enables individuals to incorporate physical activity into daily life. Furthermore, it is accessible, practical, and convenient for population groups with low levels of participation in sports and other forms of leisure-time physical activity. In countries and cities with high cycling levels, cycling for transport is undertaken by substantial proportions of the population of all ages. In addition, recreational cycling can further contribute to total population levels of physical activity.

Promoting more cycling also results in additional health benefits that flow from reduced car use. These benefits include improved air quality, reduced noise pollution, reduced greenhouse gas emissions, and reduced health risks associated with climate change. Moreover, creating people-friendly streets in support of more and safer cycling enhances urban livability and quality of life, fosters social interaction, and reduces crime.

This chapter examines the health benefits of cycling as a form of moderate to vigorous physical activity and the additional health and health-related benefits that flow from replacing private motor vehicle trips with cycling trips. We use a broad definition of health as incorporating "physical, mental, and social health and well-being" (World Health Organization 1948).

## Health Benefits of Cycling as a Form of Physical Activity

The health benefits of physical activity are well established, as documented in a recent systematic review of the relationship between physical activity and health. The review noted benefits on the risk of depression, cardiovascular disease, diabetes, blood pressure, total fat mass, abdominal circumference, weight, insulin sensitivity, breast cancer, colorectal cancer, and mortality (Weggemans et al. 2018).

For older adults, benefits were found for muscle strength, reducing fractures (especially hip fractures), fat-free mass, improving walking speed, reducing dementia and Alzheimer's disease, and preventing cognitive decline and disability. For children, the benefits were for cardiorespiratory fitness, muscle strength, insulin sensitivity, weight and fat mass, bone quality, and depressive symptoms (Weggemans et al. 2018).

In terms of quantifying a minimum level of physical activity to achieve a health benefit, the review concluded that regardless of the current level of activity, further increases in physical activity provide additional health benefits. However, in relative terms, the greatest improvements are for those currently inactive who become more active. Consistent with the World Health Organization (WHO 2010) guidelines, the review concluded that adults should "engage in physical activity of moderate intensity for at least 150 minutes every week, spread over several days. For example, walking and cycling" (Weggemans et al. 2018).

Many of the health benefits of physical activity summarized here have also been demonstrated specifically for cycling, which is consistent with the energy expenditure of most commuting cycling (5–8 METs) (metabolic equivalent of task, or five to eight times the energy expenditure at rest), placing it at the high end of the "moderate intensity" range of 3–6 MET activities for health (Ainsworth et al. 2011). Findings from cycling-specific studies are summarized in the following sections.

**Overweight and Obese**

Until recently, few studies have investigated the relationship between cycling and being overweight or obese, as most studies reported active transport in general (i.e., walking and cycling combined). Systematic reviews of these studies provide inconsistent evidence of a relationship between active transport and body mass index (BMI) (Wanner et al. 2012), possibly because of combining two travel modes with differing MET values.

More recently, a small number of studies have examined the relationship between cycling and body weight, with some longitudinal research showing that cycling was significantly and independently associated with the lowest BMI (based on objectively measured height and weight) and percentage of body fat for both men and women (Flint and Cummins 2016).

Another study analyzed data ($n$ = 2,316) from a longitudinal study on transport and health across Europe (Dons et al. 2018). The study assessed the association between transport mode and BMI as well as the change in BMI over time. BMI decreased in occasional cyclists (less than once per week) and noncyclists who increased cycling (−0.303 kg/m$^2$), while frequent cyclists (at least once per week) who stopped cycling increased their BMI (0.417 kg/m$^2$). Similar results were reported from a Danish longitudinal study of the relationship between five-year cycling habits and abdominal and general obesity and waist circumference (Rasmussen et al. 2018). These more rigorous longitudinal analyses support the findings from cross-sectional studies that cycling for transport may contribute to a lower BMI. Consistent with these findings, a recent systematic review of the health benefits of "indoor cycling" reported some evidence (based on a small number of studies) that indoor cycling together with dietary changes may improve body composition and lipid profile (Chavarrias et al. 2019).

**Cardiovascular Disease, Cancer, Type 2 Diabetes, and All-Cause Mortality**
Recently, a small number of large-scale longitudinal studies have investigated chronic disease outcomes specifically for cycling. A prospective cohort study involving 263,450 British participants aged 40–69 years investigated the association between active commuting and cardiovascular disease (CVD), cancer, and all-cause mortality (Celis-Morales et al. 2017). They found that compared with nonactive commuting, commuting by bicycle was associated with a lower risk of CVD incidence (hazard ratio = 0.54; 95% confidence interval [CI] = 0.33 – 0.88), CVD mortality (hazard ratio = 0.48; 95% CI = 0.25 – 0.92), cancer incidence (hazard ratio = 0.55; 95% CI = 0.44 – 0.69), cancer mortality (hazard ratio = 0.60; 95% CI = 0.40 – 0.90), and all-cause mortality (hazard ratio = 0.59; 95% CI = 0.42 – 0.83), with a dose-response relationship (i.e., the benefits were greater for those who cycled for a longer distance). Furthermore, commuters by bicycle were more likely than others to achieve recommended weekly levels of physical activity (90%), followed by mixed-mode cycling commuters (80%), walking commuters (54%), mixed-mode walking commuters (50%), and nonactive commuters (51%).

Similar findings for the cardiovascular health benefits of cycling have been reported from a large Danish longitudinal study, comprising 51,868 participants aged 50–65 years (Kubesch et al. 2018). This study also investigated the cardiovascular health impacts of exposure to traffic-related air pollution while cycling, based on evidence of the adverse impacts of air pollution on heart health (Franklin, Brook, and Pope 2015). Regular cycling at baseline (1993–1997) was associated with a reduced incidence of heart attacks during approximately 20 years of follow-up (hazard ratio for cycling = 0.91; 95% CI = 0.84 – 0.98) (Kubesch et al. 2018). Additionally, Kubesch et al. (2018) concluded that the benefits of cycling in reducing the risk of incident and recurrent myocardial infarction are not reduced by exposure to air pollution in cities with air pollution levels similar to Copenhagen.

Another large prospective study conducted in Sweden (23,732 men and women) found that, compared with passive travel, cycling to work at baseline was associated with reduced risks of obesity (odds ratio [OR] = 0.85; 95% CI = 0.73 – 0.99), hypertension (OR = 0.87; 95% CI = 0.79 – 0.95), hypertriglyceridemia (OR = 0.85; 95% CI = 0.76 – 0.94), and impaired glucose tolerance (OR = 0.88; 95% CI = 0.80 – 0.96). This analysis further demonstrated health benefits for participants who maintained or began bicycling to work during a 10-year follow-up relative to those who did not cycle to work at both time points or who switched from cycling to other modes of transport during follow-up. The analysis found lower odds of obesity (OR = 0.61; 95% CI = 0.50–0.73), hypertension (OR = 0.89; 95% CI = 0.80 – 0.98), hypertriglyceridemia (OR = 0.80; 95% CI = 0.70 – 0.90), and impaired glucose tolerance (OR = 0.82; 95% CI = 0.74 – 0.91) (Grontved et al. 2016).

Pooling the data from several prospective cohort studies using metaanalysis showed that six of seven studies found a beneficial association between cycling and all-cause mortality (Kelly et al. 2014). The overall effect size for individuals who cycled (mainly for transport) for a standardized level of 150 minutes per week was a 10% (95% CI = 6% to 13%) reduction in all-cause mortality.

Consistent with these studies, two recent systematic reviews found that cycling reduces cardiovascular disease incidence, physiological risk factors, and mortality (Nordengen et al. 2019a; Nordengen et al. 2019b).

### Tools to Measure Potential Cycling Benefits in Populations

A number of studies have developed metrics to quantify the population benefit of cycling. One such tool is the Impacts of Cycling Tool (ICT) for application in England (Woodcock et al. 2018). Outcome measures include transport (trip duration and trips by mode), health (physical activity levels and years of life lost), and $CO_2$ emissions from cars. Results can be calculated to estimate the potential environmental and health benefits if more people cycled regularly. The tool for the English population is available at www.pct.bike/ict.

The Health Economic Assessment Tool (HEAT) for cycling and walking developed by the WHO includes an economic assessment of the value of reduced mortality that results from specified amounts of cycling. HEAT enables the calculation of physical activity benefit, air pollution risk, crash risk, and reduced carbon emissions (World Health Organization 2017).

Application of HEAT to cycling in the Netherlands for all Dutch people 20–90 years old found that about 6,500 deaths are prevented annually and that the population has a life expectancy half a year longer because of cycling. These benefits, which apply at the population level, translate into economic benefits of €19 billion per year, representing more than 3% of the Dutch gross domestic product between 2010 and 2013 (Fishman, Schepers, and Kamphuis 2015). HEAT monetizes reduced mortality and carbon emissions based on the value of a statistical life, derived using the willingness-to-pay method (World Health Organization 2017).

### Mental Health and Cognitive Benefits of Cycling as a Form of Physical Activity

There is a wealth of systematic review and intervention evidence that physical activity is important for improving mental health and reducing anxiety and depression in clinical and general populations (Rebar et al. 2015; Mammen and Faulkner 2013; Rosenbaum et al. 2014). Physical activity can also improve cognitive function in children and older adults (Donnelly et al. 2016; Tseng, Gau, and Lou 2011).

The specific contribution to mental health from cycling as a form of physical activity is less well researched, though systematic reviews have reported evidence of mental health benefits for active transport (White et al. 2017) and for designing communities for active living (Sallis et al. 2015). A recent cross-sectional study of 1,237,194 people aged 18 years or older in the United States from three consecutive Centers for Disease Control and Prevention Behavioral Risk Factor Surveillance System surveys found that cycling as a form of exercise was associated with the second-largest reduction in the number of days of self-reported poor mental health in the past month compared to matched participants who did not exercise (21.6% lower) (Chekroud et al. 2018). In addition, physical activity that is undertaken outdoors has additional benefits from that done inside, which may be a factor for cycling for transport or recreation, which are outdoor forms of physical activity (Thompson Coon et al. 2011). Indeed, a recent randomized, controlled trial of an outdoor cycling intervention for older adults found improvements in cognitive function and well-being for both conventional cycling and e-bike cycling groups compared to a noncycling control group (Leyland et al. 2019).

A number of qualitative studies have described the joy of cycling (Whitaker 2005; Daley, Rissel, and Lloyd 2007; Zander et al. 2013). "Well-being" was the key theme to emerge from an in-depth qualitative study that explored 19 individuals' experiences of commuting (many by bicycle) in Cambridge, England, using "photovoice" and follow-up photo-elicitation interviews (Guell and Ogilvie 2015).

Conversely, research linking daily driving time with adverse health consequences has found poorer self-rated health and quality of life and increased psychological distress and time stress associated with longer driving time (Ding et al. 2014). In contrast, there is growing evidence that cycling (and walking) is associated with better quality of life scores and reduced stress. In a study of inner-city residents in Sydney, Crane et al. (2014) found that cyclists had a significantly higher quality of life score compared with walkers, public transport users, and motor vehicle users. Frequent cycling was associated with higher physical and psychological quality of life in men but not women in this sample.

Other research has also found significant associations between overall psychological well-being and (a) active travel and public transport when compared to car travel, (b) time spent (per 10 minutes) walking and driving, and (c) switching from car travel to active travel (Martin, Goryakin, and Suhrcke 2014). Similarly, an analysis based on the American Time Use Survey's well-being module showed that cyclists were the happiest commuters, followed by car passengers and then car drivers (Morris and Guerra 2015).

### Social Health Benefits of Cycling as a Form of Physical Activity

The social health benefits of cycling fall into two broad areas encompassing (a) the health-enhancing social interactions that occur when people participate in physical activities with other people, and (b) the health-enhancing lifestyles associated with living in "healthy spaces and places" that provide supportive environments for active living and social engagement (National

Heart Foundation of Australia 2018; Project for Public Spaces 2016). The health benefits of living in human-scale urban environments that support cycling and walking and discourage car use are discussed here.

In contrast to the physical and psychological health benefits of cycling, research into the social health benefits of cycling is sparse. The majority of research in this field has focused on the relationship between social isolation and ill health. This relationship is now well established (Leigh-Hunt et al. 2017).

There is also some evidence that social interaction may be a motivator for commencing and maintaining physical activities. A systematic review of qualitative studies of independently living older adults' experiences of physical activity interventions in nonclinical settings found that the enjoyment of social interaction was a key motivation for being physically active and was likely to be an important motivation for maintaining physical activity (Devereux-Fitzgerald et al. 2016).

### Cycling as a Means of Reducing Health Inequalities

Social gradients in health status exist in most countries, with the poor experiencing substantially higher mortality and morbidity than the rich (Marmot 2001). Many factors contribute to health inequalities, including inequalities in health-enhancing behaviors such as physical activity. However, physical activity through active travel is often more equitably distributed in populations than leisure-time physical activity.

In countries such as the Netherlands, which have high rates of utilitarian cycling, cycling tends to be more evenly distributed across sociodemographic groups than in countries where recreational cycling is the more dominant form of cycling (see in this volume Buehler and Pucher, chapter 2; and Martens, Golub, and Hamre, chapter 14). Moreover, where differences exist, they tend to show the opposite social gradient of that for leisure-time physical activity in countries such as Australia, where physical activity is more prevalent in high-income groups. In the Netherlands, for example, cycling rates are quite high among lower-income groups, and education levels follow a U-shaped distribution, where cycling rates are highest for individuals with the highest and lowest educational levels (Ministry of Transport, Public Works, and Water Management 2009).

In addition to socioeconomically disadvantaged population groups, other population groups that have relatively low levels of leisure-time physical activity include women, adolescent girls, and older adults. In countries such as Germany, Denmark, the Netherlands, and Japan, a high proportion of young people cycle to school, women cycle as frequently as men, and, in some cases, cycling *increases* with age (see in this volume Buehler and Pucher, chapter 2; Garrard, chapter 11; McDonald, Kontou, and Handy, chapter 12; Garrard et al., chapter 13). These diverse population groups frequently achieve adequate levels of physical activity incidentally, at low cost, and without having to find the time and money to participate in organized sports or fitness programs. The socially inclusive, population-wide participation associated with active travel may help to explain the inverse relationship observed between walking and cycling rates and obesity levels internationally (Pucher et al. 2010; Bassett et al. 2008).

These differing sociodemographic patterns for leisure-time physical activity and cycling for transport across countries indicate that the establishment of cycling as an appealing, convenient, and safe form of transport could contribute to reducing inequalities in physical activity participation and hence inequalities in health. Health is generally better and more equitably distributed when all people have access to the conditions and environments that support health and health-enhancing behaviors such as cycling to get around as part of daily life (Turrell et al. 2013).

### The Potential Impact of Increases in Cycling on Physical Activity Prevalence

Rates of cycling at the population level are much higher in many European countries than in the United States, Canada, Australia, and the United Kingdom (see Buehler and Pucher, chapter 2, this volume). The potential public health benefits of cycling promotion are therefore high in countries where the population at risk (i.e., noncyclists) is very large. The challenge is to get those with bicycles to use them on a regular basis in order to make a public health impact on the prevalence of "sufficient activity" (defined as 150 minutes or more of moderate and/or vigorous activity per week).

In this section, we model data analyzed from an Australian population dataset in which 57% of adults met the recommendation for 150 minutes per week of at least moderate-intensity physical activity and 43% were "insufficiently active." The modeling assessed what difference it would make to the population prevalence of "sufficiently active" adults if a conservatively estimated 20% of people cycled more often. The results provided estimates of the prevalence of "sufficient activity" if this subgroup of people cycled once, twice, or three times a week for 20 minutes. Then, the analysis was confined to only those who were "insufficiently active" and 20% of them adopted these amounts of cycling activity (see figure 3.1).

Given a baseline of 57% "sufficiently active," if 20% more Australian adults cycled 20 minutes once, twice, or three times per week, the percentage who meet the recommended physical activity guidelines would increase to 59.0%, 60.5%, and 64.5%, respectively. If only inactive adults were targeted, the prevalence increases would be even larger, rising to 61.0%, 67.0%, and 72.0%, respectively. Because few population interventions have achieved anywhere near a 5% absolute increase in the proportion of people meeting physical activity guidelines, the public health potential for cycling is large, even at modest amounts and frequencies of bicycle usage, especially if currently inactive people start cycling. A similar analysis conducted in the county of Scania in Sweden also estimated substantial public health benefits associated with an increase in commuting to work by bicycle (Raustorp and Koglin 2019).

These estimates are similar for other developed countries with low rates of cycling in which around half the adult population is not meeting

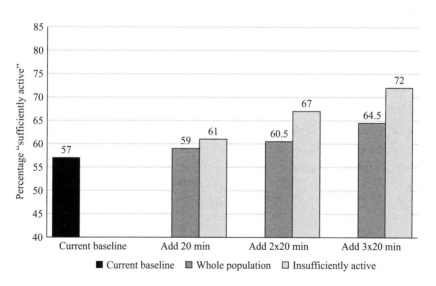

**Figure 3.1**
Effect of adding episodes of cycling each week on population level (%) of "sufficiently active" for 20% of the adult population and 20% of the adult population who are "insufficiently active." *Source*: Adapted from Armstrong et al. 2000.

minimal WHO physical activity guidelines for health. These hypothetical data are compelling: cycling is accessible, affordable, and achievable by people of all ages. Therefore, the challenge is to establish effective ways to increase cycling in the population.

### Health Benefits of Reduced Motor Vehicle Use

The health benefits of physical activity achieved when active cycling trips replace passive car trips were discussed earlier. In addition to increased physical activity, health benefits also flow from reduced motor vehicle use associated with increased bicycle travel mode share. These co-benefits include enhanced community livability (Badland et al. 2014), improved air quality, reduced noise pollution, reduced traffic injuries, and reduced greenhouse gas emissions and the health risks associated with climate change.

**Community Livability** Human-scale urban environments that support cycling and walking and discourage car use can improve social interactions and increase community attachment, livability, and amenity (Litman 2013; Litman 2017). Appleyard and Lintell's original research, which found that heavy traffic is associated with reduced street-based activities and social interactions between neighbors (Appleyard and Lintell 1980), has been replicated in other settings (Bosselmann and Macdonald 1999; Hart and Parkhurst 2011). There is also evidence that the more compact, permeable urban designs that support cycling and walking lead to crime reduction through increased street activity and "natural surveillance" (Hillier and Sahbaz 2006; Cozens and Love 2015).

**Social Inclusion** Cycling also contributes to social inclusion by providing an affordable and convenient form of personal mobility that is accessible to people who are not able to drive a motor vehicle. These groups include children and adolescents, older adults, and people on a low income. Transport costs (principally related to motor vehicles in countries such as the United States and Australia) account for a high proportion of household expenditures on goods and services. In Australia, transport costs (15%) comprise the third-largest category of household expenditures after housing (20%) and food (17%) (Australian Bureau of Statistics 2017). An investment in cycling provision is an investment in more equitable access to the multiple education, employment, economic, and community participation benefits of independent mobility, regardless of driving status. For example, building cycle paths within 5 km of all train stations and activity centers would make cycling more accessible and safer and reduce motor vehicle dependency and household transport expenditures, which would particularly benefit those experiencing housing affordability stress (Giles-Corti, Eagleson, and Lowe 2014).

**Air Pollution** Air pollution is the single largest environmental health risk in Europe, responsible for more than 400,000 premature deaths per year (IARC 2013), with motor vehicle emissions a major source of air pollution in most urban areas (European Environment Agency 2017). The health-related costs of air pollution in Europe range from €330 billion to €940 billion per year, depending on the valuation methodology (European Environment Agency 2017). A number of health impact assessment tools have been used to estimate the number of preventable premature deaths that would be achieved through compliance with international exposure recommendations for air and noise pollution in various urban settings (Mueller et al. 2017).

While a shift from motorized travel to cycling leads to an overall improvement in air quality, individual cyclists are exposed to health risks resulting from exposure to current levels of air pollution. The magnitude of these risks depends on several factors, including where and when people cycle and the intensity of effort, with more vigorous cycling increasing the respiration rate and hence the inhaled dose of air pollutants.

A number of studies have investigated differences in air pollution exposure and inhalation dose for different modes of transport, with a recent systematic review and meta-analysis reporting that while car commuters had higher exposure to all pollutants than active commuters did, active commuters had higher inhalation doses. However, overall, the benefits of physical activity resulting from active commuting were estimated to be larger than the risk from an increased inhaled dose of fine particles, with commuters using motorized transport losing up to one year of life expectancy more than cyclists (Cepeda et al. 2017). These findings are consistent with reductions in all-cause mortality and cardiovascular disease (for which air pollution is a risk factor) associated with cycling, as described earlier.

**Noise Pollution** In most urban areas, road traffic is the most important source of noise pollution, with motorized road traffic estimated to expose 40% of Europeans to daytime noise levels exceeding the WHO recommended threshold of 55 dB (Berglund, Lindvall, and Schwela 1999).

A systematic review and meta-analysis of the effects of road traffic noise exposure on the cardiometabolic systems reported that the most rigorous evidence, based on seven longitudinal studies, was for the impact of road traffic noise on ischemic heart disease, where the relative risk for the association between road traffic noise and the disease incidence was 1.08 (95% CI: 1.01–1.15) per 10 dB. A smaller number of studies reported on the association between traffic noise and stroke, diabetes, and/or obesity, and the evidence for these associations was considered less rigorous (rated from moderate to very low) (van Kempen et al. 2018).

The effects of road traffic noise on sleep have also been reviewed, with road traffic noise found to affect both objectively measured sleep physiology (awakening due to noise) and subjectively assessed sleep disturbance in adults (Basner and McGuire 2018).

The impact of air and noise pollution is greatest in dense urban centers. Cycling therefore offers significant potential to reduce these health risks, as these dense urban areas are also the most amenable to cycling because trip distances are likely to be shorter than in outer areas.

**Traffic Injuries**  Road transport injuries are responsible for a substantial burden of injury in both developing and developed countries. Motorized road transport results in 1.3 million deaths and 78.2 million nonfatal injuries requiring medical care annually, with the burden of injury increasing (Global Road Safety Facility, World Bank 2014). Over the last two decades, deaths from road crashes grew by 46%. Vulnerable road users are at increased risk of injury, with pedestrians alone accounting for 35% of road injury deaths globally and over 50% in some developing countries (Global Road Safety Facility, World Bank 2014).

As is the case for air pollution, cyclists are exposed to the risk of collision injury, but, depending on circumstances and conditions, *more* cycling (that replaces motorized trips) could either increase or decrease the risk of injury (see Elvik, chapter 4, this volume). The increased risk comes from increased exposure to motor vehicles, but a reduction in motor vehicle use because of increased cycling reduces cyclists' risk of injury by reducing their exposure to motor vehicles. The key question then becomes "What is the tipping point at which substitution of driving by cycling results in a net injury reduction benefit, including for cyclists?" This question has been addressed in a number of studies.

A review of relevant research concluded that increased cycling need not necessarily lead to increased cycling injuries, as the relationship is highly dependent on cycling conditions (see Elvik, chapter 4, this volume; Wegman, Zhang, and Dijkstra 2012). Road safety strategies, well-designed bicycle facilities, particularly physically separated networks, and reduced motorized traffic contribute to reducing injury risks for cyclists and may therefore lead to improved net safety.

Similarly, Elvik (2009) used the nonlinearity of injury risks for pedestrians and cyclists (i.e., the more pedestrians or cyclists there are, the lower the risk faced by each pedestrian or cyclist) to estimate that, for very large transfers of trips from motor vehicles to walking or cycling, a reduction in the total number of injuries is possible. Elvik states that, under these circumstances, the combined effects of "safety in numbers" for pedestrians and cyclists and a lower number of motor vehicles may lead to a lower total number of injuries. Stevenson et al. (2016) have argued, based on health impact assessment modeling, that healthier cities can be established through land-use policies that support a shift from motor vehicle travel to walking and cycling. They also stressed the importance of providing safe walking and cycling infrastructure to maximize the health benefits of a travel mode shift from passive to active transport.

While the results of these studies are not definitive and are highly context dependent, there is consistent evidence from several studies that the overall health benefits of cycling are substantially greater than the risks of both injury and exposure to air pollution (de Hartog et al. 2010; Mueller et al. 2015; Rojas-Rueda et al. 2012; Xia et al. 2015). A systematic review of comparative risk assessments and cost-benefit analyses of a mode shift from motorized to active transport (cycling and walking) estimated that health benefit-risk or benefit-cost ratios of a mode shift to active transport ranged between –2 and 360, with a median of 9. The effects of increased physical activity contributed most to the estimated health benefits, thus strongly outweighing the detrimental effects of traffic incidents and air pollution exposure on health (Mueller et al. 2015).

A recent Danish study provides additional evidence that an increase in cycling can occur without a concomitant increase in injury rates and that the health benefits of cycling substantially outweigh the injury risks. Andersen et al. (2018) reported that cycling trips increased by 10% in Denmark between 1998 and 2015, while the number of injured cyclists declined by 55%. They also estimated high benefit-hazard ratios of 238:1 for deaths and 20:1 for total cases (based on only the three common diseases of type 2 diabetes, CVD, and cancers) versus total injuries. They attributed the decline in cycling injuries to substantial improvements in cycling conditions, concluding that "the safety of cycling depends on how carefully infrastructure in all its details is planned and prioritized and is not necessarily related to traffic density" (Andersen et al. 2018).

**Greenhouse Gas Emissions and Climate Change**   Transport is a significant and growing source of the greenhouse gas emissions that contribute to climate change, with motor vehicle emissions comprising nearly one-quarter of the world's energy-related greenhouse gases (Xia et al. 2015). The environmental consequences of climate change, which include sea-level rise, degraded air quality, and extreme weather events, result in wide-ranging impacts on human health. These include the risks to human health associated with heat waves, wildfires, floods, malnutrition, and forced migration (Patz and Thomson 2018; Intergovernmental Panel on Climate Change 2014).

Cycling, as a zero-emission form of transport, offers some potential to lower emissions in the passenger transport sector. Unlike a number of high-tech options, bicycles are an equitable, off-the-shelf option that can be deployed immediately.

### Conclusions

Cycling has enormous potential to improve public health, particularly in cities and countries that currently have low levels of transport-related cycling. Active transport is consistently associated with meeting recommended levels of physical activity for health and may attenuate the commonly seen socioeconomic gradient in leisure-time physical activity participation. The potential for active travel to become one of the key solutions to the problem of physical inactivity is well recognized and deserves substantial advocacy and policy focus.

Reduced motor vehicle use also contributes to additional co-benefits in the form of improved air quality, lower noise levels, reduced road transport injuries, improved community livability, and reduced greenhouse gas emissions. When the multiple health and social benefits of a mode shift from private motor vehicle use to cycling are monetized, the economic argument for investing in increased cycling is powerful, with benefit-cost ratios substantially higher than for investments in alternative forms of transport infrastructure.

The goal of increasing cycling rates in urban areas is clearly achievable, with a number of both developed and developing countries successfully stalling or reversing unhealthy and unsustainable increases in single-occupant car travel for relatively short distances in urban areas. Cycling to get to places as part of daily life is well placed to make a substantial contribution to improving population health as well as to clean, green, and healthy cities of the future.

## References

Ainsworth, Barbara E., William L. Haskell, Stephen D. Herrmann, Nathanael Meckes, David R. Bassett Jr., Catrine Tudor-Locke, Jennifer L. Greer, Jesse Vezina, Melicia C. Whitt-Glover, and Arthur S. Leon. 2011. Compendium of Physical Activities: A Second Update of Codes and MET Values. *Medicine and Science in Sports and Exercise* 43(8): 1575–1581.

Andersen, L. B., A. Riiser, H. Rutter, S. Goenka, S. Nordengen, and A. K. Solbraa. 2018. Trends in Cycling and Cycle Related Injuries and a Calculation of Prevented Morbidity and Mortality. *Journal of Transport and Health* 9: 217–225.

Appleyard, Donald, and Mark Lintell. 1980. The Environmental Quality of City Streets: The Residents' Viewpoint. *Journal of the American Institute of Planners* 38: 84–101.

Armstrong, Tim, Adrian Bauman, and Joanne Davies. 2000. *Physical Activity Patterns of Australian Adults. Results of the 1999 National Physical Activity Survey.* Canberra: Australian Institute of Health and Welfare.

Australian Bureau of Statistics. 2017. *Household Expenditure Survey, Australia: Summary of Results, 2015–16.* Report 6530.0. Canberra: Australian Bureau of Statistics.

Badland, Hannah, Carolyn Whitzman, Melanie Lowe, Melanie Davern, Lu Aye, Iain Butterworth, Dominique Hes, and Billie Giles-Corti. 2014. Urban Liveability: Emerging Lessons from Australia for Exploring the Potential for Indicators to Measure the Social Determinants of Health. *Social Science and Medicine* 111: 64–73.

Basner, Mathias, and Sarah McGuire. 2018. WHO Environmental Noise Guidelines for the European Region: A Systematic Review on Environmental Noise and Effects on Sleep. *International Journal of Environmental Research and Public Health* 15(3): 1–45.

Bassett, David R., John Pucher, Ralph Buehler, Dixie L. Thompson, and Scott E. Crouter. 2008. Walking, Cycling, and Obesity Rates in Europe, North America, and Australia. *Journal of Physical Activity and Health* 5(6): 795–814.

Berglund, Birgitte, Thomas Lindvall, and Deitrich H. Schwela. 1999. *Guidelines for Community Noise*. Geneva: World Health Organization.

Bosselmann, Peter, and Elizabeth Macdonald. 1999. Livable Streets Revisited. *Journal of the American Planning Association* 65(2): 165–180.

Celis-Morales, Carlos A., Donald M. Lyall, Paul Welsh, Jana Anderson, Lewis Steell, Yibing Guo, Reno Maldonado, Daniel F. Mackay, Jill P. Pell, Naveed Sattar, and Jason M. R. Gill. 2017. Association between Active Commuting and Incident Cardiovascular Disease, Cancer, and Mortality: Prospective Cohort Study. *British Medical Journal* 357: j1456.

Cepeda, Magda, Josje Schoufour, Rosanne Freak-Poli, Chantal M. Koolhaas, Klodian Dhana, Wichor M. Bramer, and Oscar H. Franco. 2017. Levels of Ambient Air Pollution According to Mode of Transport: A Systematic Review. *Lancet Public Health* 2(1): e23–e34.

Chavarrias, Manuel, Jorge Carlos-Vivas, Daniel Collado-Mateo, and Jorge Perez-Gomez. 2019. Health Benefits of Indoor Cycling: A Systematic Review. *Medicina (Kaunas)* 55(8): 452.

Chekroud, Sammi R., Ralitza Gueorguieva, Amanda B. Zheutlin, Martin Paulus, Harlan M. Krumholz, John H. Krystal, and Adam M. Chekroud. 2018. Association between Physical Exercise and Mental Health in 1.2 Million Individuals in the USA between 2011 and 2015: A Cross-Sectional Study. *Lancet Psychiatry* 5(9): 739–746.

Cozens, Paul, and Terence Love. 2015. A Review and Current Status of Crime Prevention through Environmental Design (CPTED). *Journal of Planning Literature* 30(4): 393–412.

Crane, Melanie, Chris Rissel, Christopher Standen, and Stephen Greaves. 2014. Associations between the Frequency of Cycling and Domains of Quality of Life. *Health Promotion Journal of Australia* 25(3): 182–185.

Daley, Michelle, Chris Rissel, and Beverley Lloyd. 2007. All Dressed Up and Nowhere to Go? A Qualitative Research Study of the Barriers and Enablers to Cycling in Inner Sydney. *Road & Transport Research* 16(4): 42–52.

de Hartog, Jeroen Johan, Hanna Boogaard, Hans Nijland, and Gerard Hoek. 2010. Do the Health Benefits of Cycling Outweigh the Risks? *Environmental Health Perspectives* 118(8): 1109–1116.

Devereux-Fitzgerald, Angela, Rachael Powell, Anne Dewhurst, and David P. French. 2016. The Acceptability of Physical Activity Interventions to Older Adults: A Systematic Review and Meta-synthesis. *Social Science and Medicine* 158: 14–23.

Ding, Ding, Klaus Gebel, Philayrath Phongsavan, Adrian E. Bauman, and Dafna Merom. 2014. Driving: A Road to Unhealthy Lifestyles and Poor Health Outcomes. *PLoS One* 9(6): e94602.

Donnelly, Joseph E., Charles H. Hillman, Darla Castelli, Jennifer L. Etnier, Sarah Lee, Phillip Tomporowski, Kate Lambourne, and Amanda N. Szabo-Reed. 2016. Physical Activity, Fitness, Cognitive Function, and Academic Achievement in Children: A Systematic Review. *Medicine and Science in Sports and Exercise* 48(6): 1197–1222.

Dons, E., D. Rojas-Rueda, E. Anaya-Boig, I. Avila-Palencia, C. Brand, T. Cole-Hunter, A. de Nazelle, U. Eriksson, M. Gaupp-Berghausen, R. Gerike, S. Kahlmeier, M. Laeremans, N. Mueller, T. Nawrot, M. J. Nieuwenhuijsen, J. P. Orjuela, F. Racioppi, E. Raser, A. Standaert, L. Int Panis, and T. Götschi. 2018. Transport Mode Choice and Body Mass Index: Cross-Sectional and Longitudinal Evidence from a European-Wide Study. *Environment International* 119: 109–116.

Elvik, Rune. 2009. The Non-linearity of Risk and the Promotion of Environmentally Sustainable Transport. *Accident Analysis and Prevention* 41(4): 849–855.

European Environment Agency. 2017. *Air Quality in Europe—2017 Report*. Luxembourg: European Environment Agency.

Fishman, Elliot, Paul Schepers, and Carlijn Barbara Maria Kamphuis. 2015. Dutch Cycling: Quantifying the Health and Related Economic Benefits. *American Journal of Public Health* 105(8): e13–e15.

Flint, Ellen, and Steven Cummins. 2016. Active Commuting and Obesity in Mid-life: Cross-Sectional, Observational Evidence from UK Biobank. *Lancet Diabetes & Endocrinology* 4(5): 420–435.

Franklin, Barry A., Robert Brook, and C. Arden Pope III. 2015. Air Pollution and Cardiovascular Disease. *Current Problems in Cardiology* 40(5): 207–238.

Giles-Corti, Billie, Serryn Eagleson, and Melanie Lowe. 2014. *Securing Australia's Future—Sustainable Urban Mobility: The Public Health Impact of Transportation Decisions*. Melbourne: Australian Council of Learned Academies.

Global Road Safety Facility, World Bank. 2014. *Transport for Health: The Global Burden of Disease from Motorized Road Transport*. Seattle: Institute for Health Metrics and Evaluation.

Grontved, Anders, Robert W. Koivula, Ingegerd Johansson, Patrik Wennberg, Lars Ostergaard, Goran Hallmans, Frida Renstrom, and Paul W. Franks. 2016. Bicycling to Work and Primordial Prevention of Cardiovascular Risk: A Cohort Study among Swedish Men and Women. *Journal of the American Heart Association* 5(11): e004413.

Guell, Cornelia, and David Ogilvie. 2015. Picturing Commuting: Photovoice and Seeking Well-being in Everyday Travel. *Qualitative Research* 15(2): 201–218.

Hart, Joshua, and Graham Parkhurst. 2011. Driven to Excess: Impacts of Motor Vehicles on the Quality of Life of Residents of Three Streets in Bristol UK. *World Transport Policy & Practice* 17(2): 12–30.

Hillier, Bill, and Ozlem Sahbaz. 2006. *High Resolution Analysis of Crime Patterns in Urban Street Networks: An Initial Statistical Sketch from an Ongoing Study of a London Borough*. London: University College London.

Intergovernmental Panel on Climate Change. 2014. *Climate Change 2014: Impacts, Adaptation, and Vulnerability. Part B: Regional Aspects*. New York: Cambridge University Press.

International Agency for Research on Cancer (IARC). 2013. *Air Pollution and Cancer. IARC Scientific Publication No. 161*. Geneva: International Agency for Research on Cancer.

Kelly, Paul, Sonja Kahlmeier, Thomas Götschi, Nicola Orsini, Justin Richards, Nia Roberts, Peter Scarborough, and Charlie Foster. 2014. Systematic Review and Meta-analysis of Reduction in All-Cause Mortality from Walking and Cycling and Shape of Dose Response Relationship. *International Journal of Behavioral Nutrition and Physical Activity* 11(1): 132.

Kubesch, Nadine J., Jeanette Therming Jorgensen, Barbara Hoffmann, Steffen Loft, Mark J. Nieuwenhuijsen, Ole Raaschou-Nielsen, Marie Pedersen, Ole Hertel, Kim Overvad, Anne Tjonneland, Eva Prescot, and Zorana J. Andersen. 2018. Effects of Leisure-Time and Transport-Related Physical Activities on the Risk of Incident and Recurrent Myocardial Infarction and Interaction with Traffic-Related Air Pollution: A Cohort Study. *Journal of the American Heart Association* 7(15): e009554.

Leigh-Hunt, N., D. Bagguley, K. Bash, V. Turner, S. Turnbull, N. Valtorta, and W. Caan. 2017. An Overview of Systematic Reviews on the Public Health Consequences of Social Isolation and Loneliness. *Public Health* 152: 157–171.

Leyland, Louise-Ann, Ben Spencer, Nick Beale, Tim Jones, and Carien M. van Reekum. 2019. The Effect of Cycling on Cognitive Function and Well-being in Older Adults. *PLoS One* 14(2): e0211779.

Litman, Todd. 2013. *Evaluating Non-motorised Transportation Benefits and Costs*. Victoria: Victoria Transport Policy Institute.

Litman, Todd. 2017. *Community Cohesion as a Transport Planning Objective*. Victoria: Victoria Transport Policy Institute.

Mammen, George, and Guy Faulkner. 2013. Physical Activity and the Prevention of Depression: A Systematic Review of Prospective Studies. *American Journal of Preventive Medicine* 45(5): 649–657.

Marmot, Michael. 2001. Inequalities in Health. *New England Journal of Medicine* 345(2): 134–136.

Martin, Adam, Yevgeniy Goryakin, and Mark Suhrcke. 2014. Does Active Commuting Improve Psychological Wellbeing? Longitudinal Evidence from Eighteen Waves of the British Household Panel Survey. *Preventive Medicine* 69: 296–303.

Ministry of Transport, Public Works, and Water Management. 2009. *Cycling in the Netherlands*. The Hague: Ministry of Transport, Public Works, and Water Management.

Morris, Eric A., and Erick Guerra. 2015. Mood and Mode: Does How We Travel Affect How We Feel? *Transportation* 42(1): 25–43.

Mueller, Natalie, David Rojas-Rueda, Xavier Basagana, Marta Cirach, Tom Cole-Hunter, Payam Dadvand, David Donaire-Gonzalez, Maria Foraster, Mireia Gascon, David Martinez, Cathryn Tonne, Margarita Triguero-Mas, Antonia Valentin, and Mark Nieuwenhuijsen. 2017. Urban and Transport Planning Related Exposures and Mortality: A Health Impact Assessment for Cities. *Environmental Health Perspectives* 125(1): 89–96.

Mueller, Natalie, David Rojas-Rueda, Tom Cole-Hunter, Audrey de Nazelle, Evi Dons, Regine Gerike, Thomas Götschi, Luc Int Panis, Sonja Kahlmeier, and Mark Nieuwenhuijsen. 2015. Health Impact Assessment of Active Transportation: A Systematic Review. *Preventive Medicine* 76: 103–114.

National Heart Foundation of Australia. 2018. *Healthy Active by Design*. Melbourne: National Heart Foundation of Australia.

Nordengen, Solveig, Lars Bo Andersen, Ane K. Solbraa, and Amund Riiser. 2019a. Cycling Is Associated with a Lower Incidence of Cardiovascular Diseases and Death: Part 1—Systematic Review of Cohort Studies with Meta-analysis. *British Journal of Sports Medicine* 53(14): 870–878.

Nordengen, Solveig, Lars Bo Andersen, Ane K. Solbraa, and Amund Riiser. 2019b. Cycling and Cardiovascular Disease Risk Factors Including Body Composition, Blood Lipids and Cardiorespiratory Fitness Analysed as Continuous Variables: Part 2—Systematic Review with Meta-analysis. *British Journal of Sports Medicine* 53(14): 879–885.

Patz, Jonathan A., and Madeleine C. Thomson. 2018. Climate Change and Health: Moving from Theory to Practice. *PLoS Medicine* 15(7): e1002628.

Project for Public Spaces. 2016. *The Case for Healthy Places: Improving Health Outcomes through Placemaking*. New York: Project for Public Spaces.

Pucher, John, Ralph Buehler, David R. Bassett, and Andrew L. Dannenberg. 2010. Walking and Cycling to Health: A Comparative Analysis of City, State, and International Data. *American Journal of Public Health* 100(10): 1986–1992.

Rasmussen, Martin Gilles, Kim Overvad, Anne Tjonneland, Majken K. Jensen, Lars Ostergaard, and Anders Grontved. 2018. Changes in Cycling and Incidence of Overweight and Obesity among Danish Men and Women. *Medicine and Science in Sports and Exercise* 50(7): 1413–1421.

Raustorp, Johan, and Till Koglin. 2019. The Potential for Active Commuting by Bicycle and Its Possible Effects on Public Health. *Journal of Transport and Health* 13: 72–77.

Rebar, Amanda L., Robert Stanton, David Geard, Camille Short, Mitch J. Duncan, and Corneel Vandelanotte. 2015. A Meta-meta-analysis of the Effect of Physical Activity on Depression and Anxiety in Non-clinical Adult Populations. *Health Psychology Review* 9(3): 366–378.

Rojas-Rueda, D., A. de Nazelle, O. Teixido, and M. J. Nieuwenhuijsen. 2012. Replacing Car Trips by Increasing Bike and Public Transport in the Greater Barcelona Metropolitan Area: A Health Impact Assessment Study. *Environment International* 49:100–109.

Rosenbaum, Simon, Anne Tiedemann, Catherine Sherrington, Jackie Curtis, and Philip B. Ward. 2014. Physical Activity Interventions for People with Mental Illness: A Systematic Review and Meta-analysis. *Journal of Clinical Psychiatry* 75(9): 964–974.

Sallis, James F., Chad Spoon, Nick Cavill, Jessa K. Engelberg, Klaus Gebel, Mike Parker, Christina M. Thornton, Debbie Lou, Amanda L. Wilson, Carmen L. Cutter, and Ding Ding. 2015. Co-benefits of Designing Communities for Active Living: An Exploration of Literature. *International Journal of Behavioral Nutrition and Physical Activity* 12(1): 188.

Stevenson, Mark, Jason Thompson, Thiago Hérick de Sá, Reid Ewing, Dinesh Mohan, Rod McClure, Ian Roberts, Geetam Tiwari, Billie Giles-Corti, Xiaoduan Sun, Mark Wallace, and James Woodcock. 2016. Land Use, Transport, and Population Health: Estimating the Health Benefits of Compact Cities. *Lancet* 388(10062): 2925–2935.

Thompson Coon, J., K. Boddy, K. Stein, R. Whear, J. Barton, and M. H. Depledge. 2011. Does Participating in Physical Activity in Outdoor Natural Environments Have a Greater Effect on Physical and Mental Wellbeing Than Physical Activity Indoors? A Systematic Review. *Environmental Science and Technology* 45(5): 1761–1772.

Tseng, Chien-Ning, Bih-Shya Gau, and Meei-Fang Lou. 2011. The Effectiveness of Exercise on Improving Cognitive Function in Older People: A Systematic Review. *Journal of Nursing Research* 19(2): 119–131.

Turrell, Gavin, Michele Haynes, Lee-Ann Wilson, and Billie Giles-Corti. 2013. Can the Built Environment Reduce Health Inequalities? A Study of Neighbourhood Socioeconomic Disadvantage and Walking for Transport. *Health and Place* 19C(1): 89–98.

van Kempen, Elise, Maribel Casas, Goran Pershagen, and Maria Foraster. 2018. WHO Environmental Noise Guidelines for the European Region: A Systematic Review on Environmental Noise and Cardiovascular and Metabolic Effects: A Summary. *International Journal of Environmental Research and Public Health* 15(2): 379.

Wanner, Miriam, Thomas Götschi, Eva Martin-Diener, Sonja Kahlmeier, and Brian W. Martin. 2012. Active Transport, Physical Activity, and Body Weight in Adults: A Systematic Review. *American Journal of Preventive Medicine* 42(5): 493–502.

Weggemans, Rianne M., Frank J. G. Backx, Lars Borghouts, Mai Chinapaw, Maria T. E. Hopman, Annemarie Koster, Stef Kremers, Luc J. C. van Loon, Anne May, Arend

Mosterd, Hidde P. van der Ploeg, Tim Takken, Marjolein Visser, Wanda Wendel-Vos, and Eco J. C. de Geus. 2018. The 2017 Dutch Physical Activity Guidelines. *International Journal of Behavioral Nutrition and Physical Activity* 15(1): 58.

Wegman, Fred, Fan Zhang, and Atze Dijkstra. 2012. How to Make More Cycling Good for Road Safety? *Accident Analysis and Prevention* 44(1): 19–29.

Whitaker, Elizabeth D. 2005. The Bicycle Makes the Eyes Smile: Exercise, Aging, and Psychophysical Well-being in Older Italian Cyclists. *Medical Anthropology* 24(1): 1–43.

White, Rhiannon Lee, Mark J. Babic, Philip D. Parker, David R. Lubans, Thomas Astell-Burt, and Chris Lonsdale. 2017. Domain-Specific Physical Activity and Mental Health: A Meta-analysis. *American Journal of Preventive Medicine* 52(5): 653–666.

Woodcock, James, Ali Abbas, Alvaro Ullrich, Marko Tainio, Robin Lovelace, Thiago H. Sa, Kate Westgate, and Anna Goodman. 2018. Development of the Impacts of Cycling Tool (ICT): A Modelling Study and Web Tool for Evaluating Health and Environmental Impacts of Cycling Uptake. *PLoS Medicine* 15(7): e1002622.

World Health Organization (WHO). 1948. *WHO Definition of Health*. Geneva: World Health Organization.

World Health Organization (WHO). 2010. *Global Recommendations on Physical Activity for Health*. Geneva: World Health Organization.

World Health Organization (WHO). 2017. *Health Economic Assessment Tool (HEAT) for Walking and for Cycling: Methods and User Guide on Physical Activity, Air Pollution, Injuries and Carbon Impact Assessments*. Copenhagen: WHO, Regional Office for Europe.

Xia, Ting, Monika Nitschke, Ying Zhang, Pushan Shah, Shona Crabb, and Alana Hansen. 2015. Traffic-Related Air Pollution and Health Co-benefits of Alternative Transport in Adelaide, South Australia. *Environment International* 74: 281–290.

Zander, Alexis, Erin Passmore, Chloe Mason, and Chris Rissel. 2013. Joy, Exercise, Enjoyment, Getting Out: A Qualitative Study of Older People's Experience of Cycling in Sydney, Australia. *Journal of Environmental and Public Health* 2013: 6.

# 4 Cycling Safety

Rune Elvik

Cycling can significantly improve physical, mental, and emotional health (Garrard et al., chapter 3, this volume). Moreover, almost all scientific studies find that, on a population level, the health benefits of cycling far offset the health costs in terms of potential traffic injuries. Many countries and cities are seeking to encourage cycling and discourage car driving not only to promote public health but also to reduce roadway congestion, noise, air pollution, and energy use, and to slow down global warming and other harmful impacts of motor vehicle use.

There is widespread concern, however, about the actual and perceived traffic risks of cycling. As shown by Aldred and Crosweller (2015), the perceived dangers of cycling discourage it, especially among vulnerable groups (children and seniors) or risk-averse segments of the population (including most women). It is the risk not just of crashes that discourages cycling but also of near misses, where actual crashes are narrowly avoided (Fyhri et al. 2017). Improving cycling safety should be an important part of any strategy encouraging cycling.

This chapter describes current knowledge about cycling safety. It examines the following questions: How safe is cycling, and how can we know how safe it is? How serious are injuries typically sustained by cyclists, and what are their effects on health? Does the safety of cycling depend on the amount of cycling or the number of cyclists? What are the main factors related to characteristics of the cyclist and the bicycle that influence cycling safety? How can infrastructure be made safer for cycling? What are the impacts of helmets and helmet-use laws on cyclist safety?

## Cyclist Injury Risk

The safety of a certain mode of transport is often stated in terms of the number of fatalities or injured road users per million km of travel. Police reports are the most common source of data about injuries. The most common source of data about the amount of travel is household travel surveys. Estimates of the risk of injury based on these sources of data have been developed in many countries. Figure 4.1 shows fatality rates for cyclists in various countries, stated as the number of fatalities per 100 million km cycled. The rates are based on data for 2011–2016. Rates for the United States and Japan have been taken from Buehler and Pucher (chapter 2, this volume, figure 2.7). The other estimates are based on Castro, Kahlmeier, and Götschi (2018).

The risk of fatal injury varies considerably among the countries included in figure 4.1. The figure only includes countries that, according to Castro, Kahlmeier, and Götschi (2018), have data of high or very high quality about the amount of cycling and the number of fatalities. Therefore, the rates shown in figure 4.1 are comparable among countries. Rates tend to be lower in countries where the average annual distance cycled per inhabitant is longer than in countries where the average distance cycled per inhabitant is shorter. This pattern can have several explanations, some of which will be discussed later.

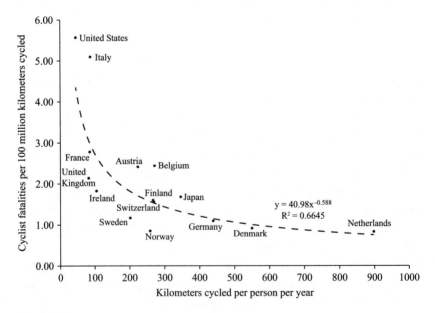

**Figure 4.1**
Cyclist fatality rates per 100 million km cycled compared with annual km cycled per capita in various countries, 2011–2016. *Sources*: Buehler and Pucher (chapter 2, this volume, figure 2.7); Castro, Kahlmeier, and Götschi (2018).

It is well known that many injuries to cyclists are not reported to the police and therefore not included in official crash statistics (Elvik and Mysen 1999; Langley et al. 2003; Watson, Watson, and Vallmuur 2015; Elvik and Sundfør 2017; Winters and Branion-Calles 2017; Shinar et al. 2018). In a study of long-term trends in bicycle fatalities in the Netherlands, Schepers et al. (2017a) show that even fatalities are not completely reported in police statistics. The reporting of slight injuries can be very low, less than 10%. Therefore, when estimating injury risk, it is essential to rely on a data source that can be assumed to have high coverage. In addition to police reports, two sources of data have been widely used: injury records kept by hospitals or other medical facilities and self-reports of injuries by cyclists.

Sweden collects data on traffic injuries both from the police and from hospitals, through the Swedish Traffic Accident Data Acquisition System. From 2010 to 2017, the average annual number of cyclists killed was 23. The estimated number of cyclists seriously injured was about 2,000 per year, almost 90 times as many as were killed (Trafikverket 2018), corresponding to a rate of one serious injury per million km cycled. In Swedish safety statistics, an injury is defined as serious if it results in a permanent medical impairment of at least 1%. The number of serious injuries is an estimated, not recorded, number. The definition of serious injury as any injury leading to permanent impairment was implemented in Sweden as a result of Vision Zero, which states that the ultimate goal is to prevent all injuries resulting in permanent impairment (Malm et al. 2008). However, the Swedish definition of serious injury differs greatly from that used in most other countries, which use the criterion of an overnight stay in the hospital. Most people would not consider an impairment level close to 1% as imposing a large restriction on daily life.

A study based on hospital discharge data in six European countries (Weijermars et al. 2018) assessed the burden of injury associated with injuries rated 3 or higher on an injury scale ranging from 1 (slight) to 6 (fatal). The burden of injury was stated in terms of years lived with disability (YLD). Years lived with disability is estimated by applying disability weights to injuries. A disability weight ranges from 0 (no disability) to 1 (dead). The larger the value, the larger the loss of function associated with an injury. For cyclists, the study found that the mean value of YLD was 0.26 for injuries that healed completely within one year and 8.8 for lifelong impacts of the injury. The latter estimate represents years lived with disability for the remaining life expectancy, which for traffic injury victims typically is around 40 years. Thus, it corresponds to an average disability weight of about 0.20–0.25 (8.8/40) for each remaining year of life. Another study (Weijermars, Bos, and Stipdonk 2016) found that the mean burden of serious injury, stated as YLD, was smaller for seriously injured cyclists than for other groups of seriously injured road users.

Table 4.1 presents estimates of cyclist injury risk based on studies made in different countries. Most studies listed in table 4.1 have relied on self-reported injuries. The studies of Blaizot et al. (2013), Bjørnskau (2017), and Nilsson et al. (2017) relied on injury records kept by hospitals or emergency clinics. The risk of injuries needing medical treatment is considerably lower than the overall risk of injuries. Most self-reported injuries are, in other words, too slight for the cyclist to seek medical treatment for them. The estimates in table 4.1 are intended to indicate the true risk when all injuries are included, even the slightest.

To indicate what these risks mean, consider a cyclist cycling 4,000 km per year, which is a little more than 10 km/day. If the risk of an injury requiring medical treatment is assumed to be 15 per million km cycled, the cyclist will, on average, sustain an injury requiring medical treatment once every 17 years. Such a risk of injury must be regarded as low. By comparison, about 10%–15% of the population seek medical treatment for any type of injury each year, corresponding to an expected incidence of one injury every 7 to 10 years (Lyons et al. 2011)—more often than most cyclists would experience.

The risk of cyclist injury varies depending on characteristics of the cyclist, the bicycle, and the traffic environment. There is huge variation around the risk estimates given in table 4.1.

### Safety in Kilometers and Safety in Numbers

The injury risks listed in table 4.1 are not constant. They depend on many factors, and this section deals with the amount of cycling performed by each cyclist and the number of cyclists in the traffic system. Both are indicators of exposure and show, as is widely accepted in modern accident research (Hauer 1995), that injury rates are not constant with respect to exposure but depend on exposure in a highly nonlinear fashion. Other risk factors, including the quality of infrastructure, are dealt with in subsequent sections.

The exposure of each cyclist can be stated in terms of the number of trips made or the number of kilometers cycled. Based on self-reported injuries

**Table 4.1**

Cyclist injury rates per million km cycled

| Study | Location | All self-reported injuries | | | Injuries requiring medical treatment | | |
|---|---|---|---|---|---|---|---|
| | | All cyclists | Females | Males | All cyclists | Females | Males |
| Aultman-Hall and Hall (1998) | Ottawa, Canada | 76 | | | 11 | | |
| Aultman-Hall and Kaltenecker (1999) | Toronto, Canada | 116 | | | 10 | | |
| Hoffman et al. (2010) | Portland, Oregon, United States | 93 | | | 24 | | |
| De Geus et al. (2012) | Brussels, Belgium | 47 | 62 | 45 | | | |
| Blaizot et al. (2013) | Rhone, France | | | | 11 | 12 | 11 |
| Palmer (2014) | Tasmania, Australia | 52 | 88 | 42 | 16 | 16 | 15 |
| Poulos et al. (2015) | New South Wales, Australia | 148 | 249 | 123 | 23 | 37 | 20 |
| Bjørnskau (2017) | Oslo, Norway | | | | 8 | 8 | 8 |
| Nilsson et al. (2017) | Stockholm, Sweden | | | | 5 | | |
| Sundfør (2017) | Norway | 62 | 54 | 65 | 16 | 14 | 17 |

and self-reported distance cycled per year, Elvik and Sundfør (2017) developed the relationships shown in figure 4.2. The curves show how the injury rate per million km cycled is related to the annual distance cycled per cyclist.

Similar results suggesting that injury risk declines as each cyclist makes more or longer trips have been reported by Rodgers (1997), Heesch, Garrard, and Sahlqvist (2011), and Schepers (2012). Published studies generally refer to these relationships as "safety in trips" or "safety in kilometers": the more trips or the more kilometers cycled, the safer each new trip or each additional kilometer becomes. The most likely explanation for this tendency is that skill improves the more you cycle.

A similar but different relationship that has attracted considerable interest from researchers recently is "safety in numbers." It refers to the tendency for the risk of injury for each cyclist to decline the more cyclists there are in a traffic system. Nearly all studies confirming the relationship of safety in numbers are multivariate crash prediction models that include terms for cyclist volume, motor vehicle volume, and one or more other

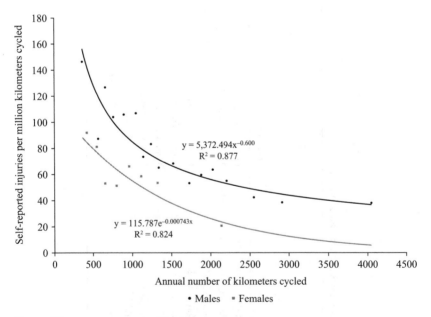

**Figure 4.2**
Declining cyclist injury rates per million km cycled as annual km cycled per cyclist increases, showing gender differences in the relationship. *Source*: Elvik and Sundfør (2017).

variables influencing the number of crashes in which both a cyclist and a motor vehicle are involved.

Some studies rely on data about bicycle volume only. This includes, for example, the classic study by Jacobsen (2003), which launched the concept of safety in numbers. In these studies, risk is often studied as a function of exposure per inhabitant.

Unfortunately, what looks like a safety-in-numbers effect in these studies may be only a statistical artifact (Elvik 2013). The potential statistical artifact arises as a result of how exposure and risk are defined. Risk is measured as injuries divided by kilometers; exposure is measured as kilometers divided by population. If the number of kilometers of cycling increases, then risk declines and exposure increases, generating precisely the negative relationship between the variables that is seen in figure 4.1.

Except for single-bicycle crashes, the number of bicycle crashes depends both on bicycle volume and on motor vehicle volume. An increase in the number of motor vehicles can increase the risk to cyclists dramatically. This can be seen clearly in historical data for the Netherlands discussed by Schepers et al. (2017b). Until the early 1960s, kilometers cycled exceeded kilometers driven by car in the Netherlands. During that period, the cyclist fatality rate (fatalities per billion km cycled) remained stable. Then, car kilometers started to overtake bicycle kilometers, and the cyclist fatality rate increased sharply until about 1975, when it stabilized. After about 1980, the cyclist fatality rate in the Netherlands declined by about 80%. Thus, cyclist injury models should include both bicycle and motor vehicle volumes as explanatory variables.

If the estimated variable coefficients for bicycle volume and motor vehicle volume have positive values between 0 and 1, that indicates a less than proportional increase in the number of crashes (e.g., traffic goes up by 40%, but crashes only go up by 15%). Coefficients with values between 0 and 1 show that there is safety in numbers. A recent meta-analysis (Elvik and Goel 2019) estimated weighted mean values of the coefficients of about 0.45 for motor vehicles and 0.40 for cyclists for crashes involving both cyclists and motor vehicles. All coefficient estimates for cyclists indicated a safety-in-numbers effect.

This means that an increase in the number of cyclists will be associated with an increase in the number of cyclist injuries but at a slower rate than the increase in the number of cyclists. A coefficient of 0.40 implies that doubling the number of cyclists would be associated with an increase of about 32% in the number of injured cyclists, all else equal. Studies of the safety in numbers hypothesis show only statistical relationships, and thus do not prove causation. Moreover, the studies are cross sectional with a single exception (Aldred et al. 2017) and do not show changes that may occur in a traffic system over time if cycling increases or better facilities are provided for cyclists. Some studies examining such changes over time are reviewed later in this chapter in the section on cycling infrastructure.

## Risk Factors Related to Cyclists and Type of Bicycle

Many risk factors influence the safety of cycling. Kilometers cycled was discussed in the section about safety in numbers and kilometers. This section reviews age, gender, experience, conspicuity, and type of bicycle. Cycling infrastructure is discussed later, in the infrastructure section of this chapter, as well as by Furth (chapter 5, this volume), whose chapter is devoted entirely to an analysis of cycling infrastructure.

The variation in cyclist injury risk by age has been studied by Heesch, Garrard, and Sahlqvist (2011), Hollingworth, Harper, and Hamer (2015), Poulos et al. (2015), Sundfør (2017), Buehler and Pucher (2017), and Feleke et al. (2018). The findings of those studies have been inconsistent, but they all show that the risk of fatal injury per kilometer cycled is highest for cyclists aged 65 years or older.

Estimates of injury risk by gender are shown in table 4.1. Most of these estimates indicate that females have a higher risk of injury than males, but they do not control for other characteristics influencing risk. Crude comparisons may be misleading. The risk of injury depends on experience and the distance cycled, and women tend to cycle shorter distances than men do. In studies controlling for experience and/or distance cycled, four out of five found that women have lower injury risk than men (Heesch, Garrard, and Sahlqvist 2011; Tin Tin, Woodward, and Ameratunga 2013; Hollingworth, Harper, and Hamer 2015; Sundør 2017). Only the study by Poulos et al. (2015) found a higher injury risk for women than for men. Thus, the evidence is inconsistent, but a majority of studies indicate a lower risk for women than for men.

The risk of injury (injuries per trip or per kilometer) is negatively correlated with the number of bike trips per week (Heesch, Garrard, and Sahlqvist 2011) and with the distance cycled per week (Hollingworth, Harper, and

Hamer 2015) or per year (Sundfør 2017). Experience, defined as the number of years one has been an active cyclist, reduces injury risk, although not all studies have found a clear dose-response relationship (i.e., the longer the experience, the lower the injury risk) (Heesch, Garrard, and Sahlqvist 2011; Hollingworth, Harper, and Hamer 2015; Poulos et al. 2015).

Cyclists on roadways can easily be overlooked by motorists. Measures enhancing their conspicuity improve safety. In two randomized controlled trials in Denmark (Madsen, Andersen, and Lahrmann 2013; Lahrmann et al. 2018), cyclists tested a permanent running bicycle light (at all times of day) and a jacket with retroreflective coating. The running bicycle light reduced multiparty injury crashes in the daytime (i.e., crashes involving both the cyclist and another road user) by 47%. The jacket reduced daytime multiparty injury crashes by 38%. These studies are unique in applying an experimental design to control for confounding factors.

In Europe, electric bikes are mostly pedelecs fitted with a battery giving support when pedaling, and the two terms are used as synonyms in this chapter. They have become quite popular in recent years. The cyclist has to pedal, but the battery gives support up to a speed of 25 km/h. That motorized support makes it considerably easier to cycle uphill. In an early study, Schepers et al. (2014) concluded that the risk of injury was higher for pedelecs than for conventional bikes. An update of that study concluded there was no difference in injury risk (Schepers, Klein Wolt, and Fishman 2018). Sundfør (2017) found a statistically nonsignificant increase in injury risk of 15% in a logistic regression model that controlled for a large number of potentially confounding factors. Based on survey data for Norway, Fyhri, Johansson, and Bjørnskau (2019) estimated that the odds ratio for a crash using a pedelec was 1.40 (95% confidence interval [CI] = 1.02 – 1.89). However, closer examination of the data found that the risk of a crash increased for women only. The odds ratio for women was 2.36 (95% CI = 1.12 – 4.60); the odds ratio for men was 0.67 (95% CI = 0.31 – 1.44).

Two studies found that injuries were, on average, more severe for pedelecs than for conventional bicycles (Weber, Scaramuzza, and Schmitt 2014; Gehlert 2017). Another study found that injury severity in pedelec crashes increased with increasing speed (Hertach et al. 2018). E-bikes travel at a slightly higher speed than conventional bikes. Based on GPS data for more than 50,000 cycle trips in Oslo, Flügel et al. (2017) developed models to explain how the speed of cycling relates to gender, trip purpose, and type

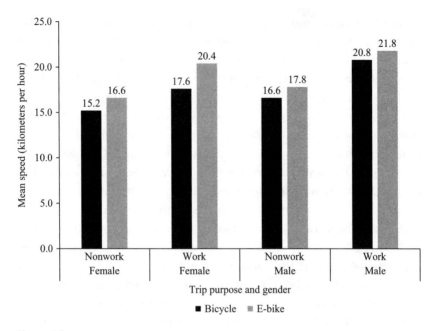

**Figure 4.3**
Average speed of cyclists in Oslo by gender, trip purpose, and type of bicycle. *Source*:
Flügel et al. (2017).

of bicycle. Figure 4.3 shows a comparison of average speeds between conventional bikes and e-bikes, by gender and trip purpose.

There is a clear, systematic pattern. Males cycle faster than females, irrespective of trip purpose and type of bicycle. Both genders cycle faster on trips to work than on trips with other purposes. E-bike riders cycle faster than conventional bike riders, irrespective of gender and trip purpose. Therefore, cyclists definitely take advantage of the support given by the battery-powered motor to increase speed.

### Infrastructure to Improve Cycling Safety

Three types of studies indicate how infrastructure influences cycling safety. (For a detailed analysis of the various kinds of cycling infrastructure, see Furth, chapter 5, this volume). The first type of study is based on data referring to individual cyclists and indicates the relative safety of cycling on different types of infrastructure. The second type of study evaluates the safety impact of new cycling infrastructure. These are often before-and-after studies, normally relying on data reported by police. The third type of study uses aggregate longitudinal data from cities that have upgraded bicycle infrastructure. Evidence from all three types of studies is reviewed here.

The risk to cyclists when riding on different types of infrastructure has been studied by Rodgers (1997), Aultman-Hall and Hall (1998), Aultman-Hall and Kaltenecker (1999), Lusk et al. (2011), Minikel (2012), Teschke et al. (2012), Harris et al. (2013), Hollingworth, Harper, and Hamer (2015), and Poulos et al. (2015). These studies have found that the injury risk is about 30%–40% lower when cycling in a bike lane, on a bike path (in some studies, referred to as a cycle track), or in traffic-calmed streets with few cars and low speed than when cycling in mixed traffic on major streets.

Three recent literature surveys (Mulvaney et al. 2015; DiGioia et al. 2017; Høye 2017) summarize studies that have evaluated the safety effects of new infrastructure. Mulvaney et al. (2015) concluded that "generally, there is a lack of high-quality evidence to be able to draw firm conclusions as to the effect of cycling infrastructure on cycling collisions." DiGioia et al. (2017) surveyed 19 studies of 22 different cycle treatments. A formal synthesis of the evidence from these studies was not feasible.

Bike boulevards are specially designed, interconnecting, bike-friendly routes on traffic-calmed neighborhood streets. Such bike boulevards, sometimes called neighborhood greenways, were associated with reduced risk of injury. Findings were mixed for unprotected but separate bike lanes, cycle tracks (protected, separate, on-street bike lanes), and bike boxes (advanced stop lines for cyclists). See Furth (chapter 5, this volume) for detailed infrastructure definitions. Raised cycle crossings (cycle tracks that are raised and act as speed humps to slow crossing traffic) were associated with reduced injury risk, but only one evaluation study was found (51% reduction; Schepers et al. 2011).

Høye (2017) reviewed several infrastructure treatments for cyclists. For bike lanes, a meta-analysis of seven evaluation studies found that they are associated with a reduction of 53% (95% CI = 66%–36% reduction) of cyclist injuries. Study findings regarding the safety effect of cycle tracks were inconsistent, but most studies indicated a reduced risk of injury. Bike boxes, where cyclists can stop in front of motor vehicles at intersections to increase their visibility and get an early start when the green traffic signal comes on, appear to reduce risk, but the decline was not statistically significant.

The third type of study about the relationship between infrastructure and risk of injury is longitudinal studies in cities or countries where extensive programs have been introduced to promote cycling. Buehler et al. (2011) studied long-term trends in cycling safety in the greater Washington, DC, area. From 1990 to 2005–2009, commuter cycling increased by a factor of 1.9 (i.e., nearly doubled). The combined length of bike lanes increased from 3 mi in 2001 to 60 mi in 2011 in Washington, DC, and from 3 mi in 1995 to 29 mi in 2011 in Arlington County. The fatality rate per 1 million bike commuters was reduced by 61% from 1998–2001 (annual average) to 2006–2009 (annual average). In the same period, the rate of serious injury per 100,000 bike commuters fell by 42%. The overall injury rate fell by 40% from 1998 to 2009. The total number of injuries was unchanged.

As reported by Pucher, Parkin, and de Lanversin (chapter 17, this volume), cycling in London increased by 173% from 2001 to 2014. The network of separated cycling facilities in London (including on-street lanes and off-street paths) increased fivefold from 2001 to 2017, from 435 km to 2,179 km. That expansion was accompanied by a dramatic change in the conceptual approach to the provision of cycle infrastructure in London, as reflected in changes to the London Cycling Design Standards—first published in 2005 and updated in 2016 (Transport for London 2016)—and the much greater commitment by the city government to promote cycling. Mayor Ken Livingstone's London Cycling Action Plan (Transport for London 2004) identified 900 km of a London Cycle Network Plus, which was designed to provide greater connectivity within the cycle network. Livingstone's successor, Mayor Boris Johnson, developed a number of "super-highways," but these were no more than cycle lanes on main roads, colored blue. These blue lanes offered only a minimal enhancement in level of service to cyclists, and their shortcomings were soon recognized. From 2001 to 2014, the risk of fatal injury was reduced by 68% and the risk of serious injury was reduced by 48%. The number of killed or seriously injured cyclists did not increase. The risk of slight injury declined by 13%; however, the number of slightly injured cyclists in London increased by 65% from 2001 to 2014. Overall, cycling in London became safer from 2001 to 2014, particularly with respect to the risk of fatal or serious injury. (For a detailed case study of cycling in London, see Pucher, Parking, and de Lanversin, chapter 17, this volume.)

Pedroso et al. (2016) report on the expansion of bicycle facilities in Boston. From 2007 to 2014, the total length of the bike lane network increased from 0.034 mi to 92.2 mi. Bicycle riding also increased markedly, from 0.9% of commuters in 2005 to 2.4% in 2014. The total number of crashes increased from 357 in 2009 to 488 in 2012 but far less than the increase in bike trips. From 2009 to 2012, the probability of getting injured when involved in a bike crash was reduced by 14% (from 82.7% to 74.6%).

Pucher and Buehler (2016) summarized longitudinal data on cycle network, amount of cycling, and cyclist risk in 10 cities in the United States. Between 2000 and 2015, the bikeway network in these 10 cities grew by a range of 17% to 381%. The number of bike trips increased by a range of 51% to 391%. The rate of fatal and serious injuries per 100,000 trips declined by a range of 43% to 79%. The decline in the injury rate most likely is a combined effect of better infrastructure and safety in numbers.

Marqués and Hernández-Herrador (2017) report on the expansion of cycle tracks in Seville, Spain. From 2006 to 2013, the total length of cycle tracks increased from 12 km to 152 km. The number of bike trips increased from 3.055 million in 2006 to 16.333 million in 2013. In the same period, injury risk was reduced by 67%. (For a detailed case study of cycling in Seville, see Geller and Marqués, chapter 19, this volume.)

New York City (Trottenberg, O'Neill, and Bassett 2017) has also promoted cycling in recent years. From 1996 to 2016, the cycle network, which consists of protected lanes (lanes that have bollards to prevent cars from encroaching on them), conventional bike lanes, and signed or marked routes expanded from 250 lane mi to 1,100 lane mi. The number of cycling trips increased from an average of 51 million per year in 1996–2000 to 134 million per year in 2011–2015. Despite this large increase in cycling, both the number of fatalities and the number of serious injuries went down. The cyclist fatality rate (per 100 million trips) was reduced by 71% from 1996–2000 to 2011–2015. The risk of serious injury (per 100 million trips) was reduced by 73%. (For a detailed case study of cycling in New York City, see Pucher, Parking, and de Lanversin, chapter 17, this volume.)

A recent study by Marshall and Ferenchak (2019) examines why cities with high bicycling rates are safer for all road users. The study was based on longitudinal data covering a period of 13 years for 12 cities. During this period, cycling for commuting increased in 9 of the 12 cities, with the largest increase a little more than 500%. The traffic fatality rate (fatalities per 100,000 inhabitants) declined in all cities, with the largest decline (64%) in the city where cycling increased the most (505%). The density of protected or separated cycling infrastructure in each city (density in hundreds of feet per square mile) had the largest influence on fatality rates. Increasing the density of cycling infrastructure from 0 to 100 ft/mi$^2$ was associated with a reduction in total fatalities (for all road user groups) of 55%, controlling for a large number of other variables influencing cycling safety. This study clearly shows the importance of providing separate or protected infrastructure to improve cycling safety. Marshall and Ferenchak concluded that the quality of infrastructure was more important for cyclist safety than the safety-in-numbers effect.

These examples consistently show that providing separate cycling facilities (lanes or tracks) increases cycling levels, while reducing the risk of cyclist injury. The reduction in risk is particularly large for fatal injury: 60%–70% in the cities examined in the study. A Dutch study (Schepers et al. 2013) confirms that the higher the degree of separation between cyclists and motor vehicles, the lower the number of killed or seriously injured cyclists.

### Effects of Bike Helmets and Bike Helmet Laws

The best available evidence on the effects of bike helmets is provided by two recent meta-analyses of studies that have evaluated these effects. Olivier and Creighton (2016) summarized evidence from 40 studies, whereas Høye (2018a) summarized evidence from 55 studies. Both studies employed state-of-the-art techniques for meta-analysis. The main findings of the studies are listed in table 4.2. Effects are stated as odds ratios. An odds ratio of 0.40 corresponds to an injury reduction of 60%.

According to these studies, wearing a bike helmet reduced fatal head injuries by 65%–70%, serious head injuries by 60%–70%, and by about 50% any head injuries (severity not specified). The reduction in facial injuries was much less, about 23%–33%. Wearing a bike helmet did not significantly reduce the number of neck injuries. Both studies (Olivier and Creighton 2016; Høye 2018a) discussed whether the findings were biased by various confounding factors but concluded that was unlikely. There is therefore no doubt that helmets protect cyclists from head injuries.

In most countries, the wearing of bike helmets is voluntary. Whether the use of helmets should be made mandatory remains controversial. A recent review by Høye (2018b) summarizes evidence on both the intended effects and possible unintended effects of bicycle helmet laws.

Table 4.2
Summary estimates of effects of cycle helmets in two recent meta-analyses

| | Summary estimates of odds ratios (95% confidence intervals in parentheses) | |
|---|---|---|
| Type of injury | Olivier and Creighton (2016) | Høye (2018a) |
| Any fatal injury | | 0.47 (0.37–0.76) |
| Fatal head injury | 0.35 (0.14–0.88) | 0.29 (0.15–0.56) |
| Serious head injury | 0.31 (0.25–0.37) | 0.40 (0.35–0.46) |
| Head injury of unspecified severity | 0.49 (0.42–0.57) | 0.52 (0.46–0.59) |
| Traumatic brain injury | | 0.47 (0.36–0.61) |
| Face injury | 0.67 (0.56–0.81) | 0.77 (0.67–0.88) |
| Neck injury | 0.96 (0.74–1.25) | 1.05 (0.87–1.26) |

For helmet laws that apply to all cyclists, the Høye (2018b) meta-analysis found that the number of head injuries to cyclists was, on average, reduced by 20%. Serious head injuries were reduced by 55%. Both effects were found to be statistically significant at the 5% level. Helmet laws that apply only to children were associated with a 20% reduction in the number of head injuries to children. Serious head injuries to children were reduced by 33%. The estimated effects were statistically significant. Although these results are based on only a few studies of impacts in some specific cities and countries, the studies suggest that bike helmet laws may achieve the intended effect of reducing head injuries to cyclists—at least among those not discouraged from cycling at all because of the helmet laws.

It has been claimed, for example, that bike helmet laws deter cycling and thus reduce the public health benefits associated with cycling (Robinson 2006). After reviewing several studies, Høye (2018b) concluded that the evidence is mixed: a helmet law may reduce the number of cyclists but does not necessarily do so. Radun and Olivier (2018) argue that the helmet-wearing law in Finland did not deter cyclists. Contradicting that claim, however, their own data actually show a 21% decline in mean cycling distance per person per day, from 0.92 km in 1998–1999 (before the helmet law) to 0.73 km in 2010–2011 (with the helmet law). Their claim that the helmet-wearing law did not deter cycling is based on the fact that only a few respondents stated in a survey that having to wear a helmet was an important obstacle to cycling. It should be kept in mind that Finland differs from other countries and that helmet wearing may be considerably less comfortable in countries with a warmer climate than Finland.

Another potential unintended effect of a helmet-wearing law is behavioral adaptation. The protection afforded by the helmet may lead cyclists to cycle faster or otherwise reduce their safety margins. Most of the studies reviewed by Høye (2018b) found no evidence of behavioral adaptation to helmets. There is, however, evidence of self-selection in helmet use that almost certainly distorts the results of most studies. For example, cyclists wearing helmets have been found to take fewer risks than cyclists not wearing helmets. For example, helmet users are less likely to cycle under the influence of alcohol or drugs, more likely to wear brighter, reflective clothing, less likely to commit traffic violations, and less likely to ride mechanically defective bicycles. Thus, when helmet wearing is voluntary, it seems highly likely that the most risk-averse cyclists will be the first to start wearing helmets and will wear them more frequently.

Helmet use is only one of many factors influencing cycling safety. Indeed, 12 of the chapters in this book provide detailed evidence and case studies of the broad range of measures needed to promote safer cycling (Buehler and Pucher, chapter 2; Furth, chapter 5; Heinen and Handy, chapter 7; Garrard, chapter 11; McDonald, Kontou, and Handy, chapter 12; Garrard et al., chapter 13; Martens, Golub, and Hamre, chapter 14; Pucher et al., chapter 15; Pardo and Rodriguez, chapter 16; Pucher, Parkin, and de Lanversin, chapter 17; Koglin, te Brömmelstroet, and van Wee, chapter 18; Geller and Marqués, chapter 19). Moreover, as noted previously in this chapter, cycling infrastructure expansion and improvements have greatly reduced fatal and serious injuries in many cases, sometimes by more than 70%. A key finding in most studies is that mixing cyclists and motor vehicles greatly increases the risk of fatal and serious cyclist injury. The more cyclists can be protected from mixing with motor vehicles, the safer cycling becomes. Thus, there is a very strong argument for providing separate and protected bicycling infrastructure.

## Discussion and Conclusions

Transport policy is changing in many parts of the world. It is no longer obvious that policy should adapt to increased use of cars by expanding road capacity. On the contrary, it is a policy objective in many countries and cities to reduce car dependence and increase cycling, walking, and travel by public transport. This raises the issue of whether increasing cycling and improving the safety of cycling can go hand-in-hand. Can both objectives be attained? More specifically, can there be a large increase in cycling without an increase in fatal and serious injuries to cyclists? The scientific evidence reviewed in this chapter suggests that the answer is yes.

There is, first of all, safety in kilometers and safety in numbers. The more often or the longer you cycle, the lower the risk per trip or per kilometer. The more bicyclists there are in traffic, the safer it is to cycle.

Second, there are some simple, cheap, and underutilized safety devices for cyclists, such as reflective clothing and a daytime running bicycle light, both of which have been found in rigorously controlled experiments to reduce the risk of injury.

Third, wearing a helmet is protective and is a wise precaution to take even if it is not mandatory. Whether it ought to be made mandatory is a political issue this chapter does not try to answer or take sides on.

Fourth, experience shows that a long-term commitment to improving infrastructure for cyclists, in particular by physically separating them from motor vehicles, improves the safety of cycling.

Fifth, in cities where cycling has increased greatly—as much as·five-fold in some cities—it is not always the case that the number of fatalities and serious injuries also increased (see Buehler and Pucher, chapter 21, this volume, table 21.1). For example, New York City, London, Paris, Copenhagen, and Amsterdam have experienced large increases in cycling but only minimal increases or even decreases in fatalities and serious injuries (see in this volume Pucher, Parkin, and de Lanversin, chapter 17; and Koglin, te Brömmelstroet, and van Wee, chapter 18).

It is therefore possible to increase cycling and improve its safety at the same time. However, to successfully combine increased cycling and increased cycling safety, there must be a long-term policy of providing safer infrastructure for cyclists. One cannot gamble on safety in numbers to do the trick by itself. There is no magic here. Cycling safety is improved as a result of infrastructure programs, and policies consisting of measures that are known to improve cycling safety, crucially including measures aimed at reducing or slowing down motor vehicle use or eliminating it altogether, such as car-free zones.

## References

Aldred, Rachel, and Sian Crosweller. 2015. Investigating the Rates and Impacts of Near Misses and Related Incidents among UK Cyclists. *Journal of Transport and Health* 2(3): 379–393.

Aldred, Rachel, Rahul Goel, James Woodcock, and Anna Goodman. 2017. Contextualising Safety in Numbers: A Longitudinal Investigation into Change in Cycling Safety in Britain, 1991–2001 and 2001–2011. *Injury Prevention*, 25(3): 236–241.

Aultman-Hall, Lisa, and Fred L. Hall. 1998. Ottawa-Carleton Commuter Cyclist On- and Off-Road Incident Rates. *Accident Analysis and Prevention* 30(1): 29–43.

Aultman-Hall, Lisa, and M. Georgina Kaltenecker. 1999. Toronto Bicycle Commuter Safety Rates. *Accident Analysis and Prevention* 31(6): 675–686.

Bjørnskau, Torkel. 2017. Sykkel i Oslo—eksponering, ulykker og risiko. Arbeidsdokument 51154. Oslo: Transportøkonomisk institutt.

Blaizot, Stefanie, Francis Papon, Mohamed Mouloud Haddak, and Emanuelle Amoros. 2013. Injury Incidence Rates of Cyclists Compared to Pedestrians, Car Occupants and Powered Two-Wheeler Riders, Using a Medical Registry and Mobility Data, Rhône County, France. *Accident Analysis and Prevention* 58(9): 35–45.

Buehler, Ralph, Andrea Hamre, Dan Sonenklar, and Paul Goger. 2011. *Trends and Determinants of Cycling in the Washington, DC Region*. Final report, contract DTRT07-G-0003. Arlington, VA: Virginia Tech Research Center, Department of Urban Affairs and Planning.

Buehler, Ralph, and John Pucher. 2017. Trends in Walking and Cycling Safety: Recent Evidence from High-Income Countries, with a Focus on the United States and Germany. *American Journal of Public Health* 107(2): 281–287.

Castro, Alberto, Sonja Kahlmeier, and Thomas Götschi. 2018. Exposure-Adjusted Road Fatality Rates for Cycling and Walking in European Countries. Discussion paper. ITF Roundtable 168. Paris: International Transport Forum.

De Geus, B., G. Vandenbulcke, L. Int Panis, I. Thomas, B. Degraeuwe, E. Cumps, J. Aertsens, R. Torfs, and R. Meeusen. 2012. A Prospective Cohort Study on Minor Accidents Involving Commuter Cyclists in Belgium. *Accident Analysis and Prevention* 45: 683–693.

DiGioia, Jonathan, Kari Edison Watkins, Yanzhi Xu, Michael Rodgers, and Randall Guensler. 2017. Safety Impacts of Bicycle Infrastructure: A Critical Review. *Journal of Safety Research* 61(6): 105–119.

Elvik, Rune. 2013. Can a Safety-in-Numbers Effect and a Hazard-in-Numbers Effect Co-exist in the Same Data? *Accident Analysis and Prevention* 60: 57–63.

Elvik, Rune, and Rahul Goel. 2019. Safety-in-Numbers: An Updated Meta-analysis of Estimates. *Accident Analysis and Prevention* 129: 136–147.

Elvik, Rune, and Anne Borger Mysen. 1999. Incomplete Accident Reporting: A Meta-analysis of Studies Made in Thirteen Countries. *Transportation Research Record* 1665(1): 133–140.

Elvik, Rune, and Hanne Beate Sundfør. 2017. How Can Cyclist Injuries Be Included in Health Impact Economic Assessments? *Journal of Transport and Health* 6(9): 29–39.

Feleke, Robel, Shaun Scholes, Malcolm Wardlaw, and Jennifer S. Mindell. 2018. Comparative Fatality Risk for Different Travel Modes by Age, Sex, and Deprivation. *Journal of Transport and Health* 8(3): 307–320.

Flügel, Stefan, Nina Hulleberg, Aslak Fyhri, Christian Weber, and Gretar Ævarsson. 2017. Empirical Speed Models for Cycling in the Oslo Road Network. *Transportation* 46(1): 1395–1419.

Fyhri, Aslak, Ole Jørgen Johansson, and Torkel Bjørnskau. 2019. Gender Differences in Accident Risk with E-bikes—Survey Data from Norway. *Accident Analysis and Prevention* 132. https://doi.org/10.1016/j.aap.2019.07.024.

Fyhri, Aslak, Hanne Beate Sundfør, Torkel Bjørnskau, and Aliaksei Laureshyn. 2017. Safety in Numbers for Cyclists—Conclusions from a Multidisciplinary Study of Seasonal Change in Interplay and Conflicts. *Accident Analysis and Prevention* 105: 124–133.

Gehlert, Tina. 2017. *Road Safety of Electric Bicycles: Compact Accident Research.* Berlin: Gesamtverband der Deutschen Versicherungswirtschaft.

Harris, M. Anne, Conor C. O. Reynolds, Meghan Winters, Peter A. Cripton, Hui Shen, Mary L. Chipman, Michael D. Cusimano, Shelina Babul, Jeffrey R. Brubacher, Steven M. Friedman, Garth Hunte, Melody Monro, Lee Vernich, and Kay Teschke. 2013. Comparing the Effects of Infrastructure on Bicycling Injury at Intersections and Non-intersections Using a Case-Crossover Design. *Injury Prevention* 19(2): 303–310.

Hauer, Ezra. 1995. On Exposure and Accident Rate. *Traffic Engineering and Control* 36: 134–138.

Heesch, Kristiann C., Jan Garrard, and Shannon Sahlqvist. 2011. Incidence, Severity and Correlates of Bicycling Injuries in a Sample of Cyclists in Queensland, Australia. *Accident Analysis and Prevention* 43(6): 2085–2092.

Hertach, Patrizia, Andrea Uhr, Steffen Niemann, and Mario Cavegn. 2018. Characteristics of Single-Vehicle Crashes with E-bikes in Switzerland. *Accident Analysis and Prevention* 117: 232–238.

Hoffman, M. R., W. E. Lambert, E. G. Peck, and J. C. Mayberry. 2010. Bicycle Commuter Injury Prevention: It Is Time to Focus on the Environment. *Journal of Trauma* 69(5): 1112–1119.

Hollingworth, Milo A., Alice J. L. Harper, and Mark Hamer. 2015. Risk Factors for Cycling Accident Related Injury: The UK Cycling for Health Survey. *Journal of Transport and Health* 2(2): 189–194.

Høye, Alena. 2017. *Trafikksikkerhet for syklister.* Rapport 1597. Oslo: Transportøkonomisk institutt.

Høye, Alena. 2018a. Bicycle Helmets—to Wear or Not to Wear? A Meta-analysis of the Effects of Bicycle Helmets on Injuries. *Accident Analysis and Prevention* 117: 85–97.

Høye, Alena. 2018b. Recommend or Mandate? A Systematic Review and Meta-analysis of the Effects of Mandatory Bicycle Helmet Legislation. *Accident Analysis and Prevention* 120: 239–249.

Jacobsen, Peter L. 2003. Safety in Numbers: More Walkers and Bicyclists, Safer Walking and Cycling. *Injury Prevention* 9(3): 205–209.

Lahrmann, Harry, Tanja K. O. Madsen, Anne V. Olesen, Jens Christian O. Madsen, and Tove Hels. 2018. The Effect of a Yellow Bicycle Jacket on Cyclist Accidents. *Safety Science* 108: 209–217.

Langley, John D., N. Dow, S. Stephenson, and K. Kypri. 2003. Missing Cyclists. *Injury Prevention* 9(4): 376–379.

Lusk, Anne C., Peter G. Furth, Patrick Morency, Luis F. Miranda-Moreno, Walter C. Willett, and Jack T. Dennerlein. 2011. Risk of Injury for Bicycling on Cycle Tracks versus in the Street. *Injury Prevention* 17(2): 131–135.

Lyons, Ronan A., Denise Kendrick, Elizabeth M. Towner, Nicola Christie, Steven Macey, Carol Coupland, and Belinda J. Gabbe. 2011. Measuring the Population Burden of Injuries—Implications for Global and National Estimates: A Multi-centre Prospective UK Longitudinal Study. *PLoS Medicine* 8(12): e1001140.

Madsen, Jens Christian O., T. Andersen, and Harry S. Lahrmann. 2013. Safety Effects of Permanent Running Lights for Bicycles: A Controlled Experiment. *Accident Analysis and Prevention* 50(1): 820–829.

Malm, Sigrun, Maria Krafft, Anders Kullgren, Anders Ydenius, and Claes Tingvall. 2008. Risk of Permanent Medical Impairment (RPMI) in Road Traffic Accidents. 52nd AAAM Annual Conference. Annals of Advances in Automotive Medicine, October 2008.

Marqués, R., and V. Hernández-Herrador. 2017. On the Effects of Networks of Cycle-Tracks on the Risk of Cycling: The Case of Seville. *Accident Analysis and Prevention* 102: 181–190.

Marshall, Wesley E., and Nicholas Ferenchak. 2019. Why Cities with High Bicycling Rates Are Safer for All Road Users. *Journal of Transport and Health* 13: 285–301.

Minikel, Erik. 2012. Cyclist Safety on Bicycle Boulevards and Parallel Arterial Routes in Berkeley, California. *Accident Analysis and Prevention* 45(3): 241–247.

Mulvaney, Caroline A., Sherie Smith, Michael C. Watson, John Parkin, Carol Coupland, Philip Miller, Denise Kendrick, and Hugh McClintock. 2015. Cycling Infrastructure for Reducing Cycling Injuries in Cyclists. *Cochrane Database of Systematic Reviews*, no. 12, 1–103.

Nilsson, Philip, Helena Stigson, Maria Ohlin, and Johan Strandroth. 2017. Modelling the Effect on Injuries and Fatalities When Changing Mode of Transport from Car to Bicycle. *Accident Analysis and Prevention* 100: 30–36.

Olivier, Jake, and Prudence Creighton. 2016. Bicycle Injuries and Helmet Use: A Systematic Review and Meta-analysis. *International Journal of Epidemiology* 46: 278–292.

Palmer, A. J., L. Si, J. M. Gordon, T. Saul, B. A. Curry, P. Otahal, and P. L. Hitchens. 2014. Accident Rates among Regular Bicycle Riders in Tasmania, Australia. *Accident Analysis and Prevention* 72: 376–381.

Pedroso, Felipe E., Federico Angriman, Alexandra L. Bellows, and Kathryn Taylor. 2016. Bicycle Use and Cyclist Safety Following Boston's Bicycle Infrastructure Expansion, 2009–2012. *American Journal of Public Health* 106 (12): 2171–2177.

Poulos, R. G., J. Hatfield, C. Rissel, L. K. Flack, S. Murphy, R. Grzebieta, and A. S. McIntosh. 2015. An Exposure Based Study of Crash and Injury Rates in a Cohort of Transport and Recreational Cyclists in New South Wales, Australia. *Accident Analysis and Prevention* 78: 29–38.

Pucher, John, and Ralph Buehler. 2016. Safer Cycling through Improved Infrastructure. *American Journal of Public Health* 106(12): 2089–2090.

Radun, Igor, and Jake Olivier. 2018. Bicycle Helmet Law Does Not Deter Cyclists in Finland. *Transportation Research Part F: Traffic Psychology and Behaviour* 58: 1087–1090.

Robinson, Dorothy. 2006. Analysis and Comment: No Clear Evidence from Countries That Have Enforced the Wearing of Helmets. *British Medical Journal* 332: 722–725.

Rodgers, Gregory B. 1997. Factors Associated with the Crash Risk of Adult Bicyclists. *Journal of Safety Research* 28(4): 233–241.

Schepers, J., P. A. Kroeze, W. Sweers, and J. C. Wüst. 2011. Road Factors and Bicycle–Motor Vehicle Crashes at Unsignalized Priority Intersections. *Accident Analysis and Prevention* 43(3): 853–861.

Schepers, J. Paul, Elliott Fishman, P. den Hertog, Karin Klein Wolt, and Arend L. Schwab. 2014. The Safety of Electrically Assisted Bicycles Compared to Classic Bicycles. *Accident Analysis and Prevention* 73: 174–180.

Schepers, J. Paul, Karin Klein Wolt, and Elliott Fishman. 2018. The Safety of E-bikes in the Netherlands. Discussion paper. Roundtable 168. Paris: International Transport Forum.

Schepers, Paul. 2012. Does More Cycling Also Reduce the Risk of Single-Bicycle Crashes? *Injury Prevention* 18(4): 240–245.

Schepers, Paul, Eva Heinen, Rob Methorst, and Fred Wegman. 2013. Road Safety and Bicycle Usage Impacts of Urban Unbundling Vehicular and Cycle Traffic in Dutch Urban Networks. *European Journal of Transport Infrastructure Research* 13(3): 221–238.

Schepers, Paul, Henk Stipdonk, Rob Methorst, and Jake Olivier. 2017a. Bicycle Fatalities: Trends in Crashes with and without Motor Vehicles in the Netherlands. *Transportation Research Part F: Traffic Psychology and Behaviour* 46: 491–499.

Schepers, Paul, Divera Twisk, Elliott Fishman, Aslak Fyhri, and Anne Mette Dahl Jensen. 2017b. The Dutch Road to a High Level of Cycling Safety. *Safety Science* 92: 264–273.

Shinar, D., P. Valero-Mora, M. van Strijp-Houtenbos, N. Haworth, A. Schrann, G. De Bruyne, V. Cavallo, J. Chliaoutaklis, J. Dias, O. E. Ferraro, A. Fyhri, A. Hursa Sajatovic, K. Kuklane, R. Ledesma, O. Mascarell, A. Morandi, M. Muser, D. Otte, M. Papadakaki, J. Sanmartin, D. Dulf, M. Saplioglu, and G. Tzamalouka. 2018. Under-reporting Bicycle Accidents to Police in the COST TU1101 International Survey: Cross-Country Comparisons and Associated Factors. *Accident Analysis and Prevention* 110: 177–186.

Sundfør, Hanne Beate. 2017. *Sykkelbruk—i trafikk og terreng. Eksponering og uhellsinnblanding*. Rapport 1665. Oslo: Transportøkonomisk institutt.

Teschke, Kay, M. Anne Harris, Conor C. O. Reynolds, Meghan Winters, Shelina Babul, Mary Chipman, Michael D. Cusimano, Jeff R. Brubacher, Garth Hunte, Steven M. Friedman, Melody Monro, Hui Shen, Lee Vernich, and Peter A. Cripton. 2012. Route Infrastructure and the Risk of Injuries to Bicyclists: A Case-Crossover Study. *American Journal of Public Health* 102(12): 2336–2343.

Tin Tin, Sandar, Alistair Woodward, and Shanti Ameratunga. 2013. Incidence, Risk, and Protective Factors of Bicycle Crashes: Findings from a Prospective Cohort Study in New Zealand. *Preventive Medicine* 57: 152–161.

Trafikverket. 2018. *Analys av trafiksäkerhetsutvecklingen 2017*. Publikation 2018:143. Borlänge: Trafikverket.

Transport for London. 2004. *Creating a Chain Reaction: The London Cycling Action Plan*. London: Transport for London.

Transport for London. 2016. *Cycle Safety Action Plan*. London: Transport for London.

Trottenberg, Polly, James O'Neill, and Mary T. Bassett. 2017. *Safer Cycling: Bicycle Ridership and Safety in New York City*. New York: New York City Department of Transportation.

Watson, Angela, Barry Watson, and Kirsten Vallmuur. 2015. Estimating Underreporting of Road Crash Injuries to Police Using Multiple Linked Data Collections. *Accident Analysis and Prevention* 83: 18–25.

Weber, T., G. Scaramuzza, and K-U. Schmitt. 2014. Evaluation of E-bike Accidents in Switzerland. *Accident Analysis and Prevention* 73: 47–52.

Weijermars, Wendy, Niels Bos, Ashleigh Filtness, Laurie Brown, Robert Bauer, Emmanuelle Dupont, Jean Louis Martin, Katherine Perez, and Pete Thomas. 2018. Burden of Injury of Serious Road Injuries in Six EU Countries. *Accident Analysis and Prevention* 111: 184–192.

Weijermars, Wendy, Niels Bos, and Henk Stipdonk. 2016. Health Burden of Serious Road Injuries in the Netherlands. *Traffic Injury Prevention* 17(8): 863–869.

Winters, Meghan, and Michael Branion-Calles. 2017. Cycling Safety: Quantifying the Under Reporting of Cycling Incidents in Vancouver, British Columbia. *Journal of Transport and Health* 7A: 48–53.

# 5 Bicycling Infrastructure for All

Peter G. Furth

Safe, convenient, low-stress, and well-connected cycling infrastructure is crucial for making the bicycle a practical way to get around cities for daily travel. In the 1890s, when cycling first became popular, the chief need of cyclists was better road pavement to ride on. Today, however, for mass cycling to occur, what is most needed is separation from the danger and stress of traffic. This chapter examines the various kinds of bike route infrastructure and how they can be used to create the connected, low-stress bike network required for cycling to become an everyday mode of transport.

## Types of Bike Route Facilities and Separation from Traffic Stress

There are four basic types of bike route facilities. One is stand-alone paths, often situated in a linear park or along an abandoned rail corridor, and sometimes shared with pedestrians. A second type is cycle tracks, also called protected bike lanes, which are bike paths or bike lanes running along or on a road but physically separated from traffic lanes by devices such as curbs, bollards, planters, concrete dividers, or a parking lane (see figure 5.1). A third type is conventional bike lanes, in which a marked stripe designates a portion of the road for bike use but without physical separation. The fourth type is roads where cyclists ride in mixed traffic, which can be considered a bike route if traffic speed and volume are low.

Only stand-alone paths and cycle tracks physically separate cyclists from motor traffic. Because opportunities for stand-alone paths in urban areas are limited, the bicycling networks of bike-friendly European cities such as Copenhagen and Amsterdam rely mainly on cycle tracks (see Koglin, te Brömmelstroet, and van Wee, chapter 18, this volume). For example, the

**Figure 5.1**
Parking-protected, one-directional cycle track on New York's 9th Avenue. *Source*:
New York City Department of Transportation.

Netherlands has 35,000 km of cycle tracks versus only 4,700 km of conventional bike lanes (Fietsersbond 2013).

In the United States, cycle tracks were essentially outlawed until 2011, with only a few exceptions, as this chapter explains. Until then, apart from stand-alone paths, the only choices American transport planners had for accommodating cyclists were conventional bike lanes and mixed-traffic routes on quiet streets. Because continuous quiet streets can be hard to find, many bike lanes and designated mixed-traffic bike routes in the United States have been implemented on roads with traffic levels far in excess of what most people will tolerate.

People vary in their tolerance for interacting with traffic. Roger Geller (2009), a bicycle planner for Portland, Oregon, found it helpful to classify the population into four groups, with size estimates as follows: the "strong and fearless" (1% of the population), who will ride in almost any traffic condition; the "enthused and confident" (6%), who demand a bit more separation but are willing to ride in a bike lane on a multilane arterial road; the "interested but concerned" (60%), who find bicycling appealing and would enjoy a chance to ride in the city, but find it too dangerous; and a group he called "no way, no how" (33%). Geller saw that American bicycle planning at the time was mainly aimed at the "enthused and confident," making it irrelevant to most people. Given the immense societal benefits of mass cycling, Geller argued that cities should focus on serving the "interested but concerned."

Building on Geller's classification, Furth coined the term "traffic stress" to describe the perceived danger that traffic imposes on cyclists and spelled out objective criteria that bike lanes and mixed-traffic segments had to meet in order to be considered a low-traffic-stress environment for cycling (Furth, Mekuria, and Nixon 2016). In the last few years, many US cities have adopted these or similar criteria, which are summarized as follows.

*Mixed traffic* Riding in mixed traffic is low stress only on streets that fit the profile of the typical local street: no centerline or marked vehicle lanes, a prevailing traffic speed of 20 mph (about 30 km/h) or less, and daily traffic volume less than 2,000 vehicles, which roughly corresponds to one car every 20 seconds during the busiest hour of the day.

*Bike lanes* Conventional bike lanes can also be low stress, but only if four requirements are met, summarized here and further discussed later in this chapter: (1) the road should have no more than one lane per direction; (2) traffic speed should be no more than 25 mph (40 km/h) wherever the bike lane is next to a parking lane and up to 35 mph (56 km/h) otherwise; (3) if next to a parking lane, the bike lane plus any marked buffer next to it should be at least 7 ft (2.3 m) wide so that one can ride far enough from parked cars to avoid being "doored" (striking a suddenly opened car door); and (4) the bike lane should not frequently be blocked by illegally parked or stopped vehicles.

The need for separation from traffic has long been understood in Europe, and in the United States it has been a cornerstone of the nation's recreational trails program. But when it comes to urban cycling, American policy strongly resisted the notion of separation from traffic until recently, promoting instead the idea that bikes should be treated as *part of* traffic. The US experience from 1975 to 2010 was a tragic failure to promote urban cycling without providing separated infrastructure. This misguided policy is illustrated by the case of Camino del Norte, a six-lane, 55 mph (90 km/h) divided highway in San Diego with conventional bike lanes. Approaching a junction where many vehicles turn right, cyclists are expected to weave across a lane of 55 mph traffic and then ride for 900 ft (270 m) in a bike lane with two lanes of traffic on their right and four lanes of traffic on their left. Needless to say, almost no one uses this bike lane.

## America's Struggle to Build Separate Cycling Infrastructure

This chapter focuses on the United States because the country provides an example of how challenging it can be to create cycling networks where cycling levels are so low that it is difficult to justify the investment and where most urban travel is by car. Similar situations exist, for example, in Canada, Australia, and New Zealand. The American example describes obstacles that have had to be overcome in the bicycling advocacy community, the engineering profession, the general public, and the political realm.

Interest in urban cycling revived in the United States in the late 1960s with the popularization of the 10-speed bike and social movements, including the environmental movement. At first, it was understandable that American city governments would resist calls to invest in separate bike infrastructure. For generations, people had known roads as having two divisions: one for motor vehicles and one for pedestrians. Adding a third division for bikes seemed like a radical and expensive idea with limited popular support. A chicken-and-egg effect was at work: Why invest in bike infrastructure when so few people are bicycling? But who will ride a bike when there is no safe infrastructure? With time, however, interest in bicycling continued to grow, spurred by external factors such as a desire for a healthy lifestyle. So why did the development of separated infrastructure lag so much?

It is impossible to understand the history of American urban cycling infrastructure without understanding the influence of John Forester's vehicular cycling (VC) theory, which posits that "cyclists fare best when they act as, and are treated as, operators of vehicles" (Forester 1992; Forester 2001). The theory asserts that the key to cycling safety is riding where drivers expect to find other vehicles. Drivers will not hit a vehicle they can see, Forester claimed, so there is no danger to riding in the middle of a traffic lane. (This is a stunning assertion, considering that there were 1.7 million rear-end collisions between motor vehicles in the United States in 2018!) According to Forester, the real danger is riding along the side of a road, where drivers aren't looking for other cars and one might be hit by a vehicle turning right. According to VC theory, cycle tracks and even bike lanes had to be avoided for the sake of bicyclist safety!

Although Forester had no empirical data to support his claims, his ardent defense of cyclists' rights gained him a devoted following. He was elected president of the League of American Wheelmen, now the League of American Bicyclists, which for over a century has been the main bicycling advocacy organization in the United States. From about 1975 to 2005, VC philosophy dominated bicycle advocacy organizations around the country, leading many of them to actively oppose bike lanes and cycle tracks. In several cities and states, VC adherents were hired as bike planners and engineers, where they used their position to prevent bike lanes and separated paths from being built.

Forester's most far-reaching influence came from getting an effective ban on separated paths written into the national handbook for bikeway design, *Guide for the Development of Bicycling Facilities*, published by AASHTO, the American Association of State Highway and Transportation Officials (AASHTO 2012; Schultheiss, Sanders, and Toole 2018). Although the AASHTO guide is not a legally binding standard, engineering officials across the United States generally treat its recommendations as standards that must be followed. The ban on separated paths, first encoded in a California manual in 1978 and then in the 1981 AASHTO guide, became self-perpetuating because the committees that control AASHTO guidelines only accept evidence from US safety studies. Without examples of separate facilities whose safety performance could be studied, the guide's recommendation against separated paths has persisted through every edition since then. (A new edition of the guide, not yet released as of May 2020, is expected to finally reverse its negative recommendations toward cycle tracks.)

Residents of European countries with extensive cycle track networks might find it astonishing that such a patently false theory that separated paths are dangerous could persist in light of decades of evidence from European cities in which millions of cyclists have ridden daily on cycle tracks, with crash rates far lower than in the United States and with far greater appeal to vulnerable populations such as children and seniors (Pucher and Dijkstra 2000; Pucher 2001; Pucher and Buehler 2008).

## The Tide Turns to Favor Separation

During the period in which vehicular cycling dominated American bicycle planning and engineering—roughly 1975 to 2010—bicycling accounted for less than 1% of daily trips in almost all American cities, clear evidence of the failure of VC theory. Davis, California, a small university town where both the university administration and city government promoted cycling, stood out as a singular exception, proving that, given the right conditions, Americans would ride bikes. As early as 1980, 28% of work trips in Davis were made by bike, and most children biked to school (Buehler and Handy 2008). Bicycling became well established in a few other university towns as well, such as Boulder, Colorado. Only in the years 2000 to 2015 did Portland, Oregon, emerge as the first large American city with an extensive cycling network, a substantial percentage of trips by bicycle, and a strong cycling culture (see Geller and Marqués, chapter 19, this volume).

In none of these places did cycling succeed based on a VC model of bikes being operated like motor vehicles; rather, these cities built extensive networks of low-stress bike routes. Davis and Boulder had long linear parks in which they built stand-alone bike paths. Davis had many wide two-lane collector roads, with far lower speeds and far less traffic than arterial roads, which they outfitted with generously wide bike lanes. Portland built traffic-protected cycle tracks on several critical highway bridges over the Willamette River to connect the two main parts of the city. The city also laid out an increasingly extensive network of local street bikeways on its nearly uninterrupted street grid using various kinds of traffic-calming measures, including diverters and speed humps (for details, see Geller and Marqués, chapter 19, this volume).

As inspiring as Davis, Boulder, and Portland became, they could not provide a general model for other US cities to follow because their special circumstances enabled them to create low-stress bike networks without dealing with the thorny issue of accommodating bikes on arterial roads. In most American cities, the only practical route between most origins and destinations involves travel along arterial roads. For low-stress bike networks to emerge there, cities would need to embrace the concept of protected bike lanes or cycle tracks.

Until 2008, there were virtually no protected bike lanes in the United States—only a few kilometers of "sidepaths" from earlier eras, many of them along beaches. In North America, only one city, Montreal, had an integrated network of modern cycle tracks, built around 1990 after a city official was inspired by a visit to Amsterdam. The success of Montreal, however, was largely ignored by engineers and planners outside Quebec.

A turning point came in 2007, when New York City, advised by consultants from Copenhagen, created a parking-protected cycle track along 9th

Avenue in Manhattan (shown in figure 5.1). More New York cycle tracks followed in 2008 and every year since then, averaging about 10 mi (16 km) per year and accelerating to 30 mi (48 km) per year by 2019 (see Pucher, Parkin, and de Lanversin, chapter 17, this volume). Those cycle tracks violated the recommendations in the AASHTO guide by placing the bike lane on the nontraffic side of a parking lane.

Across the country, bicycle planners and designers watched anxiously, but the success of New York's "experiment" was soon clear. The federal government did not penalize New York City for violating the AASHTO guidelines. Cycling levels rose dramatically on the streets with the new traffic-protected facilities. Moreover, serious cyclist injuries and fatalities fell dramatically relative to the rising number of bike trips, demonstrating the much greater safety of cycle tracks (see Pucher, Parkin, and de Lanversin, chapter 17, this volume; Waters 2018).

In the bikeway planning world, a glass ceiling had been shattered. Across the country, cities scrambled to design and implement cycle tracks like those in New York. Early adopters included Indianapolis, Austin, Washington, DC, Minneapolis, and Chicago, where cycle tracks captured the imagination of citizens and elected officials alike—at last there was a bikeway facility that "normal people" could imagine themselves using! Bicycle advocacy organizations elected new leaders unassociated with vehicular cycling; soon the organized bicycling community came to speak with a united voice in favor of separating bicycles from motor vehicle traffic (see Pucher et al., chapter 20, this volume). People for Bikes, a bicycle industry trade group, became an important promoter of cycle tracks, providing technical assistance for cities interested in creating cycle tracks and taking public officials and city transport staff on study tours to the Netherlands and Denmark. Starting in 2008, the number of kilometers of modern protected bike lanes in the United States roughly doubled every two years (see figure 5.2), reaching 684 km in 2018.

The overthrow of VC philosophy in the United States was formalized in 2011 with the publication of the *Urban Bikeway Design Guide* (NACTO 2011). To produce this manual, officials from New York and other bike-friendly cities worked through a little-known organization called the National Association of City Transportation Officials (NACTO), thereby bypassing AASHTO. By providing clear engineering guidelines for the design of cycle tracks, this manual made it easy for cities around the country to emulate

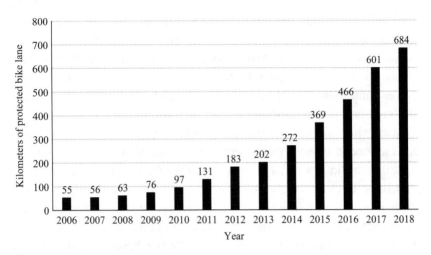

**Figure 5.2**
Growth in centerline kilometers of protected bike lanes (cycle tracks) in the United States from 2006 to 2018. *Source*: Data from People for Bikes.

the example of New York and other early adopters. The US secretary of transport endorsed the NACTO manual soon after its publication, freeing engineering officials across the nation to follow this alternative to the AAS-HTO guidelines and ending the hegemony of vehicular cycling theory over bikeway planning and design.

However, the struggle to create good bicycling infrastructure had other obstacles to overcome—taking roadway space from motor vehicles, finding economical bikeway designs, and gaining the public and political support to fund the creation of a cycling network.

## Finding Space for Bikes

There are many competing demands for space in the road right-of-way, making it a challenge to find space for bike lanes and cycle tracks. One way to make space for bikes is to resize travel lanes, parking lanes, and shoulders with appropriate widths. Since before the motorized era, the standard travel lane in the United States has been 10 ft wide (about 3 m). To this day, all vehicles except those needing special permits are required to be operable within 10 ft lanes. For example, the widest bus or truck cannot exceed a width of 8.5

ft (2.6 m). Because 12 ft (3.65 m) lanes are standard on freeways, many state and local officials believed that all safe modern roads needed lanes that wide. However, an extensive study done for the Federal Highway Administration found that on urban and suburban arterial roads, lanes 9 or 10 ft (2.75 or 3.05 m) wide are just as safe as those 11 or 12 ft (3.35 or 3.65 m) wide (Potts, Harwood, and Richard 2007). Many city streets have lanes that are up to 16 ft (4.9 m) wide; shrinking travel lanes to 10 ft (3.05 m) will not reduce a road's safety or traffic-carrying capacity, while freeing up needed space for bike lanes.

Parking lanes similarly vary surprisingly in width. The widest common personal motor vehicle in the United States is 6.6 ft (2.0 m) wide; thus, parking lanes that extend 7 ft (2.15 m) from the curb are sufficient. In addition, roads often waste space with oversized shoulders, sometimes called "edge offsets." State design guidelines sometimes call for 4 ft (1.3 m) shoulders even though the national road design manual states that, where the speed limit is 35 mph (56 km/h) or less, having no offset is acceptable (AASHTO 2018).

An even more radical way to find space for bikes is to reduce the number of travel lanes, called a "road diet." Many urban arterial roads in the United States have four lanes, two in each direction. This is inefficient because cars waiting to turn left block the inside lanes and, as a result, traffic can flow freely in only one lane per direction. A road with a three-lane layout can have the same traffic capacity: one through lane in each direction and a central zone that can be a left-turn lane where needed and elsewhere can be a raised median, making it easier for people to cross the street. San Francisco and Charlotte are examples of American cities that have each implemented more than 20 road diets, with the freed-up space usually reallocated to bike lanes.

Opposition to repurposing roadway space for bikes can be intense. In a widely publicized example, a group of prominent citizens sued the city of New York in a failed attempt to prevent a pilot road diet project on Prospect Park West from becoming permanent. Reducing the number of travel lanes on this one-way street from three to two was approved by a community board, based on traffic studies showing rampant speeding and finding that traffic would still flow smoothly with a road diet. After the road diet's implementation, data showed that neither vehicle throughput nor travel time changed, while the safety benefits were astounding. The percentage of cars exceeding 40 mph (64 km/h) fell from 47% to 2%; the percentage of cyclists riding on the sidewalk (illegal in New York City) fell from 46% to

3%; and the number of traffic injuries in six months fell from five to two. At the same time, cycling volumes doubled on weekends and tripled on weekdays. Nevertheless, the opposition group continued a draining appeal process for five more years before giving up (Sadik-Khan and Solomonow 2016).

## Stand-alone Paths and Linear Parks

Stand-alone bike paths, often built along waterways and abandoned rail corridors, enjoy strong political support because they are popular for recreation. When built in urban areas, such paths can also be valuable for utilitarian bicycle transport, particularly when they lead to a city center or other major destination.

In urban settings, dual paths—one for cyclists and another for pedestrians—are standard practice in Europe, though infrequently used in North America, where bike paths are typically shared with pedestrians. According to Dutch guidelines, bike paths should always be sized to support cyclists riding side by side (CROW 2017). In the United States, following the AASHTO guide, shared-use paths are usually 10 ft (3.05 m) wide—an arbitrary and misguided standard because this dimension is too narrow to support side-by-side cycling except where path use is very low. If paths were instead 11 ft (3.35 m) wide, the path would in effect have three lanes, enabling two people to ride side by side without one of them having to fall back every time they met a cyclist traveling in the opposite direction.

While most stand-alone paths take advantage of historical opportunities such as old rail lines and canal towpaths, new opportunities can sometimes be created. New York created the nation's busiest bike path by replacing a riverside highway with an at-grade boulevard and linear park hosting bike and pedestrian paths. Another example is Davis, California, where a private developer built a linear park with a multiuse path through its new housing development. The success of that path resulted in the city laying out plans for a connected network of linear parks and paths that successive developers were legally required to build as the city expanded (Buehler and Handy 2008). Similarly, Scottsdale, Arizona (a suburb of Phoenix), and Eagle, Idaho (a suburb of Boise), have adopted plans for bike paths following irrigation canals in undeveloped parts of those towns; developers are required to build any part of the bike path network lying in their development.

## Cycle Tracks

Cycle tracks, also called protected bike lanes, are bike paths or bike lanes running within the road right-of-way yet physically separated from motor traffic. Cycle tracks can be either at sidewalk level or at street level. At sidewalk level, separation from pedestrians is best achieved either by a row of trees or light poles or by a small, rounded curb that makes the cycle track about 1.5 in (4 cm) lower than the sidewalk, a design used widely in Denmark and the Netherlands. Dividing a sidewalk using only differing paving materials or a painted line (common in the United Kingdom, Germany, and Japan) is also possible where space is limited. However, merely designating part of a sidewalk as a bikeway can be unsatisfactory if the sidewalk lacks the space and sight lines needed for safe riding at normal bicycling speed.

For street-level cycle tracks, the traditional means of separation from motorized traffic is a raised median, common in Rotterdam and Montreal. American cities have discovered less expensive means of separation— planters, flexposts (plastic bollards that fall down when hit and then spring back up), and, most notably, parking lanes. When a parking lane is used as the barrier, a marked buffer roughly 3 ft (0.9 m) wide is needed between the parking lane and the cycle track for visibility and so that cyclists will not be injured or blocked by a car whose door is open while loading. While flexposts offer no structural resistance to vehicles, studies show that they provide the same sense of physical separation as structural barriers.

Cycle tracks can be one-way, like bike lanes; however, because they are physically separated from the street, they can also be two-way. The main advantage of two-way cycle tracks is that a single bidirectional cycle track requires less space than a pair of one-way tracks. For example, a typical pair of parking-protected one-way cycle tracks requires 16 ft (4.9 m) in total, including buffers, while a two-way cycle track requires only 11 ft (3.4 m). Moreover, during times of lighter use, two-way paths offer the possibility of riding side-by-side or passing without requiring any extra width.

One concern with two-way cycle tracks is that they complicate intersections and create safety concerns involving motor vehicles turning left. For these reasons, Copenhagen does not allow two-way cycle tracks. Amsterdam and other Dutch cities have historically favored one-way cycle tracks for the same reasons; however, Dutch guidelines allow two-way cycle tracks provided designers take appropriate measures to ensure intersection safety.

At signalized intersections, the key safety measure is separate left-turn phases for cars, a practice New York City has also followed in building its new cycle tracks (see Pucher, Parkin, and de Lanversin, chapter 17, this volume). At unsignalized intersections, the key safety measure has proven to be "side street crossing tables," also called "continuous sidewalks" (see figure 5.3); when crossing a minor street, a major street's sidewalks and cycle tracks remain elevated instead of dropping to street level. Cars entering or leaving the minor street are thereby forced to ramp up and down to cross the sidewalk and cycle track, which they do at very low speed because the ramps are dimensioned like speed humps. This treatment, which originated in Sweden (Garder, Leden, and Pulkkinen 1998), spread rapidly in the Netherlands starting around 2002. A study found that side street crossing tables halve the crash risk at unsignalized intersections, making two-way cycle tracks with this treatment as safe as one-way cycle tracks without them (Schepers et al. 2011). Thanks to this new intersection design, two-way cycle tracks have become more common than one-way tracks on Dutch arterials, and many one-way cycle tracks have been reconfigured as two-way.

**Figure 5.3**
A side street crossing table forces cars to slow down as they have to ramp up and down steeply when crossing the cycle track and sidewalk. *Source*: Peter Knoppers.

## Conventional Bike Lanes

Conventional bike lanes, separated from traffic lanes by only a painted line, are inexpensive to install and require less space than protected bike lanes (cycle tracks). That makes them an attractive option where space or funding for cycle tracks is difficult to find.

Some organizations and advocacy groups eschew conventional bike lanes altogether. Under the banner "all ages and abilities," they argue that cycle tracks are the only acceptable option on all roads except quiet local streets (NACTO 2017). However, in the appropriate setting, conventional bike lanes can provide a low-stress place for cyclists to ride, and they offer significant safety benefits compared to cycling in mixed traffic. Both the Dutch bikeway manual (CROW 2017) and the Level of Traffic Stress criteria (Furth 2017) consider conventional bike lanes an acceptable accommodation on moderate-speed roads with one lane per direction.

Riding next to a parking lane, whether in a conventional bike lane or in mixed traffic, involves a risk of being "doored." A suddenly opened car door can catch a bicycle's handlebar and throw the rider forcefully to the ground, often causing serious injuries and subjecting the victim to the risk of being run over by a motor vehicle. Typical bike lanes in US cities are too narrow to eliminate this hazard; to ride clear of car doors, cyclists in typical bike lanes have to encroach on the adjacent traffic lane. Nevertheless, studies have shown that, compared to riding in mixed traffic, painting a line to designate a bike lane reduces dooring risk, because it induces cyclists to ride farther from parked cars—presumably because that painted line makes them confident that cars approaching from behind will stay on their side of the line (Van Houten and Seiderman 2005). Still, it is preferable for bike lanes to be extra wide so that cyclists can keep a safe distance from hazards on both sides. To keep a wide bike lane from having the appearance of a parking lane, a narrow, hatched buffer can be painted on either side of a normal-width bike lane.

In commercial areas, conventional bike lanes are often blocked by illegally parked cars and delivery vehicles. One study found that almost 50% of cyclists in commercial areas had to leave the bike lane because it was blocked by a vehicle (Meng 2010). Thus, protected bike lanes are vastly preferred in commercial areas.

## Local Street Bikeways

Local streets with low traffic speed and volume, on which cyclists can comfortably ride in mixed traffic, are not only vital for access to the bicycle network; they can also be used to form main bicycle routes across the city. Bike routes that mainly follow local streets have been variously called bicycle boulevards, neighborhood greenways, and local street bikeways. In the Netherlands, such routes have proven to be safer and more popular with schoolchildren than main roads with cycle tracks.

The concept is more complex than it may first appear, because the factors that keep through traffic from using local streets—for example, being discontinuous or labyrinthine—also make them unsuitable for through bicycling. Creating a city-scale bike route using local streets can be approached in two ways. One is to take an existing long, continuous street and install infrastructure and signage to divert and slow motor vehicle traffic while allowing bikes to pass through. The West Coast cities of Berkeley, Palo Alto, and Portland in the United States and the Canadian city of Vancouver have taken advantage of their nearly uninterrupted street grid networks to turn some long local streets into local street bikeways by using partial street closures and median barriers to divert through motor traffic along with traffic circles and speed humps to slow traffic (Walker, Tressider, and Birk 2009). Older, narrow roads in or near centers of cities and towns often present a similar opportunity. As traffic demand grows, it is often impractical to widen such roads, so a bypass road is built for motor vehicles. In the Netherlands, for example, many such roads have been downgraded to local streets, using full or partial closures to force motor vehicle traffic to use the bypass road while letting bikes pass through. Such converted roads can make ideal bike routes because they follow an axis of historic urban development and generally avoid steep grades.

The second approach to developing local street bikeways is to stitch together shorter segments of local streets to form a longer route. Joining discontinuous segments can involve building connectors such as a footbridge over a creek, an underpass to cross a highway or railroad, or a path connecting nearby cul-de-sacs. In one situation in Davis, the city purchased a property that separated two quiet streets, Drexel Drive and Loyola Drive, demolished the house, and built a short path connector. Planners in Europe—and, in a few cases, in the United States as well—often incorporate short path connectors into new housing developments, creating local streets that are discontinuous for cars but continuous for bikes and pedestrians.

However, the concept of local street bikeways can be abused if the routes are too indirect. In both the United States and Europe, cities that have tried to channel cyclists onto local street routes that require large detours have found that cyclists spurn them in favor of a more direct route following main streets. A study from Portland, Oregon, found that cyclists are willing to ride only about 10% to 20% farther, depending on trip purpose, to enjoy a quieter route (Broach, Dill, and Gliebe 2012).

### Shared Road Treatments: Advisory Bike Lanes versus Sharrows

On streets lacking the space needed to create bike lanes but with more traffic than a typical local street, both Europeans and Americans have developed treatments intended to make it safe and comfortable for bikes to share a road with motor vehicles.

The American approach, marking bike silhouettes called "sharrows" in the middle of travel lanes, is both a failure and a farce. Sharrows were conceived in the days of vehicular cycling as a way to embolden cyclists to assert their right to ride in the middle of a car lane, away from the door zone, while prompting motorists to respect cyclists' right to do so. The primary study used to defend their effectiveness (Alta Planning + Design 2004) actually shows them to be quite ineffective—when cars were passing, sharrows shifted cyclists' position only 4 in (10 cm) farther from the parking lane, with most cyclists still riding in the door zone rather than in line with the sharrows. Sharrows may embolden a few of the "enthused and confident" to take the lane, but for most people sharrows do nothing to lower traffic stress. Motorists, for their part, do not understand their meaning at all.

Worse still, sharrows are a farce because cities can paint them on a road and then claim that they have created a "shared lane bicycling facility." National guidelines allow sharrows on multilane roads and on roads with speed limits up to 35 mph (56 km/h). By trying to normalize a kind of bicycling that most people eschew (riding in the middle of a travel lane on a busy road with fast traffic), sharrows make a city's cycling program appear to be out of touch with the mainstream, eroding public support for funding legitimate cycling infrastructure.

By contrast, the European treatment for road sharing, advisory bike lanes (figure 5.4), has been very effective. Dashed white lines indicate the part of the road bicyclists are expected to use. Vehicle lanes are *not* marked. Because there is no striped centerline, motorists will often drive in the middle of the road, shifting right into an advisory lane when a vehicle approaches from the opposite direction. If the advisory lane is occupied, the car will stay behind the bike until it is clear to pass.

The dashed lines of advisory lanes make the passing maneuver predictable and low stress—the bike and car each stay on their side of the line, just as they would if there were a conventional bike lane. The outcome is that advisory lanes give cyclists the same security as if they were riding in a conventional bike lane, even when there is no space for bike lanes.

**Figure 5.4**
Advisory bicycle lanes in the Netherlands. *Source*: Peter Knoppers.

Advisory bike lanes are used extensively in European countries, including the Netherlands, Germany, Denmark, and Switzerland. In the United States, they remain almost unknown. As of late 2019, only 19 streets in the entire United States had advisory lanes.

A subtle but important difference between sharrows and advisory lanes is that advisory lanes are a *shared road* treatment, while sharrows are a *shared lane* treatment. The American approach delineates motor vehicle lanes and invites cyclists to ride in them like any other vehicle. The European approach delineates the bicyclists' space and allows motor vehicles to use it when they need to. The absence of a striped centerline is critical. On streets without centerlines, it is normal for motorists to drive in the middle of the road, shifting position when encountering other vehicles. Where drivers operate with this mindset, sharing a road with bikes fits naturally. Once a centerline is marked, drivers see bikes as something blocking their lane, making the street a hostile environment for cycling.

## Planning Bicycle Networks

The requirements for bike networks can be summarized in one phrase, *low-stress connectivity*, meaning that links with low stress form a network in which origins and destinations are connected to each other without excessive detours or excessive climbs (Furth, Mekuria, and Nixon 2016). Low-stress connectivity can be decomposed into five requirements that correspond closely with those listed in the Dutch bikeway design guide (CROW 2017).

1.  *Separation from traffic stress.* As described earlier in this chapter, bike infrastructure must separate cyclists from fast and heavy traffic. That can be accomplished with stand-alone paths, cycle tracks, bike lanes (including advisory lanes) under conditions described earlier, and in mixed traffic on local streets with low traffic volume and speed.

2.  *Pleasant, well-lit, and low-crime surroundings.* Unlike people in motor vehicles, cyclists are not physically separated from their environment, so the environment around them is important. Where crime is a concern, cyclists prefer a route that is well lit and where homes with windows and active street life give it "eyes on the street." Cyclists also prefer streets with little traffic noise and with natural beauty or attractive buildings.

3. *Smooth, well-maintained pavement.* To avoid injury-inducing falls, the pavement must be well drained and well maintained, including clearing leaves, sand, and snow (or, where ice formation is not a concern, packing the snow hard). Smooth pavement improves cyclist comfort and reduces the physical effort needed to ride.

4. *Avoiding long, steep climbs.* Cycling uphill greatly increases the effort needed to propel a bicycle, so bike routes should avoid steep hills where possible.

5. *Connected and direct.* The network links that meet the first four requirements should connect people's origins and destinations without excessive detours and with safe intersection crossings. An upper limit for a detour is about 20% longer than the most direct path using *any* road or path legally open to bicycles. Because origins and destinations tend to be scattered, this requirement is tantamount to requiring that the bike network form a dense mesh (CROW 2017). The Dutch suggest a route spacing of 500 m (0.3 mi) in cities, though in practice route spacing in Dutch cities often reaches 700 m (0.4 mi), and 1 to 2 km (0.6 to 1.2 mi) outside the built-up area, where trip lengths tend to be longer.

## Funding for Bicycling Infrastructure

For a city or metropolitan region starting from scratch, creating a low-stress bicycling network is a radical initiative, akin to creating a highway or rail transit network. Bicycling infrastructure is actually very inexpensive compared to highway or rail infrastructure, but it costs more than can be accomplished without dedicated funding. While many US cities have published bike network plans, few have provided the necessary funding stream. Too often, bicycling infrastructure is funded on a piecemeal basis without any permanent budget line and at a rate at which bike network completion could take many decades.

Bicycling infrastructure costs vary depending on the specifics of each project, but here are some ballpark figures. Reconstructing a street in order to create curb-separated cycle tracks costs $10 to $20 million per mile (although full reconstruction projects can sometimes be financed by a city's road reconstruction budget). Stand-alone paths cost $1 to $2 million per mile, not counting bridges. Parking-protected cycle tracks cost around $1 million per mile, not counting traffic signal work, although at least one city (Portland, Maine) has managed to build a pair of one-way, parking-protected cycle tracks for only $100,000 per mile. Bike lane striping costs only about $15,000 per mile, and the incremental cost can be zero when part of a repaving project; however, striping has to be replaced every few years and thus requires not only an initial capital budget but also an ongoing maintenance budget. The cost of creating local street bikeways depends on the treatments involved; a typical project might cost $200,000 per mile (again, far less expensive examples can be found), plus $250,000 for every new traffic signal needed.

By comparison, highway infrastructure commonly costs $300 million to $2 billion per mile. One city's feasibility study found that the cost of building a low-stress bike network was roughly $1,000 per city resident (Götschi 2011); spread over 20 years, that would be $50 per resident per year. By comparison, per capita annual spending by Boston metropolitan area governments is $400 for public transport and $600 for roads. The Boston metro area is currently spending $2 billion to build a 4.3 mi (7 km) light rail extension that will increase the length of the rapid transit network by 5%. The same investment would suffice to create a low-stress bike network covering the entire metropolitan area.

As this comparison shows, bicycling infrastructure is clearly affordable. The issue is generating enough political and public support to make governments realign transport spending priorities so that they invest in bike network development at a meaningful pace. A few North American cities— all in Canada—provide good examples of this. Vancouver's current capital plan will invest US$34 per person per year into its bike network (City of Vancouver 2018, 79), while Ottawa and Montreal are investing US$12 per person per year (City of Montreal 2018; City of Ottawa 2017). Only a few US cities have attained this level of cycling investment, including Seattle, New York, San Francisco, and a few small university towns.

## Conclusion

Mass cycling requires bike infrastructure that separates cyclists from traffic and forms a dense, connected network. Compared to highway and rail transport, bike infrastructure is affordable and requires little space. There are many types of bicycling facilities, making it feasible to create a low-stress bicycling environment in almost every context.

Although this chapter has focused on the United States, the American experience is instructive for other car-dominated countries beginning with very low levels of cycling. Getting American cities to develop the needed bicycling network has faced four main challenges: (1) getting the engineering profession to adopt the correct design guidelines; (2) reorienting bicycling advocacy organizations to advocate for infrastructure that serves the mainstream population rather than hard-core cyclists; (3) growth in public support for bicycling; and (4) garnering political support and leadership in prioritizing investment in cycling infrastructure. In the United States, the first three have been accomplished; the last one remains a challenge.

## References

Alta Planning + Design. 2004. *San Francisco's Shared Lane Pavement Markings: Improving Bicycling Safety*. San Francisco: San Francisco Department of Parking and Traffic.

American Association of State Highway and Transportation Officials (AASHTO). 2012. *Guide for the Development of Bicycle Facilities*. Washington, DC: American Association of State Highway and Transportation Officials.

American Association of State Highway and Transportation Officials (AASHTO). 2018. *A Policy on Geometric Design of Highways and Streets*. Washington, DC: American Association of State Highway and Transportation Officials.

Broach, Joseph, Jennifer Dill, and John Gliebe. 2012. Where Do Cyclists Ride? A Route Choice Model Developed with Revealed Preference GPS Data. *Transportation Research Part A: Policy and Practice* 46(10): 1730–1740.

Buehler, Ted, and Susan Handy. 2008. Fifty Years of Bicycle Policy in Davis, California. *Transportation Research Record* 2074: 52–57.

City of Montreal. 2018. *2019–2021 Three-Year Capital Works Program Highlights*. City of Montreal. https://ville.montreal.qc.ca/portal/page?_pageid=44,2243535&_dad=portal &_schema=PORTAL.

City of Ottawa. 2017. *Budget*. https://ottawa.ca/en/city-hall/budget.

City of Vancouver. 2018. *Final 2019–2022 Capital Plan & Plebiscite Questions*. RTS 12408. July 12, 2018. https://council.vancouver.ca/20180724/documents/regurr1.pdf.

CROW. 2017. *Design Manual for Bicycle Traffic*. Ede, Netherlands: CROW.

Fietsersbond. 2013. *Hoeveel Kilometer Fietspad Is Er in Nederland?* [How Many Kilometers of Bike Path Are There in the Netherlands?]. https://www.fietsersbond.nl /nieuws/bijna-35-000-km-fietspad-in-nederland/.

Forester, John. 1992. *Effective Cycling*. Cambridge, MA: MIT Press.

Forester, John. 2001. The Bikeway Controversy. *Transportation Quarterly* 55(2): 7–17.

Furth, Peter G. 2017. Level of Traffic Stress Criteria, version 2.0. http://www.north eastern.edu/peter.furth/criteria-for-level-of-traffic-stress/.

Furth, Peter G., Maaza C. Mekuria, and Hilary Nixon. 2016. Network Connectivity for Low-Stress Bicycling. *Transportation Research Record* 2587: 41–49.

Garder, Per, Lars Leden, and Urho Pulkkinen. 1998. Measuring the Safety Effect of Raised Bicycle Crossings Using a New Research Methodology. *Transportation Research Record* 1636: 64–70.

Geller, Roger. 2009. *Four Types of Cyclists*. Portland, OR: Portland Bureau of Transportation. https://www.portlandoregon.gov/transportation/44597?a=237507.

Götschi, Thomas. 2011. Costs and Benefits of Bicycling Investments in Portland, Oregon. *Journal of Physical Activity and Health* 8(s1): S49–S58.

Meng, Dun, 2012. Cyclist Behavior in Bicycle Priority Lanes, Bike Lanes in Commercial Areas and at Traffic Signals. MS diss., Northeastern University, Department of Civil and Environmental Engineering.

National Association of City Transportation Officials (NACTO). 2011. *Urban Bikeway Design Guide*. Washington, DC: National Association of City Transportation Officials.

National Association of City Transportation Officials (NACTO). 2017. *Designing for All Ages & Abilities: Contextual Guidance for High-Comfort Bicycle Facilities*. Washington, DC: National Association of City Transportation Officials.

Potts, Ingrid B., Douglas W. Harwood, and Karen R. Richard. 2007. Relationship of Lane Width to Safety for Urban and Suburban Arterials. *Transportation Research Record* 2023: 63–82.

Pucher, John. 2001. Cycling Safety on Bikeways vs. Roads. *Transportation Quarterly* 55(4): 9–12.

Pucher, John, and Ralph Buehler. 2008. Making Cycling Irresistible: Lessons from the Netherlands, Denmark, and Germany. *Transport Reviews* 28(4): 495–528.

Pucher, John, and Lewis Dijkstra. 2000. Making Walking and Cycling Safer: Lessons from Europe. *Transportation Quarterly* 54(3): 25–50.

Sadik-Khan, Janette, and Seth Solomonow, 2016. *Streetfight: Handbook for an Urban Revolution*. New York: Viking.

Schepers, J. P., P. A. Kroeze, W. Sweers, and J. C. Wüst. 2011. Road Factors and Bicycle–Motor Vehicle Crashes at Unsignalized Priority Intersections. *Accident Analysis and Prevention* 43(3): 853–861.

Schultheiss, William, Rebecca L. Sanders, and Jennifer Toole. 2018. A Historical Perspective on the AASHTO Guide for the Development of Bicycle Facilities and the

Impact of the Vehicular Cycling Movement. *Transportation Research Record* 2672(13): 38–49.

Van Houten, Ron, and Cara Seiderman. 2005. How Pavement Markings Influence Bicycle and Motor Vehicle Positioning: Case Study in Cambridge, Massachusetts. *Transportation Research Record* 1939: 1–14.

Walker, Lindsay, Mike Tressider, and Mia Birk. 2009. *Fundamentals of Bicycle Boulevard Planning and Design*. Portland, OR: Center for Transportation Studies, Portland State University.

Waters, Carlos. 2018. Why Protected Bike Lanes Are More Valuable Than Parking Spaces. Video. Vox. https://www.vox.com/videos/2018/9/12/17832002/nyc-protected-bike-lanes-janette-sadik-khan.

# 6 Bicycle Parking

## Ralph Buehler, Eva Heinen, and Kazuki Nakamura

Local and national governments have increasingly encouraged cycling through various initiatives and greater investments in bikeway infrastructure (see in this volume Buehler and Pucher, chapter 2; Furth, chapter 5; Pucher, Parkin, and de Lanversin, chapter 17; Koglin, te Brömmelstroet, and van Wee, chapter 18; Geller and Marqués, chapter 19). While the number of bicycling-related peer-reviewed publications—in particular about bikeway infrastructure—has soared over the past decade, most studies ignore bicycle parking. This is remarkable, given that bicycles are parked the majority of the time. For example, in Germany and the United States, cyclists report riding for about 40 minutes a day—leaving more than 23 hours of stationary time for their bicycle per day (USDOT 2018; German Ministry of Transport 2012). Bicycles owned by infrequent cyclists are parked even longer.

Bicycles are parked at different locations, typically near trip destinations: at home, at work or school, at public transport stops, and at many other places, such as restaurants, shops, and on sidewalks. Bike parking facilities and the design of racks and locks used for bike parking vary between cities and countries. Parked bicycles are at risk of being vandalized or stolen—either the entire bicycle or just parts, such as saddles, lights, or tires. Bike locks used by riders include heavy chain locks and U-locks to secure bicycles to racks and street furniture, but also simple O-locks bolted to the back wheel of the bicycle just to block the back wheel from turning. Bicycle parking practices vary by city as well (e.g., Aldred and Jungnickel 2013). Larsen (2017) observed that whereas in some cities, such as New York, bicycles were typically attached to something when parked, it was common in Copenhagen and Amsterdam to park bikes unattached to fixed objects. Two studies noted that many elements of the built environment were used for parking counter to their primary purpose, sometimes referred to as "informal" parking or "fly-parking" (Gamman, Thorpe, and Willcocks 2004; Larsen 2017). This may be an indicator that current bicycle parking provisions are mismatched to user requirements and a result of higher user demand (Nakamura and Abe 2014). An alternative explanation is that when informal parking is prevalent, it becomes the norm (Abou-Zeid et al. 2013).

The remainder of this chapter briefly summarizes the scientific literature on bicycle parking and highlights bike-parking practices in 11 bicycle-friendly cities from three continents. A better understanding of the effect of the quality and quantity of bike parking can help estimate bicycle demand at home, work, businesses, and public transport stops and stations. A supply of bicycle parking that meets quantity and quality demands can encourage cycling.

## Literature Review

### Parking at Public Transport Stops and Stations

Providing bicycle parking at public transport stops can increase the demand for both modes (Krizek and Stonebraker 2011). Because of its speed, bicycling can help enlarge the public transport catchment area and reduce the need to operate feeder (bus) services. The vast majority of papers on bicycle parking at public transport stops find a positive relationship between bicycle parking supply and public transport ridership, cycling levels, or the stated likelihood to cycle to access public transport (e.g., Harveyet al. 2016). The type of bicycle parking available at rail stations ranges from simple bicycle racks without weather protection to full-service bicycle parking garages with video surveillance, repair services, and bicycle rentals (Pucher and Buehler 2008). Studies suggest that bicyclists value covered bike parking and bike lockers more than simple bike parking without weather protection. While cyclists appreciate full-service bike parking garages, fees charged for bike parking in such garages reduce the likelihood of parking bicycles there (Geurs, La Paix, and Van Weperen 2016; Molin and Maat 2015).

Generally, cyclists prefer to park close to public transport and in areas with public or video surveillance (e.g., de Souza et al. 2017). For example, in New South Wales, Australia, 80% of bicycles were parked in bike racks and lockers within 30 m of the rail station entrances, which also had greater public surveillance from public transport passengers passing by (Arbis et al. 2016). Several studies also report that abandoned bicycles parked at stations hinder bicycling to stations by occupying needed parking spots or by creating an impression of an unsafe location to park, thereby deterring usage (Harvey et al. 2016).

## Parking at Work, Universities, and Schools

Virtually all studies indicate that the presence of bicycle parking at work increases the likelihood of commuting to work by bicycle (e.g., Bopp et al. 2016; Maldonado-Hinarejos, Sivakumar, and Polak 2014). For example, Hunt and Abraham (2007) found that cyclists value secure parking at work as much as commuting an additional 26.5 minutes in mixed traffic. Some studies concluded that the combination of facilities at work is important—with commuters valuing bicycle parking and cyclist showers more than just bicycle parking without cyclist showers (Buehler 2012).

Other research indicates that better-quality parking facilities may increase ridership (Mrkajic, Vukelic, and Mihajlov 2015). Heinen, Maat, and van Wee (2013) showed that indoor (instead of outdoor) bicycle parking increased the likelihood of cycling to work and the frequency of cycling to work. McDonald et al. (2013) showed that providing covered bicycle parking was associated with an 11% increase in cycling to work. Studies on student commuting to school or university also show a positive relationship with bike parking (e.g., Wang, Akar, and Guldmann 2015). In addition, inadequate bicycle parking is often mentioned as a reason for not cycling to school.

## Parking at Home

The location for bike parking at home varies. In China, parking sheds are most commonly used (~40%) and are the most preferred option to park the bicycle (~60%) (Lusk, Wen, and Zhou 2014). In Singapore, cyclists also prefer to park their bicycle inside, perhaps because of a lack of adequate parking outside (e.g., Lusk, Wen, and Zhou 2014). Parking locations also differ between cities and suburbs. In the Vienna, Austria, region, urban cyclists were more likely to park on staircases (10% vs. 4% for suburban cyclists), in living rooms (8% vs. 3%), or on the street (6% vs. 3%) but less likely to park in a garage (4% vs. 15%) (Pfaffenbichler and Brezina 2016). Those who parked in bicycle storage rooms with easy access or in gardens were more satisfied with their storage facilities. These differences may partly result from variations in crime levels. Fear of crime was mentioned as a reason why more blacks and Hispanics in the United States preferred to park their bicycle inside (52% and 47%, respectively) compared to whites (28%) (Lusk et al. 2017). However, Van Lierop et al. (2015) showed that residents in Montreal, Canada, tend to underestimate the risk of theft at their own residential location.

## Parking in the City and Other Locations

Bicycle parking in cities can be a challenge from a transport and urban planning perspective. Different cities have dissimilar levels and quality of bicycle parking infrastructure. Many Dutch, Danish, and German cities have large supplies of high-quality bicycle parking (Pucher and Buehler 2008), which are sometimes absent elsewhere—although their supply may be increasing in the United States (Hirsch et al. 2016).

A limited number of studies investigated parking at specific locations and often revealed the lack of high-standard parking. For example, Thompson (2006) showed that only half the libraries in Pennsylvania had any bicycle parking. Van Lierop et al. (2015) revealed that only 30% of Montreal respondents were happy with the bicycle parking facilities downtown or at the grocery store. There is typically more bicycle parking in countries with higher cycling rates, but shopkeepers and authorities there often struggle with informal parking and abandoned bicycles (Van der Spek and Scheltema 2015).

Several studies discuss the importance of safe parking and protection from theft for public parking. Main obstacles to buying and using a private bicycle in Spain are fear of theft (46%) and the lack of proper parking (57%) (Castillo-Manzano, Castro-Nuño, and López-Valpuesta 2015). The latter is particularly important to women. In contrast, protection from theft and proper parking are not very important variables for mode choice in India (Majumdar and Mitra 2015).

## Case Studies

To capture some of the current practices and variability of bike parking, we provide a summary overview of bike parking in 11 cities on three continents (see table 6.1). City size ranges from small towns, such as Groningen, the Netherlands, and Freiburg, Germany, with about 200,000 inhabitants, to large cities such as London, United Kingdom, or Tokyo, Japan, with several million inhabitants. All our case-study cities actively promote bicycling, and several are among the most bike-friendly cities in their countries, such as Groningen; Strasbourg, France; Seville, Spain; or Kyoto, Japan. Shares of trips made by bicycle range from 2.0% in London and 4.5% (of commuters) in Washington, DC, and Minneapolis, Minnesota, to 34% of trips by bicycle in Freiburg and 46% in Groningen.

**Table 6.1**
Overview of bike parking supply and policies in select cities in Europe, the United States, and Japan (sorted by city size)

Groningen, the Netherlands (200,000 inhabitants, 46% of trips <7.5 km by bicycle)

There are three covered, guarded bike parking garages in the city center for 2,230 bicycles, all offering free parking and additional services such as air pumps and bicycle repairs. Facilities are open at least between 8:00 a.m. and 9:30 p.m., longer at many locations, and bicycles can be parked up to one month, after which they will be removed.
All three train stations offer free covered parking.
The main train station offers parking for 9,900 bicycles, including free guarded parking for 5,500 bicycles.
All five Park and Ride locations have bicycle boxes or bicycle racks.
Since April 1, 2012, Dutch building regulations have required residences to have storage of sufficient size to park several bicycles.
Red carpet is being used to prevent bicycle parking at locations in the city center to allow space for pedestrians.
In cases of expected high demand, boxes are drawn in which bicycles can be parked for a short moment or racks are placed temporarily.

Freiburg, Germany (220,000 inhabitants, 34% of trips by bicycle)

There are 12,000 parking spaces in the city (7,000 in the city center, 1,000 in a bike parking garage at the main train station, and 4,000 at public transport stops), about 3,000 of which are weather protected or sheltered.
Bike parking supply in the city center increased from 2,200 in 1987 to 7,000 in 2018.
There is a bike parking garage with secure and sheltered bike parking at the main train station, with 1,000 parking spots for €1 per day, €10 per month, or €80 per year.
Parking is banned at Bertoldsbrunnen in the city center because parked bikes forced pedestrians to walk on or close to light rail tracks.
The city removes abandoned bicycles and has created a special email address for reporting abandoned bicycles.
Parking in the city center and at the main train station is not sufficient. The goal is to increase the bike parking supply, improve its quality, and require bike parking by building ordinance.

Eurométropole de Strasbourg, France (490,000 inhabitants, 11% of trips by bicycle)

The city had 34,612 parking racks in 2017.
There are 21 garages or sheds at light rail stops, with a total of 2,056 parking spaces.
The city building code requires bike parking for any new building.
There are 736 bike parking spaces in car parking garages throughout the city.
Special e-bike parking and charging are available at a few locations.

*(continued)*

**Table 6.1 (continued)**
Overview of bike parking supply and policies in select cities in Europe, the United States, and Japan (sorted by city size)

Minneapolis, Minnesota, United States (410,000 inhabitants, 4.1% of workers commute by bicycle)

The city has 20,000 parking spots in 4,500 racks and 350 lockers.
The city allows bike parking at all signposts.
The city provides bike lockers throughout downtown and at the university, including car parking garages with cyclist showers.
A 50% cost share program for businesses that request bike corrals or racks is offered.
Off-street parking regulations require bike parking for new offices, schools, and residential buildings with multiple apartments.
The city provides a detailed map showing bike parking spaces.
The city has guidelines for designing bicycle parking.
Metro Transit provides bike parking at most light rail stops.

Washington, DC, United States (660,000 inhabitants, 5.4% of workers commute by bicycle)

A parking garage for 100+ bicycles is available at the main train station, with a changing room and day lockers, including bike rental and bike repair, for $96 per year or $12 per month.
There are an estimated 3,000 bike parking spots in the city.
The city requires bike parking in all new buildings with car parking (the minimum number of bike parking spaces is 5% of car parking spots).
The city provides guidance on bike parking design and placement.
The city waives permitting fees for businesses that install bike racks.
Metrorail offers 4,800 bike parking spots in bike racks and 1,300 bike lockers ($120 per year) at its stations.
The parking shed at the College Park Metrorail stop is accessible free of charge after registration.
There is an ongoing expansion of the bike parking supply in the city.

Seville, Spain (700,000 inhabitants, 5% of trips by bicycle)

There were 5,700 parking places installed along newly built cycle tracks between 2006 and 2010.
There is a policy to expand parking alongside newly constructed cycle tracks.
City regulations explicitly allow bike parking at traffic signs, traffic signals, streetlights, and other street furniture.
The new bike master plan (2017) provides city grants for neighborhoods, companies, and schools to build bike parking.
There are plans to build bike parking sheds or garages at public transport stops.
There is plentiful bicycle parking at university buildings spread throughout the city.

---

Kyoto, Japan (1,500,000, 21% of trips by bicycle)

---

The city has 35,000 public bike parking spots and 10,000 private spots, most of which are off-street.

Informal bike parking is banned around station areas; this practice has been expanded to the whole city since 2016.

The city council provides an online map of bike parking for both public and private spaces.

Most stations have bike parking, especially in the city center and around large stations, universities, and sightseeing places.

Of the 2,100 parking spots in the city center and 600 spots at Kyoto central station, 1,000 parking spots are in automated machines. The spots in Kyoto central station are available only for monthly use.

Public parking requires daily (¥150) or monthly (¥2,700) passes, with private parking costing more, price varying by location.

---

Nagoya, Japan (2,300,000, 16% of all trips by bicycle)

---

Informal on-street bike parking is banned around 102 stations. Bike parking is provided around most stations with parking control.

Public parking operated by the city council has a total capacity of 90,000 spaces in 363 locations.

In Aichi prefecture, which includes Nagoya, 70% of the parking is public and 30% is private.

While many parking spaces are on-street in racks, 25% of bike parking spaces are off-street, such as underground. Some underground and indoor bike parking spaces are adjacent to stations with direct access during times when trains are in service.

Parking fees range from ¥100 per day to ¥2,000 per month.

The city center has 87 bike parking spaces, of which 73 are for daily use and 16 are for monthly use. The Sakae district of the city center has many free on-street bike parking spots.

The main station area of the city center has two large commercial buildings with bike parking, one of which is automated underground parking, with a capacity of 952 spots.

---

Berlin, Germany (3,500,000 inhabitants, 13% of trips by bicycle)

---

There are 26,600 parking places at U-Bahn and S-Bahn stations.

There is a full-service bike parking garage in Bernau, with 566 spaces for bikes (500 free of charge and 66 bike boxes for €10 per month or €95 per year).

A zoning ordinance requires bike parking for commercial and office buildings and for residential buildings with three or more apartments.

Since the mid-1990s, the supply of parking at public transport stations and in certain central areas of neighborhoods has increased.

There have been upgrades to parking facilities since the 1990s (e.g., installation of U-shaped bike stands).

The city has a goal of building 100,000 additional bike parking spaces by 2025.

---

*(continued)*

**Table 6.1 (continued)**
Overview of bike parking supply and policies in select cities in Europe, the United States, and Japan (sorted by city size)

London, United Kingdom (8,790,000 inhabitants, 2% of all trips by bicycle)

An interactive online map indicates the presence and amount of bicycle parking in the city.
London Kings Cross station offers bicycle parking at several locations—mainly unguarded at U-shaped racks.
The type and quantity of parking vary by borough. The City of London offers 364 parking places where bicycles can be parked for free, 24 hours a day, in four staffed car parking facilities.
Secured bicycle parking is limited but is installed at Finsbury Park station and at stations in the borough of Waltham Forest, and secured on-street residential parking is being delivered by boroughs. The London borough of Hackney has installed the most.
Bicycle parking requirements for new buildings are set out in the London Plan. The new policy sets higher parking requirements for some uses and specific geographical areas.
London currently developing a cycle parking strategy.

Tokyo, Japan (9,300,000 inhabitants, 14% of all trips by bicycle)

Informal on-street parking is banned in 417 areas around rail stations.
Tokyo prefecture has 920,000 bike parking spots, of which 50% are operated by ward offices and 50% are private. Private parking has doubled in the last decade.
There are many on-street parking spots. Most off-street parking spots are located underneath elevated expressways and railways.
There are a range of parking fees for hourly use, all-day use, monthly use, and yearly use. The spaces for temporary use are free for the first couple of hours. For regular use, public parking offers a discount for ward residents.
In the ward of Chiyoda, a central area covering Tokyo station, many bike parking spots are available only for regular use, for which ward residents pay ¥3,000 per year and others ¥6,000 per year.
There are 23 locations of "mega-size" public bike parking with more than 2,000 spots per location, of which nine locations are underground.

*Source*: Data collected by the authors directly from local governments.

## Total Supply of Bike Parking

None of our case study cities collected data on the total amount of bike parking available in the entire city. Several cities provided estimates for public bike parking. For example, Strasbourg reports about 3,500 public bicycle parking racks and 2,100 spaces in bike parking garages. In Washington, DC, an online-based crowdsourcing project estimated 3,000 public bike parking spots in the city. Estimates typically exclude privately supplied bicycle parking—either in privately operated bike parking garages, private offices, or commercial and residential buildings. In Japan, there is a trend toward privatization of bike parking. Private companies increasingly operate off-street parking garages and on-street parking with the city's permission to manage these areas. The city of Kyoto, for example, operates 35,000 public bike parking spots, but there are an additional 10,000 privately operated bike parking spots. In Tokyo, about 50% of bike parking spots at rail stations are privately operated.

## Bicycle Parking Facilities

Several cities have information about bike parking in specific areas or specific facilities. For example, Freiburg increased bike parking in its city center from 2,200 bicycle parking spaces in 1987 to 7,000 in 2018. Between 2006 and 2010, Seville built 5,700 bicycle parking spaces along its newly constructed cycle track network (see Geller and Marqués, chapter 19, this volume). Many cities report the number of bike parking spots close to public transport stops and stations, such as Tokyo and Berlin, with 920,000 and 26,600 bike parking spaces at rail stops, respectively. Moreover, urban campuses of universities (e.g., in Seville and Kyoto) are often cited as providing a large (but often unknown) number of bike parking spaces.

Cities also have different practices for counting bike parking. For example, Minneapolis and Seville specifically encourage bicycle parking on traffic signs and street signs throughout the city—counting each signpost as bicycle parking greatly inflates the number of reported bicycle parking spaces.

Most cities have large garages for bike parking that are safe, secure, and weather protected. Except for Japan, these bike parking garages frequently also offer bike repair, air pumps, bike accessories for sale, bike rentals, and other services. Bicycle parking garages are typically in the city center or at rail stations. Size, pricing, and operations of bike parking garages vary. Washington's only bike parking garage, at Union Station, accommodates slightly more than 100 bicycles, compared to Groningen's large bike parking garage at the main train station, which holds 5,500 bicycles. Freiburg, Strasbourg, Washington, and the Japanese cities charge for parking in their bicycle parking garages. Bike parking garages are free in Groningen. While bike parking garages typically offer short-term as well as monthly or annual parking passes, central Tokyo and Kyoto only provide garage parking spaces for regular users. Several cities offer discounted bicycle parking at train stations for individuals with monthly or annual public transport passes. Most bicycle parking garages are supervised by a parking attendant and/or CCTV. Kyoto, Tokyo, and Nagoya also provide fully automated bicycle parking garages without an attendant. Several cities, including Strasbourg, additionally provide smaller bike cages with restricted access and CCTV surveillance at public transport stops to facilitate bike and transit integration. Figures 6.1–6.3 show examples of the various bike parking facilities discussed.

### Dealing with High Demand

With the exception of the Japanese cities, most cities offer free (on-street) bike parking close to bike parking garages to accommodate additional bike parking demand. For example, roughly 10,000 bicycle parking spaces are available

**Figure 6.1**
Simple, uncovered bike parking in (from left to right) Berlin, Nagoya, Groningen, and Minneapolis. *Sources*: Uwe Kunert, Kazuki Nakamura, Eva Heinen, and Elmer Lovrien.

**Figure 6.2**
Bike parking garage in Groningen (left) and entrance to fully automated bike parking garage in Nagoya (right). *Sources*: Eva Heinen and Kazuki Nakamura.

**Figure 6.3**
Bike and public transport integration: bike parking on the train platform in the United Kingdom (left) and bike parking garages at main train stations in Freiburg (center) and Washington, DC (right). *Sources*: Eva Heinen and Ralph Buehler.

at Groningen's main train station—5,500 inside the main parking garage and 4,500 outside (all are often entirely full). As at Groningen's main station, even at locations with a large supply of bicycle parking, excess bike parking demand can result in bicycles being parked informally on sidewalks and becoming a nuisance for pedestrians. For example, Freiburg banned informal bicycle parking in its city center (Bertoldsbrunnen) because parked bicycles forced pedestrians to walk dangerously close to light rail tracks. Tokyo and Nagoya also ban informal bike parking around 417 and 102 rail stations, respectively. Since 2016, Kyoto has banned informal bike parking in the entire city. Groningen uses red carpets on the pavement to maintain spaces for pedestrian movement, where bicycle parking (and use) is prohibited.

In bike-friendly cities, large public events typically result in spikes in bike parking demand. Several cities have responded to this. Groningen paints temporary bike parking locations to provide additional short-term parking facilities on its streets. Minneapolis and Washington, DC, provide "valet parking" for bicycles, where volunteers securely park bikes for event visitors (sometimes for a fee).

### Regulations and Subsidies for Installing Bicycle Parking

Several of our case study cities require bike parking in newly constructed buildings. In the Netherlands, this has been a national requirement for residential buildings since 2012 (it had already been required for office buildings). Typically, bicycle parking spots are required for commercial buildings and large, multiunit residential buildings. The specific number of bike parking spots and quality of bike parking vary by city and building type—often based on square footage, the number of workers or residents, and the number of car parking spots provided.

Most cities provide extensive design guidelines on how to build adequate bicycle parking. Some cities additionally support bike parking through grants and partnerships. For example, Seville provides grants to schools, neighborhoods, and companies interested in building bike parking. Minneapolis offers up to 50% cost sharing for businesses installing new bike corrals or bike parking. Washington, DC, waives permitting fees for businesses that install bike parking. Since 2009, Tokyo has offered subsidies for the development of private bike parking around rail stations.

**Other Initiatives to Promote Bike Parking**

London, Kyoto, and Minneapolis provide online maps that allow cyclists to identify specific bicycle parking locations. Kyoto's map also includes bike parking provided by private companies. Most cities have plans to improve the quality and expand the quantity of bike parking over the coming years. Berlin's plan may be the most ambitious at this point, calling for 100,000 additional bike parking spaces by 2025. Seville is set to continue with its tight coordination between bike parking supply and further expansion of the cycle track network.

**Conclusions**

Bicycle parking supply appears to be a determinant of cycling for current and potential cyclists. Conversely, a lack of bicycle parking and/or inadequate bicycle parking discourages cycling. Both cyclists and potential cyclists prefer higher-quality (e.g., weather-protected) and more convenient bicycle parking facilities over lower-quality facilities or no bicycle parking. Convenience includes easy access to bicycles (e.g., no stairs, and short distances between bicycle parking and actual trip origins or destinations). However, preferences, quality, and convenience vary by user group, and the higher user cost for quality parking garages may discourage the use of those facilities. Current and potential cyclists prefer bicycle parking facilities that increase personal security and safety from bicycle theft and vandalism. Finally, the practice of parking bicycles varies by country, city, and even rail station, including the prevalence of illegal parking outside designated spaces.

References

Abou-Zeid, M., J. -D. Schmoecker, P. F. Belgiawan, and S. Fujii. 2013. Mass Effects and Mobility Decisions. *Transportation Letters* 5(3): 115–130.

Aldred, Rachel, and Katrina Jungnickel. 2013. Matter in or out of Place? Bicycle Parking Strategies and Their Effects on People, Practices and Places. *Social and Cultural Geography* 14(6): 604–624.

Arbis, David, Taha Hossein Rashidi, Vinayak V. Dixit, and Upali Vandebona. 2016. Analysis and Planning of Bicycle Parking for Public Transport Stations. *International Journal of Sustainable Transportation* 10(6): 495–504.

Bopp, M., D. Sims, J. Colgan, L. Rovniak, S. A. Matthews, and E. Poole. 2016. An Examination of Workplace Influences on Active Commuting in a Sample of University Employees. *Journal of Public Health Management and Practice* 22(4): 387–391.

Buehler, Ralph. 2012. Determinants of Bicycle Commuting in the Washington, DC Region: The Role of Bicycle Parking, Cyclist Showers, and Free Car Parking at Work. *Transportation Research Part D: Transport and Environment* 17(7): 525–531.

Castillo-Manzano, Jose I., Mercedes Castro-Nuño, and Lourdes López-Valpuesta. 2015. Analyzing the Transition from a Public Bicycle System to Bicycle Ownership: A Complex Relationship. *Transportation Research Part D: Transport and Environment* 38: 15–26.

Gamman, Lorraine, Adam Thorpe, and Marcus Willcocks. 2004. Bike Off! Tracking the Design Terrains of Cycle Parking: Reviewing Use, Misuse and Abuse. *Crime Prevention and Community Safety* 6(4): 19–36.

German Ministry of Transport. 2012. *National Household Travel Survey: Mobility in Germany 2008/2009*. Berlin: German Ministry of Transport.

Geurs, Karst T., Lissy La Paix, and Sander Van Weperen. 2016. A Multi-modal Network Approach to Model Public Transport Accessibility Impacts of Bicycle-Train Integration Policies. *European Transport Research Review* 8(4): 8–25.

Harvey, Elizabeth, Charles T. Brown, Stephanie DiPetrillo, and Andrew Kay. 2016. Bicycling to Rail Stations in New Jersey. *Transportation Research Record* 2587: 50–60.

Heinen, Eva, Kees Maat, and Bert van Wee. 2013. The Effect of Work-Related Factors on the Bicycle Commute Mode Choice in the Netherlands. *Transportation* 40(1): 23–43.

Hirsch, J. A., K. A. Meyer, M. Peterson, D. A. Rodriguez, Y. Song, K. Peng, J. Huh, and P. Gordon-Larsen. 2016. Obtaining Longitudinal Built Environment Data Retrospectively across 25 Years in Four US Cities. *Frontiers in Public Health* 4. https://doi.org/10.3389/fpubh.2016.00065.

Hunt, John Douglas, and J. E. Abraham. 2007. Influences on Bicycle Use. *Transportation* 34(4): 453–470.

Krizek, Kevin J., and Eric W. Stonebraker. 2011. Assessing Options to Enhance Bicycle and Transit Integration. *Transportation Research Record* 2217: 162–167.

Larsen, Jonas. 2017. The Making of a Pro-Cycling City: Social Practices and Bicycle Mobilities. *Environment and Planning A: Economy and Space* 49(4): 876–892.

Lusk, Anne C., Albert Anastasio, Nicholas Shaffer, Juan Wu, and Yanping Li. 2017. Biking Practices and Preferences in a Lower Income, Primarily Minority Neighborhood: Learning What Residents Want. *Preventive Medicine Reports* 7: 232–238.

Lusk, Anne C., Xu Wen, and Lijun Zhou. 2014. Gender and Used/Preferred Differences of Bicycle Routes, Parking, Intersection Signals, and Bicycle Type: Professional Middle Class Preferences in Hangzhou, China. *Journal of Transport and Health* 1(2): 124–133.

Majumdar, Bandhan Bandhu, and Sudeshna Mitra. 2015. Identification of Factors Influencing Bicycling in Small Sized Cities: A Case Study of Kharagpur, India. *Case Studies on Transport Policy* 3(3): 331–346.

Maldonado-Hinarejos, Rafael, Aruna Sivakumar, and John W. Polak. 2014. Exploring the Role of Individual Attitudes and Perceptions in Predicting the Demand for Cycling: A Hybrid Choice Modelling Approach. *Transportation* 41(6): 1287–1304.

McDonald, Noreen C., Yizhao Yang, Steve M. Abbott, and Allison N. Bullock. 2013. Impact of the Safe Routes to School Program on Walking and Biking: Eugene, Oregon Study. *Transport Policy* 29: 243–248.

Molin, Eris, and Kees Maat. 2015. Bicycle Parking Demand at Railway Stations: Capturing Price-Walking Trade Offs. *Research in Transportation Economics* 53: 3–12.

Mrkajic, Vladimir, Djordje Vukelic, and Aandjelka Mihajlov. 2015. Reduction of $CO_2$ Emission and Non-environmental Co-benefits of Bicycle Infrastructure Provision: The Case of the University of Novi Sad, Serbia. *Renewable and Sustainable Energy Reviews* 49: 232–242.

Nakamura, Hiroki, and Naoya Abe. 2014. Evaluation of the Hybrid Model of Public Bicycle-Sharing Operation and Private Bicycle Parking Management. *Transport Policy* 35: 31–41.

Pfaffenbichler, Paul Christian, and Tadej Brezina. 2016. Estimating Bicycle Parking Demand with Limited Data Availability. *Engineering Sustainability* 169(2): 76–84.

Pucher, John, and Ralph Buehler. 2008. Making Cycling Irresistible: Lessons from the Netherlands, Denmark and Germany. *Transport Reviews* 28(4): 495–528.

Souza, Flavia de, Lissy La Paix Puello, Mark Brussel, Romulo Orrico, and Martin van Maarseveen. 2017. Modelling the Potential for Cycling in Access Trips to Bus, Train and Metro in Rio de Janeiro. *Transportation Research Part D: Transport and Environment* 56: 55–67.

Thompson, Samantha Hypatia. 2006. Bicycle Access to Public Libraries: A Survey of Pennsylvania Public Libraries and Their Accessibility to Patrons Arriving via Bicycle. *Library Philosophy and Practice* 9(1).

US Department of Transportation (USDOT). 2018. *National Household Travel Survey 2016/2017*. Washington, DC: US Department of Transportation, Federal Highway Administration.

Van der Spek, Stefan Christiaan, and Noor Scheltema. 2015. The Importance of Bicycle Parking Management. *Research in Transportation Business and Management* 15: 39–49.

Van Lierop, Dea, Michael Grimsrud, and Ahmed El-Geneidy. 2015. Breaking into Bicycle Theft: Insights from Montreal, Canada. *International Journal of Sustainable Transportation* 9(7): 490–501.

Wang, Chih-Hao, Gulsah Akar, and Jean-Michel Guldmann. 2015. Do Your Neighbors Affect Your Bicycling Choice? A Spatial Probit Model for Bicycling to the Ohio State University. *Journal of Transport Geography* 42: 122–130.

# 7   Programs and Policies Promoting Cycling

Eva Heinen and Susan Handy

Over the past decade, levels of cycling have soared worldwide (see in this volume Pucher and Buehler, chapter 1; Buehler and Pucher, chapter 2). Governments have used a variety of approaches to achieve this increase. Most attention in policy and research has been directed toward capital expenditures such as cycling-specific infrastructure (see also Furth, chapter 5, this volume). Noninfrastructural policies and programs also have an important role to play, despite the fact that they may not receive the attention they require. Such programs may complement infrastructure investments, ensuring that such investments pay off to their fullest, or may serve as the main initiative itself. Indeed, the cities that are most successful at promoting cycling have done so through an integrated package of infrastructure investments combined with policies and programs (Pucher, Dill, and Handy 2010).

This chapter focuses on programs and policies that aim to increase cycling. It builds on several (systematic) reviews that have evaluated the effectiveness of noninfrastructural or what are sometimes called "soft" interventions (Scheepers et al. 2014; Ogilvie et al. 2004; Pucher, Dill, and Handy 2010). Given the large number of initiatives worldwide, we will not be able to discuss all of them but will instead aim to include representative or exceptional examples. We focus on those programs and policies that are bicycle-specific or have potential to increase cycling.

A diverse set of policies and programs may encourage cycling (for an overview, see table 7.1). Some of these have been adopted with the explicit aim of increasing cycling, such as laws that protect vulnerable road users. Other policies and programs do not aim to promote cycling specifically but nevertheless may contribute to higher levels of cycling as a secondary aim or even as an unintended consequence. Some policies and programs directly encourage or improve conditions for cycling, while other policies and programs

Table 7.1
Overview of programs and policies to promote cycling

| | |
|---|---|
| Promotions | Education and training programs<br>Promotional events<br>Individualized marketing<br>Social marketing<br>Information |
| Incentives | Payments<br>Tax credits<br>Other rewards<br>Driving costs |
| Laws | Rules of the road<br>Liability laws<br>Laws for cyclists<br>Enforcement |
| (Built) environment | Land-use planning<br>Network connectivity<br>Traffic calming<br>Public transport integration<br>Smaller design features |

help to promote cycling by slowing or discouraging motor vehicle traffic. Policies and programs can be initiated by and be the responsibility of various actors, including but not limited to governments, employers, schools, and voluntary organizations; advocacy by community groups often provides an important push for their implementation. As we discuss their effects and present examples in the sections that follow, we group these strategies into four categories: promotions, incentives, laws, and environment.

## Promotions

Whether someone bicycles depends not only on the quality of bicycling infrastructure but also on their own awareness of, comfort with, and enjoyment of bicycling (Schneider 2013). These psychological factors are shaped over one's life by bicycling experience and other factors, but interventions of various sorts can enhance them. Programs designed to promote cycling by addressing these psychological factors include education and training, community events, social marketing, individualized marketing, and information.

Education and training programs are perhaps most important. Cycling training programs for children can have substantial and lasting effects on their cycling skills. In Belgium, for example, a four-session course for children in the fourth grade (ages nine and ten) effectively increased basic cycling skills, and those improvements persisted for at least five months (Ducheyne et al. 2014). Cycling training in schools is common in some European countries, such as the Netherlands and Germany (Pucher and Buehler 2010). Training programs for adults are also effective in increasing cycling skills and comfort levels. Adults who completed a six-hour cycling proficiency training program in Sydney felt considerably more confident and comfortable with bicycling after the training and continued to feel that way two months after the training (Telfer et al. 2006).

Psychological barriers to bicycling also declined following a 12-week intervention involving bicycling instruction and weekly group rides for overweight adults in low-income Latino and black communities in Milwaukee, Wisconsin (Schneider et al. 2018). A six-week bicycle proficiency class in Columbia, South Carolina, led to a significant increase in bicycling (Thomas et al. 2009), while participants in an adult cycle training program in London were bicycling more frequently and for longer periods three months after the program (Johnson and Margolis 2013). Training programs for adults are less common than programs for children, but their availability is increasing.

Promotional events are another way to increase awareness of, comfort with, and enjoyment of bicycling. The universe of such events is large, including both formal and informal events and events focused on recreational cycling as well as utilitarian cycling. These events generally aim to build a sense of community around bicycling by getting people on their bicycles in a safe and fun environment. The "granddaddy" of such events is Bogota, Colombia's Ciclovía, established in 1974 (figure 7.1). For seven hours every Sunday, 76 mi of the city's streets are closed to traffic, and some 1.7 million people, about a quarter of the city's population, bicycle, walk, or otherwise propel themselves around the city. As many as 400 cities worldwide now hold some version of a Ciclovía, though not always on a weekly basis. At "slow-up" events in Switzerland, for example, long stretches of roads in rural areas are closed to motor vehicles, and food stands and other concessions are set up along the route; 20 such events were planned in 2018 (Bike Switzerland 2018). Evidence of the effectiveness of Ciclovía-style events on bicycling is sparse, although at least one study shows a connection between participation in Ciclovía and utilitarian cycling (Gómez et al. 2005). Bike to Work Days, another common promotional event, aim to

**Figure 7.1**
Ciclovía event in Bogota, Columbia. *Source*: Carlosfelipe Pardo.

increase bicycle commuting with enticements such as free breakfasts, give-aways, contests, and other activities offered over a day, week, or month. Evidence suggests that such events increase bicycling beyond the event by attracting first-time bicycle commuters (Pucher, Dill, and Handy 2010). Other types of promotional events, such as bicycle film festivals, help to increase awareness of bicycling as a mode of transport and can enhance social norms around bicycling.

Social marketing is a strategy widely used in the public health field to encourage healthier behaviors, such as smoking cessation, increased exercise, and healthier eating. Social marketing campaigns in transport have generally targeted behaviors related to safety, including seat-belt use and drunk driving, but examples of social marketing campaigns focused on cycling are on the rise. The grassroots Love London, Go Dutch campaign in 2012, for example, pushed mayoral candidates to support cycling and involved endorsements from prominent athletes and politicians, petitions and letters from thousands of residents, and a parliamentary debate (London Cycling Campaign 2018). Research provides some guidance as to how best to design such campaigns. Studies show that campaigns that emphasize health might be more effective than campaigns that focus on safety (Gamble, Walker, and

Laketa 2015). Social networks in the form of family, friends, peers, and colleagues appear to have a mostly tacit influence on bicycling but could be harnessed in conjunction with social marketing campaigns to improve their effectiveness (Sherwin, Chatterjee, and Jain 2014).

Targeted marketing is another strategy for promoting cycling. Individualized marketing programs such as TravelSmart and SmartTrips involve one-on-one consultation with households about their current travel patterns and the alternatives available to them, including cycling. A review of 16 studies of individualized marketing programs concluded that such programs produced eight additional cycling trips per person per year (Yang et al. 2010). Employer-based trip reduction programs similarly aim to shift commuters from driving to alternative modes, though in a less individualized way. In these programs, employers provide information on driving alternatives to employees, offer incentives for choosing modes other than driving, and put on promotional events. Studies show that such programs have often resulted in increases in bicycling (Pucher, Dill, and Handy 2010). Information about bicycling routes in the form of printed bike maps and smartphone apps are an important part of such efforts. Some cities provide cycling guides or have established "cycling embassies" to assist potential cyclists.

Evidence and experience point to important synergies among these different kinds of promotional programs. In Sydney, Australia, for example, a promotional effort combined information provision, cycle training, free bike hire, and a Ride to Work Day campaign as a strategy for increasing the use of existing cycle paths. This modest program produced significant increases in counts on cycle paths in the intervention areas but no measurable increase in cycling across the system (Rissel et al. 2010). More commonly, promotional programs are combined with improvements in infrastructure, making it difficult to isolate the effect of the promotional programs themselves. A multifaceted program in Odense, Denmark, for example, produced an increase in the share of trips by bicycle from 22.5% to 24.6% over the three-year period of the program, with a net increase in the distance cycled of 100 m per person per day (as described in Yang et al. 2010). Results were also promising in six Cycling Demonstration Towns in England, where infrastructure improvements were combined with promotional programs such as training (Goodman et al. 2013). Indeed, cities seeing the greatest increases in bicycling share have combined promotional programs with infrastructure improvements (Pucher, Dill, and Handy 2010), pointing to the strong complementarity between these strategies.

## Incentives

Costs are often assumed to be a key determinant of mode choice. The fact that cycling is inexpensive is one of the reasons why people may choose to cycle (Heinen et al. 2010). Using that line of reasoning, paying people to cycle may further encourage cycling. Research suggests that if people in Britain were to receive £2 each day to cycle to work, the level of cycling would almost double (Wardman, Tight, and Page 2007). However, few long-term examples are available in practice, and many suggest that the use of such short-term incentives will not last, and that individuals will return to their old behavior when such incentives are removed (Scheepers et al. 2014).

In addition to direct financial payments, many other programs exist to support cyclists financially. Several countries have policies in place that offer financial support for the purchase of a bicycle for commuting. For example, in the Netherlands, until 2015, the "fietsplan" allowed employers to offer employees the opportunity to buy a bicycle with their income before taxation, which resulted in a 42% reduction on average in the cost of purchase. This was later replaced with a different scheme that offered comparable benefits (CNV 2018). In the United Kingdom, employers can provide their employees with access to a "cycle to work" scheme. This scheme was introduced in a 1999 government initiative in which employers bought bikes and leased them to employees to cycle to work, resulting in savings between 32% and 42% for the lease compared to the purchase price (Department for Transport 2011). Financial contributions to commute journeys by bicycle are also provided in some countries. For example, in Belgium, an employer may offer to pay €0.23 per kilometer tax-free (Belgium Federal Ministry of Finance 2018). Similarly, in Germany, employers can make a contribution to commuting expenses independent of transport mode (Bundesverfassungsgericht 2008). It is important to highlight that these contributions were often already in place for other transport modes, with the ability to use them for cycling added on later. The effectiveness of the preceding policies has not been thoroughly researched.

Various smaller local initiatives exist. Many of those are reward schemes. The idea behind these schemes is that they monitor cycling levels of individuals or groups (e.g., companies), who collect points based on the distance or number of trips cycled. These points can then be used for rewards, sometimes in competition with other individuals or groups. For example, the Dutch scheme "trappers" (Trappers 2018) measures actual cycling behavior, for which the participant is rewarded with points that can be used at an online retailer. Other reward programs offer discounts at supporting businesses. For example, Bicycle Benefits is a program that rewards cyclists who shop by bike and show their Bicycle Benefits helmet sticker at participating businesses (Bike Benefits 2018).

Costs of traveling by other modes of transport may also affect bicycle use. At least one study found a significant relationship between bicycle use and gasoline prices when comparing 61 states and provinces in the United States and Canada (Pucher and Buehler 2006). In an earlier study, however, Dill and Carr (2003) did not find a significant relationship when comparing 42 American cities. Other costs attached to car use may also deter driving and at the same time encourage cycling. Toll fees have been studied to determine whether they result in a modal shift from the car to the bicycle (Meland, Tretvik, and Welde 2010; Scheepers et al. 2014), but no effect on cycling was found. In contrast, restricted parking and the need to pay for parking have been shown to increase the likelihood of cycling and may even stimulate a modal shift away from the car toward the bicycle (Heinen et al. 2014; Brockman and Fox 2011; Panter et al. 2013).

## Laws

Laws can have conflicting effects on bicycling levels. Most bicycling-related laws aim to increase the safety of bicycling, which helps to encourage bicycling. Some laws, however, tend to make bicycling less convenient or enjoyable, thereby discouraging cycling even if they make it safer.

Countries, states, and local governments have different traffic regulations and enforcement levels. The legal position of cyclists differs between places. In the United States, for example, only 30 states consider bicycles to be vehicles, and fewer still consider them motor vehicles (Bopp, Sims, and Piatkowski 2018). In other countries, such as the Netherlands, a clear distinction is made between cyclists (i.e., those using nonmotorized bicycles) and drivers of motorized vehicles. The growing popularity of electric bicycles of various sorts is complicating the legal situation in many places, however. These differences in classification can affect the rights and duties of cyclists.

Differences between countries and states exist as to where cyclists are allowed to cycle. In some countries, such as the United Kingdom, cyclists are allowed to cycle on car segments of roads even if dedicated cycling infra-structure is available. In about half of US states, bicyclists are required to use bicycle-specific facilities wherever available, following mandatory-use laws (Bopp, Sims, and Piatkowski 2018). In most states, this requirement only holds if the facilities are adequate. In the Netherlands, the use of bicycle facilities is also the norm, but the quality of the facilities is better than in most countries.

Right-of-way regulations also differ greatly. Uncertainty about the right-of-way may increase the chances of crashes resulting from different views of road users, which in turn may discourage cycling. Especially with turning traffic, uncertainty of right-of-way may cause crashes and thus discourage cycling. In the Netherlands, turning traffic is required to yield to traffic going straight, independent of whether traffic is a bicycle or a car. As a consequence, car driv-ers need to yield for cyclists if they turn. In the highway code of the United Kingdom, the right-of-way depends on the positions of the car and bicycle. If the bicycle is ahead of the car, the car should not overtake the bicycle and turn; however, if the car is in front of the cyclist, the car has the right-of-way. In the United States, vehicle codes require that vehicles turning right do so from the rightmost part of the road. Cyclists often use this part of the road, and motor vehicles are normally not supposed to use bicycle facilities. As a result, motor vehicles often do not move to the far right of the road when turning, even when they are supposed to. The general confusion over rules increases the probability of right-turn crashes, and liability is often uncertain.

It has frequently been argued that the legal position of cyclists when they are involved in crashes with a motor vehicle may explain higher cycling levels in some countries (e.g., Pucher and Dijkstra 2000). Many European countries have a regulation in place that "protects" cyclists from liability in cases of a crash with a motor vehicle, although European liabil-ity legislation has not yet been fully synchronized. For example, in the Netherlands, a car driver is automatically liable in the case of a collision between a car and a cyclist (or pedestrian), independent of whether they are responsible for the crash, unless one can prove circumstances beyond one's control (Dutch National Government 1994). Most European countries have similar legislation, but the United Kingdom, Cyprus, Malta, Romania, and Ireland, as well as most countries outside the European Union, including the United States, do not have such a presumed liability system in place (Gaffney 2015). In this case, the cyclist would need to prove that the car driver was at fault, which is a difficult procedure because there are often no independent witnesses. The exact influence of these differences in liability on cycling is unknown. Countries with relative high levels of cycling, such as Denmark and the Netherlands, have legislation in place that offers greater protection to cyclists, but the introduction of such legislation often occurred after cycling rates were already high or increasing.

An often-discussed issue in the United States is legislation regarding how cyclists are expected to behave at stop signs. In most states, the bicycle is required to behave like a car and therefore come to a full stop when approaching an intersection with a stop sign. In 1982, Idaho passed a law that permits cyclists to treat stop signs like yield signs. It requires cyclists to slow down to a reasonable speed and, if required for safety, stop. If there is no other traffic, the cyclist can cross without coming to a full stop. Attempts to adopt a similar law in other states have failed, except in some smaller places and in Delaware, where, with support from the police, a "stop = yield" law was adopted in 2017. In recent years, Arkansas, Oregon, and Washington also passed laws allowing cyclists to treat stop signs as yield signs. Such "stop = yield" laws probably help increase cycling by reducing the time and energy to get places, although this effect has not been documented.

Another example, related to perhaps one of the most heated topics in cycling, is legislation on the use of bicycle helmets (See Elvik, chapter 4, this volume). Although health officials agree that, for some crashes, wearing a helmet would reduce injuries, its effects on cycling and its overall benefits are still a topic of debate (Robinson 2006; Radun and Oliver 2018) and outside the scope of this chapter. Countries differ in their requirements and practices for wearing a helmet. In Australia, most states and territories have mandatory helmet laws. In Canada, several provinces require helmets, while other provinces do not. In the United States, no state requires adult cyclists to wear helmets, but young cyclists are required to wear them in 21 states. Similar differences exist in Europe, where countries such as Sweden and France require children to wear a helmet, but it is voluntary in most other countries. It has been suggested that helmet laws may serve as a deterrent to bicycle use. However, it is difficult to research their effect in isolation (see Elvik, chapter 4, this volume). Even in countries that introduced or eliminated helmet laws, other changes often occurred at the same time and, as a result, the independent effect is difficult to determine.

A third example of a difference in regulations for cyclists regards substance intoxication. In many countries, cycling is treated similarly to driving a car, so the legal blood-alcohol level for cyclists is the same as for drivers. However, the consequences differ significantly between countries. For example, in the United Kingdom, the maximum penalty for cycling while under the influence of alcohol or drugs is a £1,000 fine, but one would not receive points on one's license. In Germany, the limit for operating a bicycle is higher than for driving a car (Hagemeister and Kronmaier 2017), but if enforced and ticketed, the penalty will include a fine as well as points on the license. In the Netherlands, similar limits on alcohol are applied to both cyclists and car drivers, but these laws are rarely enforced on cyclists. Such laws—and their enforcement—could discourage cycling in certain situations.

It is reasonable to assume that stricter enforcement of laws for motor vehicles could improve bicycle safety and thereby increase ridership. Few studies compare differences in enforcement between countries and regions. Pucher and Dijkstra (2000) noticed that, compared to the United States, enforcement of traffic regulations is stricter for motorists in Germany and the Netherlands. Types of regulations for which enforcement is particularly important include speeding and behavior around cyclists. Recently, several police forces in England started to actively enforce and fine motorists who overtake cyclists without providing them with sufficient space. Driver training could also improve traffic safety. The length of training and focus on the rights and behavior of pedestrians and cyclists in the theoretical and practical exams differs to a large extent between countries.

Enforcement on cyclists also differs between countries and may discourage cycling, depending on the focus and approach, even if it helps to make cycling safer. Some countries predominantly enforce regulations that would increase bicycle safety directly, such as having lights on at night. Other countries may focus on regulations that may have no benefit for cyclists or other road users, or even put cyclists in greater danger (e.g., location on the road). Concerns may go beyond the issue of enforcement of cycling regulations: black and Hispanic cyclists in the United States cite fear of racial profiling by the police as a deterrent to bicycling (Brown and Sinclair 2017).

**Built Environment**

Other chapters in this volume discuss interventions specific to cycling, such as infrastructure (e.g., bicycle paths) (Furth, chapter 5), bicycle sharing programs (Fishman and Shaheen, chapter 10), and bicycle parking (Buehler, Heinen, and Nakamura, chapter 6). In this section, we will discuss policies that influence the bicycling environment more indirectly but will also include examples of roadway and roadside infrastructure that influence cycling directly.

Distance is one of the most fundamental factors influencing the individual's decision to bicycle. Distances to potential destinations are largely determined by the density of development and mix of land uses in a community. Land-use planning is thus essential to the success of efforts to increase bicycling. Zoning regulations are especially important in determining what kinds of land uses are allowed where, and they often put limits on the density of development. Many cities in the United States are modifying their zoning to allow greater densities and more mixing of land uses around transit stations, an approach that should also help to increase the viability of bicycling. Such patterns of development are common in European countries where bicycle ridership is high.

The connectivity of the network and the directness of routes are also important in determining distances to potential destinations. Connectivity is largely determined by the street network, but in communities with many off-street bike paths, the bicycle network has greater connectivity than the street network available to motorized vehicles (Tal and Handy 2012). On the other hand, bicyclists may face physical barriers where drivers do not—for example, when bicyclists are prohibited from using bridges. To increase cycling, planners have used a strategy of providing connections between locations that are more direct for cyclists than for cars. Sometimes called "filtered permeability," it can be applied at both a city level and on a smaller scale in neighborhoods. Many cities have implemented one-way streets for cars that can be used bidirectionally by cyclists, as well as shortcuts for cyclists that are inaccessible to cars in selected locations. Good examples of its application at a city level are found in Houten and Groningen in the Netherlands. Traveling by car between two locations within Houten via the city center is not possible; drivers must use the ring road instead. In contrast, cyclists can travel directly from one side of the community to the other.

A related policy is the establishment of car-free or car-light zones, which are found widely in European city centers but rarely in the United States. The goal of this approach is to create a vibrant commercial district where pedestrians are given priority in public spaces. This approach can improve conditions for bicyclists as well, although bicycle riding is sometimes restricted in these areas as a measure to ensure the safety of pedestrians, and the quality of bicycle routes from residential areas to the city center is also critical. Case studies show that the implementation of car-free zones does indeed reduce driving and can increase bicycling along with walking and public transport (Topp and Pharoah 1994). Although several case studies have shown that restricting car access does not necessarily harm the local economy, resistance from local shop owners is often encountered, and cities may want to introduce car-restrictive measures incrementally to first demonstrate their benefits (Lambe, Murphy, and Bauman 2017). The impact of restricting car access may be enhanced with the implementation of car parking restrictions and pricing policies.

Short of banning cars altogether, cities often adopt traffic-calming measures that aim to reduce the volume and speed of motor vehicles to create a space better suited for cycling. Reducing the number of lanes for motor vehicles may create space for bicycle facilities and at the same time reduce the speed of traffic to levels suitable for cyclists. Cities use speed reduction measures such as speed humps and devices that horizontally deflect traffic to directly reduce the speed of motor vehicles, with the aim of reducing the severity of crashes, given that speed is the most important determinant of the severity of consequences to vulnerable road users when they collide with a car. However, current evidence is inconclusive as to whether traffic calming encourages cycling (Brown, Moodie, and Carter 2017).

One study showed that traffic calming was not correlated with cycling in most cities, although in Ghent, Belgium, individuals living in a neighborhood with more traffic-calming features were more likely to cycle (Mertens et al. 2017). One explanation for such differences in findings is that traffic-calming efforts vary widely from community to community. Some communities specifically design their traffic-calming measures so they do not inadvertently discourage bicycling. For example, speed bumps and humps can be installed with narrow pass-throughs for cyclists, but which are still unavoidable by motor vehicles and thus slow them down (figure 7.2).

Public transport policies and investments also play a critical role in encouraging cycling behavior. The bicycle is often used in combination with public transport journeys, and countries offer many facilities at stations, such as parking, to encourage this (see Buehler, Heinen, and Nakamura, chapter 6, this volume).

Cities can implement many smaller design features that serve the needs of cyclists (see also Furth, chapter 5, this volume). Many cities have installed

**Figure 7.2**
Bicycle-friendly speed bumps. *Source*: Eva Heinen.

bicycle route signage to provide guidance to cyclists as to the location of attractive or safe routes. Bicycle boxes and advance stop lines, which allow cyclists to wait in front of motorized traffic and reduce conflicts with turning vehicles, are used in various countries, especially in places without separated cycling infrastructure. Dedicated bicycle traffic lights (figure 7.3) are common in locations with higher levels of cycling. These lights often also allow cyclists to start before motorized traffic or offer a separate signal phase only for cyclists. Many cities have policies that require businesses to provide bicycle facilities, such as parking and showers for employees (see Buehler, Heinen, and Nakamura, chapter 6, this volume). The degree to which the presence of such facilities increases cycling is uncertain.

## Conclusions

A wide variety of programs and policies have the potential to increase cycling, including promotions, incentives, laws, and policies regarding the built environment. Some policies and programs do not specifically aim to encourage cycling but instead focus on deterring the use of other modes of transport, driving specifically. Policies and programs that discourage

**Figure 7.3**
Bicycle traffic light. *Source*: Susan Handy.

driving are an essential complement to those that encourage bicycling, and policies and programs of both types are an essential complement to investments in cycling infrastructure.

The policies and programs discussed in this chapter are often not rigorously evaluated, and therefore the independent effect of each policy and program is not well known. One complicating issue is that, in practice, policies and programs are often used in combination with one another, making it difficult for researchers to determine the independent effect of each program or policy (Scheepers et al. 2014; Pucher et al. 2010). As a result, it is not easy to compare the effectiveness of these policies and programs with one another or with other strategies, such as investments in infrastructure. However, researchers and practitioners alike have concluded that an integrated combination of policies and programs together with suitable infrastructure is essential to encourage cycling over an extended time.

It is essential to differentiate among three kinds of policies and programs: those that encourage individuals who do not cycle to begin cycling; those that encourage current cyclists to cycle more frequently; and those that make cycling safer and more enjoyable for both current and potential cyclists. Heinen et al. (2011) concluded that the decision to cycle and the frequency of cycling are correlated with different policies and schemes. For example, they concluded that having bicycle storage, changing facilities, and travel compensation schemes available at work encouraged individuals to cycle but did not influence cycling frequency. Interventions using the programs and policies discussed should be tailored to specific populations to increase their effectiveness, but the policies and programs that encourage cycling for some groups will also likely encourage cycling for the general population (Pucher and Buehler 2008).

Policies and programs aiming to reduce car use through restrictions or pricing are among those most commonly evaluated. Evidence suggests that these policies and programs are effective at encouraging cycling, and they may come with additional benefits, such as improved urban livability and reductions in air pollution. Despite the fact that these policies and programs are not necessarily expensive, they may not be politically viable everywhere.

This chapter focused on existing policies and programs that may encourage cycling and provided examples from multiple countries that may serve as good examples for practitioners from elsewhere or cases for further investigation. Studies that isolate each policy and program may provide essential information on their unique effectiveness. This could most easily be done by introducing or removing individual policies or programs within a particular community. To effectively encourage cycling, practitioners should explore combining a variety of programs and policies in addition to infrastructure provision to reach their policy aims in relation to cycling.

### References

Belgium Federal Ministry of Finance. 2018. Fietsvergoeding. https://financien.belgium.be/nl/particulieren/vervoer/aftrek_vervoersonkosten/woon-werkverkeer/fiets#q1.

Bike Benefits. 2018. How the Program Works. http://www.bicyclebenefits.org/wp#/how-the-program-works.

Bike Switzerland. 2018. What Is a Swiss Slow-up? https://bikeswitzerland.com/what-is-a-swiss-slow-up/.

Bopp, Melissa, Dangaia Sims, and Daniel Piatkowski. 2018. Policy and Law Approaches to Bicycling. In *Bicycling for Transportation, an Evidence-Base for Communities*, 165–191. Amsterdam: Elsevier.

Brockman, Rowan, and Kenneth Fox. 2011. Physical Activity by Stealth? The Potential Health Benefits of a Workplace Transport Plan. *Public Health* 125: 210–216.

Brown, Charles, and James Sinclair. 2017. Removing Barriers to Bicycle Use in Black and Hispanic communities. Paper no. 17–03327, presented at the Annual Meeting of the Transportation Research Board, January, Washington, DC.

Brown, Vicky, Marj Moodie, and Rob Carter. 2017. Evidence for Associations between Traffic Calming and Safety and Active Transport or Obesity: A Scoping Review. *Journal of Transport and Health* 7(A): 23–37.

Bundesverfassungsgericht. 2008. *Leitsatz zum Urteil des Zweiten Senats vom 9 Dezember 2008*. https://www.bundesverfassungsgericht.de/SharedDocs/Entscheidungen/DE/2008 /12/ls20081209_2bvl000107.html.

Christelijk Nationaal Vakverbond (CNV). 2018. *Werkkostenregeling fiets van de zaak*. https://www.cnvvakmensen.nl/diensten/kennisbank/werkkostenregeling-fiets-van -de-zaak.

Department for Transport. 2011. *Cycle to Work Scheme Implementation Guidance*. https://www.gov.uk/government/publications/cycle-to-work-scheme-implementation -guidance.

Dill, Jennifer, and Theresa Carr. 2003. Bicycle Commuting and Facilities in Major US Cities: If You Build Them, Commuters Will Use Them—Another Look. *Transportation Research Record* 1828(1): 116–123.

Ducheyne, Fabian, Ilse De Bourdeaudhuij, Matthieu Lenoir, and Greet Cardon. 2014. Effects of a Cycle Training Course on Children's Cycling Skills and Levels of Cycling to School. *Accident Analysis and Prevention* 67: 49–60.

Dutch National Government. 1994. Artikel 185 Wegenverkeerswet. http://wetten .overheid.nl/BWBR0006622/2018-03-15#HoofdstukXII_Artikel185.

Gaffney, Richard. 2015. Cycling Accidents and Presumed Liability: UK vs Europe. Slater and Gordon. https://www.slatergordon.co.uk/media-centre/blog/2015/08/cycling -accidents-and-presumed-liability-uk-vs-europe/.

Gamble, Tim, Ian Walker, and Aleksandra Laketa. 2015. Bicycling Campaigns Promoting Health versus Campaigns Promoting Safety: A Randomized Controlled Online Study of "Dangerization." *Journal of Transport and Health* 2(3): 369–378.

Gómez, Luis, Olga Sarmiento, Diego Lucumí, Gladis Espinosa, Roberto Forero, and Adrian Bauman. 2005. Prevalence and Factors Associated with Walking and Bicycling for Transport among Young Adults in Two Low-Income Localities of Bogotá, Colombia. *Journal of Physical Activity and Health* 2(4): 445–459.

Goodman, Anna, Jenna Panter, Stephen Sharp, and David Ogilvie. 2013. Effectiveness and Equity Impacts of Town-wide Cycling Initiatives in England: A Longitudinal, Controlled Natural Experimental Study. *Social Science and Medicine* 97: 228–237.

Hagemeister, Carmen, and Markus Kronmaier. 2017. Alcohol Consumption and Cycling in Contrast to Driving. *Accident Analysis and Prevention* 105: 102–108.

Heinen, Eva, Kees Maat, and Bert van Wee. 2010. Commuting by Bicycle: An Overview of the Literature. *Transport Reviews* 30(1): 59–96.

Heinen, Eva, Kees Maat, and Bert van Wee. 2011. The Role of Attitudes toward Characteristics of Bicycle Commuting on the Choice to Cycle to Work over Various Distances. *Transportation Research Part D: Transport and Environment* 16(2): 102–109.

Heinen, Eva, Jenna Panter, Alice Dalton, Andy Jones, and David Ogilvie. 2014. Socio-spatial Patterning of the Use of New Transport Infrastructure: Walking, Cycling and Bus Travel on the Cambridgeshire Guided Busway. *Journal of Transport and Health* 2(2): 199–211.

Johnson, Rebecca, and Sam Margolis. 2013. A Review of the Effectiveness of Adult Cycle Training in Tower Hamlets, London. *Transport Policy* 30: 254–261.

Lambe, Barry, Niamh Murphy, and Adrian Bauman. 2017. Smarter Travel, Car Restriction and Reticence: Understanding the Process in Ireland's Active Travel Towns. *Case Studies on Transport Policy* 5(2): 208–214.

London Cycling Campaign. 2018. Love London, Go Dutch. https://lcc.org.uk/pages /go-dutch.

Meland, Solveig, Terje Tretvik, and Morton Welde. 2010. The Effects of Removing the Trondheim Toll Cordon. *Transport Policy* 17(6): 475–485.

Mertens, Lieze, Sofie Compernolle, Benedikte Deforche, Joreintje Mackenbach, Jeroen Lakerveld, Johannes Brug, Celina Roda, Thierry Feuillet, Jean-Michel Oppert, Ketevan Glonti, Harry Rutter, Helga Bardos, Ilse De Bourdeaudhuij, and Delfien Van Dyck. 2017. Built Environmental Correlates of Cycling for Transport across Europe. *Health and Place* 44: 35–42.

Ogilvie, David, Matt Egan, Val Hamilton, and Mark Petticrew. 2004. Promoting Walking and Cycling as an Alternative to Using Cars: Systematic Review. *BMJ* 329(7469): 763.

Panter, Jenna, Simon Griffin, Alice Dalton, and David Ogilvie. 2013. Patterns and Predictors of Changes in Active Commuting over 12 Months. *Preventive Medicine* 57(6): 776–784.

Pucher, John, and Ralph Buehler. 2006. Why Canadians Cycle More than Americans: A Comparative Analysis of Bicycling Trends and Policies. *Transport Policy* 13(3): 265–279.

Pucher, John, and Ralph Buehler. 2008. Making Cycling Irresistible: Lessons from the Netherlands, Denmark and Germany. *Transport Reviews* 28(4): 495–528.

Pucher, John, and Ralph Buehler. 2010. Walking and Cycling for Healthy Cities. *Built Environment* 36(4): 391–414.

Pucher, John, and Lewis Dijkstra. 2000. Making Walking and Cycling Safer: Lessons from Europe. *Transportation Quarterly* 54(3): 25–50.

Pucher, John, Jennifer Dill, and Susan Handy. 2010. Infrastructure, Programs, and Policies to Increase Bicycling: An International Review. *Preventive Medicine* 50(Suppl): S106–S125.

Radun, Igor, and Jake Olivier. 2018. Bicycle Helmet Law Does Not Deter Cyclists in Finland. *Transportation Research Part F: Traffic Psychology and Behaviour* 58: 1087–1090.

Rissel, Chris, Carolyn New, Li Wen, Dafna Merom, Adrian Bauman, and Jan Garrard. 2010. The Effectiveness of Community-Based Cycling Promotion: Findings from the Cycling Connecting Communities Project in Sydney, Australia. *International Journal of Behavioral Nutrition and Physical Activity* 7(1): 8.

Robinson, Dorothy. 2006. Analysis and Comment: No Clear Evidence from Countries That Have Enforced the Wearing of Helmets. *British Medical Journal* 332: 722–725.

Scheepers, Eline, Wanda Wendel-Vos, Lea den Broeder, Elise van Kempen, Pieter van Wesemael, and Albertine Schuit. 2014. Shifting from Car to Active Transport: A Systematic Review of the Effectiveness of Interventions. *Transportation Research Part A: Policy and Practice* 70: 264–280.

Schneider, Robert. 2013. Theory of Routine Mode Choice Decisions: An Operational Framework to Increase Sustainable Transportation. *Transport Policy* 25C: 128–137.

Schneider, Robert, Jennifer Kusch, Anna Dressel, and Rebecca Bernstein. 2018. Can a Twelve-Week Intervention Reduce Barriers to Bicycling among Overweight Adults in Low-Income Latino and Black Communities? *Transportation Research Part F: Traffic Psychology and Behaviour* 56: 99–112.

Sherwin, Henrietta, Kiron Chatterjee, and Juliet Jain. 2014. An Exploration of the Importance of Social Influence in the Decision to Start Bicycling in England. *Transportation Research Part A: Policy and Practice* 68: 32–45.

Tal, Gil, and Susan Handy. 2012. Measuring Nonmotorized Accessibility and Connectivity in a Robust Pedestrian Network. *Transportation Research Record* 2299(1): 48–56.

Telfer, Barbara, Chris Rissel, Jeni Bindon, and T. Bosch. 2006. Encouraging Cycling through a Pilot Cycling Proficiency Training Program among Adults in Central Sydney. *Journal of Science and Medicine in Sport* 9(1–2): 151–156.

Thomas, Ian, Stephen Sayers, Janet Godon, and Stacia Reilly. 2009. Bike, Walk, and Wheel: A Way of Life in Columbia, Missouri. *American Journal of Preventive Medicine* 37(6): S322–S328.

Topp, Hartmut, and Tim Pharoah. 1994. Car-Free City Centres. *Transportation* 3: 231–247.

Trappers. 2018. Fietsen is verdienen! https://site.trappers.net/.

Wardman, Mark, Myles Tight, and Matthew Page. 2007. Factors Influencing the Propensity to Cycle to Work. *Transportation Research Part A* 41(4): 339–350.

Yang, Lin, Shannon Sahlqvist, Alison McMinn, Simon Griffin, and David Ogilvie. 2010. Interventions to Promote Cycling: Systematic Review. *British Medical Journal* 341: c5293.

# 8 Evaluation of Cycling Policies and Projects

Bert van Wee

This chapter discusses challenges for the evaluation of bicycle projects and highlights the strengths and weaknesses of three evaluation tools: cost-effectiveness analysis (CEA), multicriteria analysis (MCA), and cost-benefit analysis (CBA).

Cycling projects and policies cost money and may impact travel times for cyclists and other road users, the safety and health of cyclists, and the environment. This raises the question of how to evaluate the benefits relative to the costs. Such evaluations are helpful in making decisions on the overall budgets available for cycling policies, prioritizing policy options within a given budget, and making decisions about road space allocation.

There is some guidance for the evaluation of cycling policies. For example, the Health Economic Assessment Tool (HEAT) for cycling and walking (WHO 2014) provides support on how to estimate the value of reduced mortality. In addition, the Centre for Diet and Activity Research (CEDAR) provides the Integrated Transport and Health Impact Modelling Tool (ITHIM) (Woodcock, Givoni, and Morgan 2013; see also Cedar n.d.) providing practical guidance to assess health benefits of transport scenarios and policies at the urban and national levels, including cycling investments. These tools provide guidance for partial evaluations, mainly health benefits, but do not aim to evaluate the overall costs and benefits of cycling (and walking) policies.

The chapter focuses on the evaluation of candidate policy options for cycling (sometimes labeled as ex ante policies, as opposed to ex post evaluations, which evaluate policies already implemented).

The next section discusses whether we should evaluate cycling policies at all and is followed by a section that gives an overview of challenges for evaluations of candidate cycling policy options and a section that elaborates on the pros and cons of evaluation tools (CEA, MCA, CBA). Then the chapter presents a few real-world applications of the evaluation of candidate policy options, followed by a section summarizing the main conclusions of this chapter.

## Should You Evaluate at All?

The first and very fundamental question is, should one evaluate candidate cycling policies at all? I interpret *evaluation* as carrying out an assessment of the effects and costs of effects, and in some cases as a way of integrating multiple effects to come to an integrated assessment of all the important effects.

In several cases, such an assessment is not needed. This is at least the case when it is clear that the policies improve cycling safety and stimulate cycling in an economically effective way. This might apply to many relatively inexpensive cycling policies, such as placing (additional) signs to show cyclists the right direction to specific destinations. The assessment costs can easily be relatively high compared to the costs of the policies. In addition, in several cases, a brief estimation of the effects can probably be based on previous experiences in the same (or a comparable) town, city, or region. Often the conclusion will be that it is a good idea to implement the policy and that no advanced evaluation is needed. So, an alternative for a relatively expensive assessment could be to do a brief literature, document, and web search focusing on experiences elsewhere.

Guidelines for cycling infrastructure can save time and costs as well as improve infrastructure quality. In the Netherlands, perhaps the world's premier cycling country, regularly updated national guidelines provide detailed recommendations for virtually every aspect of infrastructure design and engineering. Local municipalities "just do it"; they provide cycling infrastructure in all cities and towns as well as in rural areas. Borrowing from the Dutch saying "better well stolen than badly invented," it is particularly useful for other countries to examine the well-developed and time-tested guidelines from the Netherlands and other countries that have them. Such guidelines can be used as a point of reference, but they must be customized and adapted to the local conditions specific to each particular city and neighborhood.

## Challenges for Evaluations for Cycling Projects

This section discusses several challenges to being able to better evaluate the pros and cons of cycling projects and policies. These challenges are relevant for all three evaluation methods—CEA, MCA, and CBA.

First, cycling is generally poorly included in transport models. Because many of the benefits of cycling policies depend on the levels of cycling that they induce, and models are often used to estimate these levels, this can be problematic. Thus, it is a major challenge to develop better models to forecast the effects of cycling policies on travel behavior. Such models need to be country- or even region-specific, including cultural factors such as the status of cycling. Data availability is often a problem, but opportunities based on new information and communication technologies (e.g., big data collected via mobile apps) might provide solutions. Sharing datasets and using parameters from other countries or regions might also provide a solution. At least carrying out a sensitivity analysis using different parameter settings is recommended.

Second, safety effects, or at least the fear of increased bicyclist injuries and fatalities, seem to be a barrier for the implementation of cycling policies, at least in cities and towns where cycling is not (yet) common (see in this volume Buehler and Pucher, chapter 2; Elvik, chapter 4). Unfortunately, the way in which safety effects are estimated is often incorrect. Several studies use aggregated average risk factors per mode, but these averages do not need to apply to a specific case or project. Aggregate average risk factors for cars include the relatively safe motorways, but cycling is rarely an alternative for long-distance car trips, for which motorways are used. Another methodological problem is that cycling becomes safer if cycling levels increase: the "safety in numbers" effect (see Elvik, chapter 4, this volume). HEAT advises including this effect by recommending reductions in (default) risk factors up to 45%, and ITHIM also proposes to use nonlinear risk factors.

Third, because health benefits are often a dominant benefit category of cycling policies, it is very important to adequately assess the health benefits. But this is not easy. To quote Woodcock, Givoni, and Morgan (2013): "Methods to estimate the health impacts from transport-related physical activity and injury risk are in their infancy." One reason is that cycling is only one of multiple forms of exercise. People might substitute other forms of exercise for cycling, but it is also possible that a person who feels fit because of cycling takes the stairs more frequently as opposed to the elevator. What is needed for evaluation is the *change* in health benefits resulting from a *change* in physical activity. Decisio and Transaction Management Centre (2012) assume that for trips of up to 3.7 km, 75% of health benefits from cycling are additional, and this percentage goes down to 25% for trips over 15 km. Estimating the health benefits of cycling is also difficult because of potential self-selection effects: maybe the people with better than average health are the first to start cycling or increase cycling levels. To the best of my knowledge, there is no tool (including HEAT and ITHIM) that explicitly includes self-selection effects and how to deal with them. Also, the air pollution intake of people cycling is higher than for people driving or traveling by public transport, and concentrations do not have to be the same for places where one cycles compared to the roads a driver uses. The time of exposure also is not necessarily the same. A what-if study comparing making shorter trips by bike rather than by car shows that the positive health impact of exercise resulting from cycling is much more important than the negative impact of the higher intake of pollutants and the risks associated with cycling (de Hartog et al. 2010).

Van Wee and Ettema (2016) conceptualize the complex relationship between travel behavior and health and explicitly include travel-related physical activity, the intake of pollutants, and travel incidents (see figure 8.1). Tools

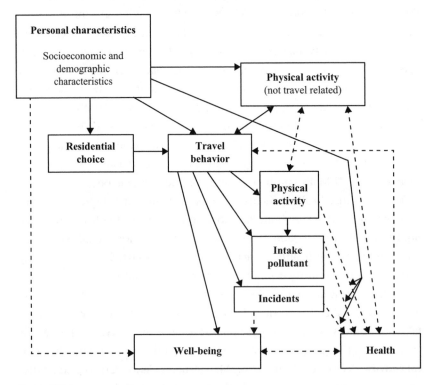

**Figure 8.1**
A conceptual model for the relationships between travel behavior and health. *Source*: van Wee and Ettema (2016).

such as HEAT and ITHIM do include some of the factors and relationships but not all, which is understandable because there are no data and models for the full causal model yet. It is beyond the aim of this chapter to discuss the figure in detail. The conceptual model is included to provide some guidance for evaluation purposes. More specifically, the model makes clear which factors could matter and how they are interrelated.

Fourth, cycling-friendly cities are often attractive in general (see in this volume Koglin, te Brömmelstroet, and van Wee, chapter 18; Geller and Marqués, chapter 19); for example, because of a reduction in the negative environmental impacts of cars. Thus, an ex ante evaluation should include an increase in the attractiveness of cities, but it is very difficult to quantify, let alone monetize, such benefits.

Fifth, people value higher levels of accessibility, and cycling policies can increase these levels. Benefits relating travel time and cycling level are generally included in ex ante assessments. In addition, people might value having options available, even if they do not use them (the option value). Ex ante evaluations should preferably include such option values, but doing this is not at all easy. Besides, it is unclear if and how an increase in options to combine travel and other activities in the case of passive travel (e.g., trains and self-driving cars) will affect the appreciation for cycling (and related accessibility) and mode choice.

Sixth, in many rural areas and some urban areas, social exclusion is a problem. Some people, especially poor people without a car, cannot fully participate in societal activities. Public transport is often seen as the solution, but cycling can be an alternative as well (see Martens, Golub, and Hamre, chapter 14, this volume). There is a long way to go to quantify (and monetize) social exclusion effects, so including such effects in ex ante evaluations is problematic.

### Which Evaluation Framework Should Be Used: CBA, MCA, or CEA?

If policy makers have some estimates for the separate effects of policies, the next question is how to integrate those effects. In some cases, CEA may be an option, especially if policies aim to have a specific effect and several alternatives are available. CEAs generally look at the monetary cost per unit of change in a policy-relevant variable, examples being costs of avoiding one fatality or one ton of $CO_2$ emitted.

The effect of cycling policies could be an increase in cycling levels. That increase then has several effects (e.g., health, accessibility, and urban quality), but not all of these have to be quantified (or valued). The question then becomes, how many (expected) additional cycling kilometers per euro or dollar (or other currency) do the policy options generate? Another application of CEA can be to estimate the price per (expected) reduction in the number of fatalities and injuries. CEA is much less data-intensive than a CBA and still allows measures to be selected based on efficiency. An even simpler form of CEA is to use threshold values: policies are "sound" if the cost per additional kilometer cycled or fatality reduced is below a certain threshold. Then, the evaluator only needs to determine whether a candidate policy is inexpensive enough.

Focusing on only one effect, CEAs ignore other effects, such as increased access to destinations, safety improvements, or emission reductions. If those other effects are substantial, a multicriteria analysis (MCA) or a cost-benefit analysis (CBA) might be the preferred method, especially if such other effects are not highly correlated with cycling levels. In such cases, MCA and CBA are the two methods most often used to evaluate alternative policy options.

MCA is like a table summarizing the evaluation of consumer products. All relevant decision criteria are listed, some in monetary terms (such as the cost of a bike parking facility), some in their own unit (e.g., the capacity expressed in the number of bikes that can be stalled or parked), some in rank order (e.g., rating the ease of finding an available place to stall one's bike as – for below average, 0 for average, + for above average, or ++ for much better than average), and some verbally (e.g., is there paid parking, yes or no?). Based on scores and weight factors, incomparable effects are aggregated to come to the "best" option. MCA is not very data-intensive, because it does not need to quantify all effects. Some effects can be estimated based on expert judgment, a method that can be used to derive scores ranging from – to ++.

The most data-intensive evaluation method is CBA. It is an overview of all relevant pros and cons of alternative policy options. These options are quantified and expressed in monetary terms to the extent possible. Costs of building and maintaining infrastructure are often available in monetary terms. Unit costs for travel time savings are also generally available (e.g., Abrantes and Wardman 2011), and some countries (such as the Netherlands) even prescribe values of time to be used for CBAs. The impact of changes in noise levels on the value of dwellings is also addressed in the literature (e.g., Wilhelmsson 2000). Estimating changes in travel times and noise levels resulting from transport policies requires a model to estimate changes in travel behavior. However, in the case of cycling policies, most models yield unreliable and controversial estimates, as discussed earlier. Unit costs are based on market prices and other consumer evaluations, an important exception being the changes in emissions in greenhouse gases such as $CO_2$. Costs and benefits in future years must be converted to the price level of one common year. To do so, a so-called discount rate is used to convert values from different years to the same year. Such a discount rate is similar to an interest rate expressing to what extent people prefer money and benefits now compared to in the future. Discount rates also reflect a risk premium, expressing people's preference for lower risk and uncertainty. Finally, CBA expresses the results of its calculations in one or a few key indicators such as total benefits minus total costs or, more frequently, the ratio of total benefits to total costs.

CBA needs quantitative data for all relevant factors and unit costs, such as projected changes in emission, noise, congestion, and safety levels, as well as estimates of the value of the travel time saved. Despite numerous technically sophisticated studies aimed at estimating unit costs, there remain many debates about their appropriate estimates. That is especially the case for those benefits and costs for which there is no market price, such as fatalities, injuries, noise, and quality of life and the general attractiveness of the area where a project is implemented. Alberini and Ščasný (2013) provide a detailed discussion of alternative methodologies for estimating the monetary costs of fatalities.

The choice of discount rate is controversial, with a wide range of possible values that have important impacts on the estimated overall benefit-cost ratio of a proposed project. Most investment projects incur costs at the outset and generate benefits over decades in the future. Such long-term investment projects are disadvantaged by high discount rates, which reduce the estimated value of future benefits while giving full weight to the value of current costs. Conversely, low discount rates generally favor long-term projects.

CBAs allow comparisons across projects, based on benefit-cost ratios (BCRs), just as projects can be compared based on CEAs. Experience shows that BCRs for cycling projects generally are much higher than for public transport or road infrastructure projects, especially in larger cities. Despite the many inputs required for the analysis, CBAs for cycling are not always expensive. Experiences in the Netherlands show that these CBAs can be based largely on two workshops with experts and a set of default unit costs.

An important decision is which evaluation method to choose for each specific case. Economists generally tend to prefer CBA, because of the low risk of counting the same effects twice (double counting). For example, if lower noise emissions (directly) and lower noise impacts on nearby dwellings (indirectly) are both included, the overlap is large. Another reason why economists prefer CBA is that weighting is based on consumer preferences. Therefore, possibilities for manipulating results are limited, especially compared to an MCA.

By comparison, planners often prefer MCA because more effects can be included, especially those for which there are no unit costs or that cannot be quantified at all. Another advantage of MCA is that it can be easily linked to different stakeholders with different preferences, using weights, as will be demonstrated.

How do these three methods relate to health impact assessments (HIAs)? First, HIAs only focus on health effects, whereas MCA and CBA have a broader scope. Second, HIA results can be inputs for CEAs, MCAs, and CBAs. For CEA evaluations, they can focus on the cost in euros or US dollars per health benefit unit (e.g., estimated mortality rates or life years gained). By comparison, monetary values are needed for health impacts in a full CBA, but placing a monetary value on health impacts is controversial. For example, it can require assigning a value to human life, pain and suffering, and disability.

## Real-world Cases

This section presents a few real-world examples of all three methods, starting with CEA. Several CEAs are carried out for cycling projects, focusing on health impacts, examples being the cost per disability-adjusted life-year (DALY) or quality-adjusted life-year (QALY). An alternative is to calculate the number of people being more physically active because of cycling facilities.

Wang et al. (2004) provide an example of the latter approach by calculating the cost-effectiveness of bicycle and pedestrian trails. They estimated annual trail costs based on construction and maintenance expenses. Based on a questionnaire among users of the trails, three indicators were selected for physical activity levels: (1) the number of users who increased their physical activity because of the trail; (2) the number of users who became physically active to improve their general health; and (3) the number of users who became physically active in order to lose weight (Wang et al. 2004, 239). Cost-effectiveness ratios were derived "by dividing the costs of trail development and maintenance (total cost of the four trails) by the selected physical activity-related outcomes/items related to using trails" (Wang et al. 2004, 239). In addition, Wang et al. carried out sensitivity analyses by changing the number of users and the life cycle of the trails. Table 8.1 presents the results.

Beale et al. (2012) provide another example of a CEA of cycling infrastructure, focusing on health expressed in costs per QALY. Cost-effectiveness estimates range between £100 and £10,000 per QALY. The range of values results from different assumptions about the intervention's effectiveness and the proportion of the total intervention cost attributable to health (5% or 100%). This hundredfold range of values demonstrates the importance of such assumptions. For the comparison of cycling options, and keeping the assumptions constant, this is not much of a problem, because the relative ranking will not change. This can be a major problem, however, for the comparison of cycling projects with other projects aimed at increasing QALYs.

Keseru et al. (2016) provide an example of a multiactor multicriteria analysis (MAMCA). A multiactor MCA has the advantage of showing

**Table 8.1**
Costs, number of users, and cost-effectiveness of trails in promoting or supporting physical activity and public health (2003 US$)

| Item | Value |
|---|---|
| Trail development costs (US$/year) | |
| Construction of trails | 210,019 |
| Maintenance of trails | 79,016 |
| Total costs | 289,035 |
| Number of persons using the trails | |
| Users who are more physically active | 2,950 |
| Users who are physically active to improve general health | 2,037 |
| Users who are physically active to lose weight | 327 |
| Cost-effectiveness ratio (US$/user) | |
| Cost per user who is more physically active | 98 |
| Cost per user who is physically active to improve general health | 142 |
| Cost per user who is physically active to lose weight | 884 |

*Source*: Wang et al. (2004).

synergies or conflicts between stakeholders in terms of their preferences. The example provided here evaluates five alternatives for a cycle super-highway of about 14 km between Waalwijk and Tilburg, two cities in the southern Netherlands. The alternatives differ with respect to the urban and natural environment, level- and split-level junction design, and design of infrastructure (width, lighting, curves, and pavement choices). The selection of objectives, criteria, and weighting factors was based on inputs from local experts, a stakeholder workshop, and an online questionnaire that was filled in by stakeholders. The alternatives are compared with the BAU scenario (no cycle superhighway). Most indicator values were based on a traditional transport model. Other indicators were evaluated qualitatively by local experts. Table 8.2 shows the indicator values.

Based on these values, the alternatives were ranked for the MCA in general (not stakeholder-specific) and according to stakeholder group preferences for the MAMCA (table 8.3). Table 8.3 shows that public transport operators have very different preferences than the other stakeholders. This is because new cycling infrastructure would divert some public transport passengers to cycling instead, resulting in a loss of passenger revenue.

CBA is regularly used to evaluate cycling policies. Cavill et al. (2008) reviewed numerous CBAs that evaluate cycling and walking projects. Focusing on health implications, they concluded that there was a need for standardization of methods, especially in the evaluation of health effects. Perhaps their most important conclusion was that all except one of the cases they examined had benefits exceeding costs. Health-related benefits are often a dominant benefit category in CBAs for cycling. Since the Cavill et al. study, much progress has been made in improving health impact analysis. HEAT and ITHIM now provide much better and more detailed guidance for real-world applications.

Table 8.4 provides an overview of a real-world CBA case study to illustrate a typical CBA for cycling projects. The case study is of a bicycle bridge that crosses a motorway, a railway line, and a canal in Utrecht, the Netherlands. It shows that travel time savings dominate the benefits. Unlike several other CBAs for bicycle projects, health benefits are limited, because the bridge would reduce measured kilometers of cycling by providing a shorter route.

Table 8.4 highlights the important impact of uncertainty in the large range of values estimated for the three scenarios. Benefits minus costs range

**Table 8.2**

MAMCA: Values for different stakeholder categories, types of effects, and alternatives

| Stakeholder category and effect type | Category of effects (Env = environmental, Soc = social, Econ = economic) | Weight of the effect, as used in the multicriteria evaluation | Business as usual | Alternative | | | | |
|---|---|---|---|---|---|---|---|---|
| | | | | A | B | B1 | C | A- |
| **Citizens** | | | | | | | | |
| Air quality (BAU = 1) | Env | 0.38 | 1 | 0.996 | 0.981 | 0.980 | 0.976 | 0.986 |
| Noise (BAU = 1) | Env | 0.23 | 1 | 0.982 | 0.989 | 0.989 | 0.987 | 0.992 |
| Equity (BAU = 1) | Soc | 0.16 | 1 | 1.008 | 1.005 | 1.005 | 1.006 | 1.003 |
| Health of citizens (modal share of active travel modes in %) | Soc | 0.23 | 0.49 | 0.5056 | 0.4988 | 0.4990 | 0.5008 | 0.4965 |
| **Government** | | | | | | | | |
| Air quality (BAU = 1) | Env | 0.19 | 1 | 0.996 | 0.981 | 0.980 | 0.976 | 0.986 |
| Noise (BAU = 1) | Env | 0.34 | 1 | 0.982 | 0.989 | 0.989 | 0.987 | 0.992 |
| Health of citizens (modal share of active travel modes in %) | Soc | 0.17 | 0.49 | 0.506 | 0.499 | 0.499 | 0.501 | 0.497 |
| Sociopolitical acceptance (qualitative*) | Soc | 0.19 | 0 | 2 | −1 | −1 | 1 | 1 |
| **Public transport operators** | | | | | | | | |
| Cost-effectiveness of public transport operation (number of passengers on bus services) | Econ | 0.68 | 28,200 | 27,567 | 27,843 | 27,833 | 27,760 | 27,937 |
| Public funding of transport (level of public funding to operate bus services) (qualitative*) | Econ | 0.07 | 0 | 0 | 0 | 0 | 0 | 0 |
| Accessibility of stops (qualitative*) | Soc | 0.26 | 0 | 3 | 2 | 1 | 2 | 1 |

*(continued)*

**Table 8.2 (continued)**

MAMCA: Values for different stakeholder categories, types of effects, and alternatives

| Stakeholder category and effect type | Category of effects (Env = environmental, Soc = social, Econ = economic) | Weight of the effect, as used in the multicriteria evaluation | Business as usual | Alternative | | | | |
|---|---|---|---|---|---|---|---|---|
| | | | | A | B | B1 | C | A- |
| Employers | | | | | | | | |
| Economic activity (number of jobs in the region) | Econ | 0.26 | 23,673 | 23,734 | 23,713 | 23,714 | 23,720 | 23,699 |
| Accessibility (qualitative*) | Soc | 0.33 | 0 | 2 | 2 | 2 | 2 | 1 |
| Health of employees (modal share of active travel modes in %) | Soc | 0.28 | 0.49 | 0.5056 | 0.4988 | 0.4990 | 0.5008 | 0.4965 |
| Livability (qualitative*) | Soc | 0.12 | 0 | 1 | 1 | 1 | 1 | 1 |

*Note:* Alternatives are a combination of (1) municipalities to be connected, (2) type of land use to be connected, and (3) level of service of the bicycle infrastructure.

*Qualitative scale between very negative (−3) and very positive (+3).

*Source:* Based on Keseru et al. (2016), with modifications of labels made by the author.

Table 8.3

MAMCA: Ranking of the alternatives by MCA order and by the order preferred by the stakeholder groups

| | | MAMCA order: ranking by stakeholder group | | | |
|---|---|---|---|---|---|
| Alternatives | MCA order | Citizens | Government | Public transport operators | Employers |
| BAU | 6 | 6 | 6 | 1 | 6 |
| A | 1 | 1 | 1 | 6 | 1 |
| B | 4 | 4 | 3 | 3 | 4 |
| B1 | 1 | 5 | 3 | 2 | 4 |
| C | 2 | 2 | 4 | 5 | 2 |
| A– | 3 | 5 | 5 | 2 | 5 |

from −16.3 + Pro Memory to 65.0 (excluding Pro Memory—not quantified/ monetized effects). Uncertainties in the estimates of investment costs and travel time benefits differ most among the alternatives, although by much less than the factor of 100 noted in the CEA example earlier. For travel times, the uncertainty results from the poor quality of the traffic model used. As indicated in table 8.4, researchers were not able to quantify spatial quality / environmental impacts at all. Moreover, option values and social exclusion effects were not included in the analysis. The unit costs used in the estimation procedure are similar to those shown in table 8.5.

In order to compare costs and benefits in monetary terms, standardized values for unit costs are needed. A recent Dutch report presents some default unit costs aimed at providing guidance for cycling CBAs (see table 8.5). This chapter presents a few examples. Providing default values for unit costs makes it easier to conduct ex ante assessments and increases the consistency across such evaluations. The marginal values of travel time savings for car, bus, and train travel were based on a stated-preference survey. The value of travel time savings for cycling was based on the judgment of an expert group, which advised using the same time value as for car drivers. Effects on labor productivity only include the difference in working days of cyclists and others. Effects on life-years were based on de Hartog et al. (2010). Effects of the two other health categories were based on a combination of assumptions and empirical data. Health benefits of additional cycling depend on the type of project. Infrastructure projects often improve the health of people who already cycle. Cycling promotion campaigns

**Table 8.4**

CBA results for a bicycle bridge in Utrecht, the Netherlands, for three scenarios (net present value, € millions)

| | Scenario | | |
|---|---|---|---|
| | Pessimistic | Neutral | Optimistic |
| Costs | | | |
| Investments | −25.5 | −21.5 | −13.9 |
| Maintenance | −2.5 | −2.5 | −2.5 |
| Total | −28.0 | −23.9 | −16.4 |
| Direct benefits | | | |
| Benefits for school | Pro Memory | | Pro Memory |
| Travel time gains for cyclists | 9.9 | 29.1 | 58.4 |
| Travel time gains for car users | 0.1 | 5.4 | 2.7 |
| Travel cost reductions for cyclists | 1.0 | 1.8 | 2.7 |
| Total | 10.9 | 36.3 | 63.8 |
| Indirect effects | | | |
| Labor productivity | −0.6 | −0.8 | −0.4 |
| Life-years | −0.2 | −0.3 | −0.1 |
| Levies on car fuels | −0.1 | −0.3 | −0.2 |
| Subsidies for public transport | 0.2 | 0.7 | 12.1 |
| Spatial quality / environment | Pro Memory | Pro Memory | Pro Memory |
| Total | −0.7 | −0.7 | 11.5 |
| External effects | | | |
| Emissions of pollutants | 0.1 | 0.4 | 1.2 |
| Noise | 0.0 | 0.2 | 0.6 |
| Road safety | 1.3 | 2.4 | 4.3 |
| Total | 1.4 | 2.9 | 6.1 |
| Total benefits | 11.6 + Pro Memory | 38.5 + Pro Memory | 81.5 + Pro Memory |
| Benefits minus costs | −16.3 + Pro Memory | 14.6+ Pro Memory | 65.0 + Pro Memory |
| Benefit-cost ratio | 0.4 | 1.7 | 5.7 |

*Note*: In this study, "Pro Memory" indicates where an impact could not be quantified or monetized because of the lack of available data. Thus, "Pro Memory" impacts were not included in the benefit-cost ratio but were simply indicated as relevant impacts to consider.
*Source*: Decisio and Transaction Management Centre (2012).

**Table 8.5**
Selected unit costs for valuations to be used for cycling projects

Marginal value of time, averages for all trip purposes,
by mode, in 2010 euros

| Unit | Bike | Car | Bus | Train |
|---|---|---|---|---|
| Euros per hour | 9.00 | 9.00 | 6.75 | 9.25 |

Health benefits per kilometer cycled (eurocents per kilometer cycled)

| | Type of policy instrument increasing cycling levels | |
|---|---|---|
| Benefit type | Cycling infrastructure | Campaigns to stimulate people to cycle |
| Increased labor productivity | 4 | 8 |
| Reduction of health costs | 3 | 5 |
| Reduced burden of illness | 2–4 | 8–12 |
| Life-years gained | 4–5 | 7–10 |
| Total | 13–16 | 28–35 |

Costs of air pollution and $CO_2$ (eurocents per kilometer)

| | Area type | | | |
|---|---|---|---|---|
| Mode of travel | City | Built-up area—other | Rural areas | Average |
| Car—vehicle km | 3.34 | 2.08 | 1.20 | 1.56 |
| Car—passenger km | 1.80 | 1.35 | 0.78 | 1.01 |
| Train | | | | 0.26 |
| Bus | 2.33 | 1.55 | 1.41 | 1.61 |
| Bike | 0.02–0.04 | 0.02–0.04 | 0.02–0.04 | 0.02–0.04 |

*Source*: Decisio (2017).

often target people who do not cycle, such as obese persons, who therefore have higher benefits per kilometer cycled (Decisio 2017). The cost of pollutants per kilometer is based on published research that estimates these impacts (VU/CE Delft 2014). Carbon dioxide costs per ton are based on the distance-to-target method, which estimates the most expensive measured cost needed to reach the lower emission goal. Costs for the bike included additional $CO_2$ emissions based on the study's assumption that cyclists require additional food to support their physical activity.

Alternative assumptions can have an important impact on the results, yet the estimates of the marginal value of travel time (MVOT) savings and emission levels are within the range of values found in the extensive literature on this topic. Because health benefits are a dominant category in CBAs for cycling, the uncertainty in quantifying and monetizing such effects can have a significant impact on CBA results.

Table 8.5 shows that default unit costs are sometimes available and can be used for the assessment of cycling projects. It also shows that cyclists value travel time reductions about the same as car drivers or train passengers and significantly more than bus users. It also shows that the health benefits per kilometer cycled are substantial, amounting to 13–35 eurocents per kilometer. The measured costs of air pollution and $CO_2$ are much lower for bicycling than for other modes, but there are some costs as well, because producing bikes requires energy and natural resources. Moreover, cyclists burn more energy and thus probably need to eat more food than they would if they did not cycle. That would not be true, however, for people cycling in order to lose weight. The uncertainty regarding the food intake by cyclists explains the range of values estimated for cycling.

## Conclusions

The first conclusion that can be drawn from the analysis in this chapter is that it is important to develop and implement guidelines for designing cycling policies. That saves time and money and can lead to less expensive, safer, and more consistently designed cycling infrastructure networks. It also reduces the need for ex ante evaluations.

Second, advanced ex ante (before implementation) evaluations of candidate cycling policy options are not always needed. In some cases, especially in the case of relatively inexpensive policy options, the best and simplest advice is: Just do it! That is especially true if there are good reasons to believe—even without formal analysis—that the benefits are much larger than the costs.

Third, regardless of the evaluation framework, assessing the expected effects of candidate cycling projects in advance of their implementation remains difficult. One reason for that difficulty is that cycling is often excluded or inadequately included in formal transport planning models. In addition, it is difficult to estimate the risk of cyclist injury for a specific situation or project evaluation and even more difficult to estimate the projected decline in injury risk as cycling levels increase. Finally, there is limited understanding, let alone quantification, of self-selection effects that could influence the results of project evaluations. For example, if mainly healthy people cycle, the health benefits of cycling projects might be overestimated.

Fourth, CEA, MCA, and CBA vary in their ability to integrate the various costs and benefits into the analysis. If the main purpose is to select among several options with roughly the same types of effects that are strongly correlated with levels of cycling (number of kilometers or trips), a simpler CEA is probably sufficient. However, if there are multiple effects that are not strongly correlated with cycling levels, then MCA or CBA might be preferable. In the case of a full CBA, the main problem is that the necessary data often are not available. In that case, expert judgment might provide an alternative solution. If the main benefits and costs can be expressed in monetary terms and there is consensus on cost and benefit estimates, then a full CBA would be preferable. CBA guidelines for methodologies and unit costs, as well as independent supervision and quality assessment, are necessary to reduce the potential danger of manipulation that favors or rejects a particular project. If such guidelines and standards are not available, MCA might be the preferred method. The weight setting can importantly influence the resulting outcome of an MCA because weights determine the importance of an effect type (e.g., emissions or health) in the overall evaluation of alternatives. Therefore, the methodology for determining weights should be made explicit. That would reduce the risk of manipulation (i.e., setting weights to favor the alternative preferred by the person or agency setting the weights). One option is to set specific weight factors to explicitly take into account the different valuations by the stakeholders of the various costs and benefits in the project, as demonstrated by the MAMCA methodology in the earlier example.

This chapter has focused exclusively on ex ante evaluations. In addition, however, ex post evaluations (after implementation) can be helpful. Researchers and policy makers can learn from real-world implementations of cycling projects. For example, researchers could evaluate their assumptions, expert judgments, and statistical models (if available). Well-analyzed impacts of cycling projects that have actually been implemented are more reliable than ex ante evaluations, because they are based on actual impacts as opposed to predicted impacts. They may also be more useful and convincing for policy makers elsewhere who are considering similar projects.

## References

Abrantes, Pedro, and Mark Wardman. 2011. Meta-analysis of UK Values of Travel Time: An Update. *Transportation Research Part A: Policy and Practice* 45(1): 1–17.

Alberini, Anna, and Milan Ščasný. 2013. Exploring Heterogeneity in the Value of a Statistical Life: Cause of Death v. Risk Perceptions. *Ecological Economics* 94: 143–155.

Beale, Sophie, Matthew Bending, Paul Trueman, and Bhash Naidoo. 2012. Should We Invest in Environmental Interventions to Encourage Physical Activity in England? An Economic Appraisal. *European Journal of Public Health* 22(6): 869–873.

Cavill, Nick, Sonja Kahlmeier, Harry Rutter, Francesca Racioppi, and Pekka Oja. 2008. Economic Analyses of Transport Infrastructure and Policies Including Health Effects Related to Cycling and Walking: A Systematic Review. *Transport Policy* 15(5): 291–304.

Centre for Diet and Activity Research (CEDAR). n.d. Integrated Transport and Health Impact Modelling Tool (ITHIM). Cambridge: University of Cambridge. https://www .cedar.iph.cam.ac.uk/research/modelling/integrated-transport-and-health-impact -modelling-tool-ithim/.

de Hartog, Jeroen Johan, Hanna Boogaard, Hans Nijland, and Gerard Hoek. 2010. Do the Health Benefits of Cycling Outweigh the Risks? *Environmental Health Perspectives* 118(8): 1109–1116.

Decisio. 2017. *Waarderingskengetallen MKBA Fiets: State-of-the-Art* [Prefixes SCBA Bicycle: State-of-the-Art]. Amsterdam: Decisio.

Decisio and Transaction Management Centre. 2012. *Maatschappelijke kosten-batenanalyse van de fiets* [Social Cost Benefit Analysis of the Bicycle]. Amsterdam: Decisio / The Hague: Transaction Management Centre.

Keseru, Imre, Jeroen Bulckaen, Cathy Macharis, and Joost de Kruijf. 2016. Sustainable Consensus? The NISTO Evaluation Framework to Appraise Sustainability and Stakeholder References for Mobility Projects. *Transportation Research Procedia* 14: 906–915.

van Wee, Bert, and Dick Ettema. 2016. Travel Behaviour and Health: A Conceptual Model and Research Agenda. *Journal of Transport and Health* 3(3): 240–248.

Vrije Universiteit Amsterdam (VU) and CE Delft. 2014. *Externe en infrastructuurkosten van verkeer* [External and infrastructure costs of transport]. Amsterdam and Delft: Free University Amsterdam and CE Delft.

Wang, Guijing, Caroline Macera, Barbara Scudder-Soucie, Tom Schmid, Michael Pratt, and David Buchner. 2004. Cost Effectiveness of a Bicycle/Pedestrian Trail Development in Health Promotion. *Preventive Medicine* 38: 237–242.

Wilhelmsson, Mats. 2000. The Impact of Traffic Noise on the Values of Single-Family Houses. *Journal of Environmental Planning and Management* 43(6): 799–815.

Woodcock, James, Moshe Givoni, and Andrei Scott Morgan. 2013. Health Impact Modelling of Active Travel Visions for England and Wales Using an Integrated Transport and Health Impact Modelling Tool (ITHIM). *PLOS One* 8(1). https://doi .org/10.1371/journal.pone.0051462.

World Health Organization (WHO). 2014. *Health Economic Assessment Tools (HEAT) for Walking and for Cycling: Methods and User Guide.* http://www.euro.who.int/en /health-topics/environment-and-health/Transport-and-health/activities/guidance -and-tools/health-economic-assessment-tool-heat-for-cycling-and-walking.

# 9 E-bikes in Europe and North America

Christopher R. Cherry and Elliot Fishman

Electric bikes (e-bikes) are a transport technology that has emerged in the last couple of decades, going from a technological novelty to a mainstream transport mode that has revolutionized the transport systems in some countries. Today, almost all major bicycle companies offer e-bike models, and some companies only sell e-bikes. The boost in e-bikes in Europe and North America over the past decade has been the result of the evolution of the technology as well as market demands. E-bikes are aimed at reducing barriers to cycling. They can reduce terrain challenges and enable cycling among individuals with physical challenges that preclude cycling, such as older adults. E-bikes are seen as a way to maintain moderate levels of activity and prolong time on a bicycle. Recently, the diversity of e-bike designs has been growing. Cargo bikes have become more popular, allowing households to replace cars for many of their transport errands, such as shopping and traveling with children. This chapter will discuss the growth of e-bikes in Europe and North America, starting with a short history of the technology, then addressing policy frameworks, and finally discussing the impacts e-bikes are having on the transport system.

Contemporary e-bike growth started in China, where basic motors, controllers, and lead-acid batteries enabled production of vehicles that were low cost and provided high utility. China's e-bikes proliferated early this century, with the market growing exponentially each year. This growth led to industrial economies of scale and major technological advances. Indeed, many of the e-bikes sold today in China rely on the same fundamental drivetrain technologies as 20 years ago. In China, more people commute by low-speed electric two-wheelers than by personal car in the United States. There are about 200 million to 250 million electric two-wheelers on the streets of China today.

The development of China's manufacturing capacity and basic technology opened up e-bikes to Western markets. At first, late in the first decade of this century, many Western e-bike brands, online retailers, and others began sourcing e-bikes from China. Many of the e-bikes in that first market were similar to the e-bikes in China, which in many cases resembled scooters. The early market for imported e-bikes bred confusion about vehicle classifications. In the past decade, a large shift in the e-bike market has come from technological advances driven by large manufacturers and component makers. Well-known firms such as Shimano and Bosch jumped into the e-bike market and brought with them important technological improvements and capacity that increased the performance of e-bikes while also increasing their price. Almost all major bicycle companies have a line of e-bikes that are driven by sophisticated drive systems. Relative to earlier-generation e-bikes, these e-bikes improve the cycling experience by providing true pedal assistance, where the motor engagement is proportional to the pedaling energy from the rider. Almost all the modern e-bikes in the West use lithium-ion batteries. In some cities in Europe, e-bikes make up about half of new bicycle sales (though they represent a smaller fraction of the total fleet of bicycles).

### Let's Define an E-bike … If We Can

There is a lot of debate on what an e-bike is. Most agree that a bicycle is a human-powered vehicle with two wheels and two pedals that drive the wheels. By extension, an e-bike is a bicycle with two wheels and two pedals that requires both human power and an electric motor that can help drive the wheels. Most people tend to be less prescriptive on physical design elements of bicycles.

Much of this debate about e-bikes is motivated by an attempt to understand how to govern their use within the existing transport system. For example, should they be allowed to use bicycle infrastructure? The first e-bikes in Europe and North America were imported directly from China and generally followed Chinese performance characteristics. The earliest standards in China were developed in association with the China Bicycle Association, which steered performance toward that of conventional bicycle design. The key factors in China's first standard included a relatively low top speed, 25 km/h, operable pedals, and a maximum weight of 45 kg. Based on these standards, e-bikes are regulated as bicycles. China's earlier regulation did not mandate pedal assistance, a more complicated technology. Subsequently, even today, almost all e-bikes in China are controlled by a throttle, and a class of Western e-bikes follows this design. In China, the design has mutated to include more utilitarian styling, essentially resembling a motor scooter, which allows transporting multiple people and goods. These types of e-bikes have not proliferated in Europe or North America. The *Electric Bike World Report* calls this type of vehicle a "scooter style electric bike"; one that resembles a bicycle is a "bicycle style electric bike" (Jamerson and Benjamin 2016). Figure 9.1 shows samples from a continuous spectrum of "e-bikes." Each of these vehicles has the same underlying technology—the same drive system and the same modes of operation. Most readers will probably consider the top left vehicle to be an e-bike but not the bottom right vehicle. Readers will probably disagree on which of the intermediate variations are e-bikes. The "X" reflects the various perceptions of what may define an e-bike among most readers.

Many suggest that e-bikes must have human assistance and that the motor should only provide power when the rider is actively pedaling. These pedal-assist (often referred to as pedal+electric=pedelec in Europe) bikes are indeed closer to what most would consider an e-bike. The most recent technological advancement is the rise of sophisticated pedal-assistance systems where rider effort seamlessly interfaces with electrical power. Advanced sensors and controllers can instantly identify how much human power is being generated and provide proportional assistance to the rider. The proportion is not prescribed, and some simple pedal-assist e-bikes have sensors that

**Figure 9.1**
Design variations of e-bikes noted in the literature. *Source*: Mojdeh Azad.

detect whether a rider is pedaling and provides power but does not measure rider effort. For example, a rider can pedal but provide very little effort and still call their vehicle a pedal-assist e-bike. The European Union and many states in the United States have adopted relatively nuanced definitions of e-bikes. The US model definitions that have been adopted by the Bicycle Product Supplier Association, along with People for Bikes (2015), include a three-class system for e-bikes (see figure 9.2) that is making its way into state legislation. The simple system is distinguished primarily by speed and pedal assistance, as follows.

Class 1: Pedal assist only, maximum assisted speed 20 mph (32 km/h)

Class 2: Throttle controlled, maximum assisted speed 20 mph (32 km/h)

Class 3: Pedal assist only, maximum assisted speed 28 mph (45 km/h)

The model legislation enables the definition of a speed pedelec, a class 3 e-bike that has different rules regarding where it can be ridden (e.g., not on shared-use paths). This is similar to Europe's type-approval approach, which places class 3 e-bikes in a different category. Notably, pedal assistance technology and assistance levels are not prescribed in typical legislation. To complicate matters, some e-bikes have pedal assist like those in class 1 but can switch modes and be throttle controlled like those in class 2, making their inclusion in the legal and practical definition of *bicycle* more challenging.

| VEHICLE TYPE | | VEHICLE | | USER | | | | BIKEWAY ACCESS | | | |
| --- | --- | --- | --- | --- | --- | --- | --- | --- | --- | --- | --- |
| | | PEDAL OPERATED | MAXIMUM MOTOR-ASSISTED SPEED (MPH) | MINIMUM AGE (YEARS) | DRIVER'S LICENSE | LICENSE PLATE | HELMET | CLASS I BIKE PATH | CLASS II BIKE LANE | CLASS III BIKE ROUTE | CLASS IV PROTECTED LANE |
| BICYCLE | | YES | N/A | N/A | NO | NO | 17 AND UNDER | YES | YES | YES | YES |
| TYPE 1 E-BIKE | | YES | 20 | N/A | NO | NO | 17 AND UNDER | YES | YES | YES | YES |
| TYPE 2 E-BIKE | | NO | 20 | N/A | NO | NO | 17 AND UNDER | YES | YES | YES | YES |
| TYPE 3 E-BIKE | | YES | 28 | 16 | NO | NO | YES | NO | YES | YES | NO |
| MOPED | | NO | N/A | 16 | YES | YES | YES | NO | YES | YES | NO |

**Figure 9.2**
Three classes of e-bikes compared with pedal-assist bicycles and mopeds. *Source*: People for Bikes (2015).

## What Is Not an E-bike

It is also useful to clarify what is not an e-bike. There are many other technologies that have emerged in the "micromobility" space that blur definitions even more than those presented here. Stand-on two-wheeled electric scooters with two tandem wheels, like kick scooters, are an emerging and popular mode. They are not e-bikes. Some variations include a seat, so they begin to resemble some of the e-bikes presented in figure 9.1, but they still lack any ability to be pedaled. Other one- and two-wheeled vehicles are being invented and adopted, and they are following some of the early adoption patterns of e-bikes. Moreover, large electric scooters, such as those popular in China, are generally excluded from this chapter's definition of e-bikes. Lastly, electric motorcycles, whose operating speeds exceed the 20–30 mph threshold, are excluded from this analysis.

For the purposes of this chapter, e-bikes are two- or three-wheeled vehicles that the user has the ability to practically pedal. Generally, e-bikes follow the same form and function as conventional bicycles, including reasonable top speeds and sizes. Some types of cargo bikes are also e-bikes. This definition includes e-bikes of classes 1–3 in the United States or electric bicycle vehicle classes up to speed pedelecs (equivalent to class 3 e-bikes in the United States) in Europe. We do not imply that excluding other vehicles from this chapter's definition precludes them from being included in policy discussions surrounding e-bikes. Other typologies could include those vehicles.

## Chapter Scope

This chapter will focus primarily on European and North American experiences with e-bikes. Pucher et al. (chapter 15, this volume) has a brief discussion on the impact of electric two-wheelers on cycling in China and India. However, the Asian e-bike experience is difficult to translate to Western cycling. This chapter also focuses on transport e-bike use and does not describe the debate on natural-surface trails (i.e., mountain bikes). E-bike researchers have been studying the role of e-bikes for about a decade in Europe and North America, aiming at the main policy aspects to identify how people use them and their effect on two common research domains of cycling that dominate impact assessment methods—safety and health. Importantly, these metrics are balanced against the overall mobility and accessibility goals of a transport system. This chapter describes the growth of e-bikes in Europe and North America and their impacts.

## Policy Frameworks for E-bikes

E-bikes have largely been embraced by bicycling user groups, advocacy organizations, and industry. In the United States and Europe, primary advocacy organizations (e.g., People for Bikes in the United States and the European Cyclists Federation) have highlighted the role of e-bikes in growing cycling and have devoted significant effort and resources toward understanding and promoting e-bike use. E-bike growth in Europe has outpaced e-bike growth in North America. Figure 9.3 shows growth levels in the United States and the five largest e-bike markets, estimated by the *Electric Bike World Report* (Jamerson and Benjamin 2016). At the high end, about one-half of bicycles sold in the Netherlands are e-bikes. Germany dominates the European market, in part because of the size of its market and its aggressive marketing of e-bikes, resulting in rapid growth, among other reasons. In contrast, the US market is about one-quarter the size of the German market, despite the United States having four times Germany's population. Figure 9.3 omits some other important bicycle markets (e.g., Denmark) because the size of the market is relatively low though per capita adoption is still relatively high.

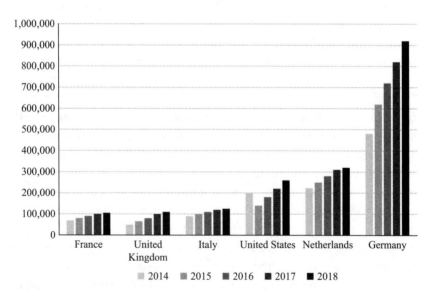

**Figure 9.3**
Growth levels in the United States and the five largest e-bike markets, estimated by the *Electric Bike World Report*. *Source*: Jamerson and Benjamin (2016).

The conventional policy approach toward e-bikes has centered on identifying how e-bikes are functionally and practically the same as bicycles and therefore should be regulated as bicycles. For example (as we discuss later in this chapter), if they are no more dangerous than a traditional bicycle, should they be mandated to adopt more safety equipment than a bicycle, such as compulsory (motorcycle) helmets, turn signals, or brake lights? The European and North American philosophy has been to force e-bikes to conform to the form and function of bicycles through policy. One of the policy challenges that is manifesting as emerging personal mobility technology evolves is the sanctity of the "bicycle lane" being reserved for nonmotorized vehicles only or any vehicle that fits the performance envelope of a bicycle (i.e., wheeled, human scaled, and slow).

**Behavior Matters**

In the United States and Europe, most of the debate surrounding e-bike use is aimed at how e-bikes are used in the transport system. For example, if e-bikes just replace bicycles, should they be encouraged? Researchers are tasked with identifying an e-bike's relative mobility gains, safety performance, and net physical activity impacts. E-bike users' travel behavior is more complex, and the overall result of e-bike use tends to result in e-bike users riding more often and taking longer trips, often replacing motorized transport modes.

The most recent and comprehensive study (Macarthur et al. 2018) in the United States, a nationwide survey of e-bike users ($n$=1,740) in 2017 and 2018, found that e-bike users tend to take more and longer trips. E-bike riders are white (85%) and male (70%). They tend to be older than the average cyclist, with more than half the e-bike riders being over 45. About one-quarter of the respondents stated that they had a physical limitation that diminished their ability to ride a conventional bicycle. The median household income of US e-bike riders is about $100,000 per year, compared to the US average of about $60,000. E-bike riders are not necessarily new to cycling; almost all respondents have ridden a bicycle as an adult.

One challenge of analyzing e-bike use in the United States is disentangling on- and off-road recreational use from utilitarian use. In the United States, about twice as many respondents reported using an e-bike for utilitarian purposes as for recreational purposes. Focusing on transport, e-bikes can remove some of the main barriers to cycling. The survey asked users to identify barriers to bicycling for transport trips and to give reasons for using e-bikes. As expected, long distances and hills are a barrier that e-bikes overcome. Other common responses included riding with cargo or kids, affordability, health, and physical limitations. Perhaps the most striking finding from the survey was that more than half of all vehicle miles traveled for utilitarian purposes replaced car-based modes. Two-thirds of riders use their e-bikes to access different destinations or take different routes than with a conventional bicycle.

In Europe, several studies have investigated e-bike use, and the findings are largely similar to those for the United States. E-bike use varies across Europe, and several studies have taken country-specific approaches to analyze how e-bikes are used. E-bike trips tend to be longer and more frequent. The highest uptake of e-bikes per capita has been in the Netherlands, with more than one million e-bikes in their fleet. A few studies have investigated e-bike use either through focus groups or interviews, surveys, or e-bike use trials.

One of the larger efforts to study e-bike behavior was in Norway and included paired trials that tracked e-bikes loaned to prospective owners, following their behavior over several weeks and comparing it to that of a control group. E-bike users increased their travel distance and frequency. E-bikes compete better with car-based modes and do not diminish normal cycling. Moreover, those with e-bikes also changed travel patterns over time, increasing their use of e-bikes in terms of both trip rate and length, and substituting car trips. Participants also expressed a higher willingness to pay for e-bikes, implying the increased utility and marginal benefit of the e-bike (Fyhri et al. 2017; Fyhri and Fearnley 2015).

Detailed interviews of e-bike users in the Netherlands and United Kingdom found that about half the users switched from bicycles to e-bikes and the others switched from car- or transit-based modes (Jones, Harms, and Heinen 2016). In the Netherlands, e-bike users travel about 50% farther by e-bike than by bicycle. Longer travel distances and declining physical ability to bicycle among older adults were motivating factors for purchasing and using e-bikes (see Garrard et al., chapter 13, this volume). Many riders still focused on e-bikes as a way to sustain a moderately physically active lifestyle. These results are confirmed by GPS-validated studies focusing on the actual trip and route choice behavior (Plazier, Weitkamp, and van den Berg 2017).

To date, almost all studies published on the behavior and mode shift of e-bikes rely on surveys, interviews, or relatively limited experiments.

Still, regardless of the methodological approach, consistent themes emerge. Compared to conventional bicycles, e-bikes are used for trips that tend to be longer, more frequent, and more complex. E-bikes substitute for a large number of car trips, even in bicycle-friendly areas. Some of the main motivations for e-bike adoption include the ability to traverse more difficult grades without high levels of exertion and to carry goods or children. Early adopters of e-bikes included the elderly and, despite recent anecdotes about growing numbers of younger users, the demographic still skews older and more affluent (Leger et al. 2018; Macarthur et al. 2018). One of the main motivators for e-bike purchase was the desire to stay active while acknowledging a diminished physical ability to bicycle long distances.

## Health

In evaluating nonmotorized projects, health and safety tend to be among the largest external impacts a mode of travel has on the transport system and for public health outcomes. Garrard et al. (chapter 3, this volume) describe the health impacts of cycling. HEAT, a commonly used active transport project evaluation tool, focuses on the health implications of active transport modes. In the case of e-bikes, it has been an open question whether they are an "active" mode of transport that is capable of generating positive health outcomes (Castro et al. 2019; Bourne et al. 2018). Several experimental and longitudinal studies have been performed on pedal-assist e-bikes (class 1), most in Europe but a couple in North America. Most of the experiments focused on measuring physical activity metrics (e.g., energy expenditure), while others measured other health indicators. In summary, e-bikes still tend to provide moderate levels of physical activity (i.e., 3–6 metabolic equivalents of task, MET), generally falling between walking and conventional cycling. Generally, the largest difference between modes is on uphill segments, where conventional cycling tends to exceed 6 METs (i.e., vigorous physical activity). E-bikes truly are physical activity levelers, and one of their main appeals is that they reduce terrain barriers and allow riders to arrive at their destinations without too much exertion, thus maintaining moderate physical activity levels throughout the trip.

## Safety

Safety is the primary focus of policy debate surrounding e-bikes in the transport system. It has been a difficult challenge to address. Most Western countries either do not have enough e-bike crashes for statistical analysis or e-bikes are not distinguished from conventional bicycles in the crash records. In general, there have been three main domains in which safety effects of e-bikes have been analyzed: perceived safety derived from surveys or interviews, crash records from police reports or hospital databases, and naturalistic data from e-bikes equipped with measuring instruments. The fundamental questions tend to focus on differences in safety metrics compared to conventional bicycles. This focus tends to omit the role of secondary safety impacts of shifts from other modes, such as personal vehicles.

Safety perception is an important aspect for increasing rates of cycling, especially among some groups of potential cyclists (see Elvik, chapter 4, this volume). There have been several e-bike user surveys or interviews that have included questions on perceived safety (Macarthur et al. 2018; Jones, Harms, and Heinen 2016; Popovich et al. 2014). While not definitive in their findings, they lend some insight. Specifically, people who use e-bikes, especially women, report in surveys and interviews that they feel safer on them than on conventional bicycles. Some of this safety perception could stem from higher-quality bicycle components, better speed management through difficult conflict zones, or keeping up with traffic. Paradoxically, despite higher safety perceptions, e-bike users acknowledge that they travel at higher speeds on e-bikes, which is confirmed by observational speed studies. Still, one study in Denmark found that 30% of e-bike survey respondents reported a conflict or other risk that would not have occurred had they been on a bicycle (Haustein and Møller 2016).

Naturalistic studies use data, often from onboard GPS devices, sensors, or video, to observe real-world behavior of e-bike riders to gather objective data on safety behavior. In most cases, naturalistic studies are conducted by deploying fleets of vehicles or placing measuring instruments on personal vehicles. Speed is the  metric reported most often. While e-bikes are able to sustain relatively higher speeds than conventional bicycles, it is important to measure how people actually use e-bikes in real-world environments. Most naturalistic studies confirm that e-bikes tend to have higher average speeds, by about 2.6 km/h (mean speed 16.5 km/h), compared to conventional bicycles (Cherry and Macarthur 2017).

A few studies have investigated class 3 e-bikes (speed pedelecs with a maximum assisted speed of 45 km/h) and found average speeds around 30 km/h, or about twice the operating speed of bicycles and e-bikes (Cherry and Macarthur 2017). There is some evidence that the gains in average speed are amplified on uphill segments, where e-bikes are able to sustain higher speeds, and that the speed differential is smaller on flat or downhill segments.

One of the consequences of higher speeds is that conflicts with car drivers may be magnified. For example, some studies found that cars allow smaller gaps when crossing in front of e-bike riders than for conventional bicycle riders (Petzoldt, Schleinitz, Krems, et al. 2017; Dozza, Bianchi Piccinini, and Werneke 2016). This is thought to be the manifestation of a driver expectancy problem where drivers, seeing someone cycling at a seemingly leisurely cadence or posture, expect a slower speed, yet e-bikes are traveling faster. E-bikes appear much like bicycles but have different performance characteristics that could result in higher conflicts, particularly at intersections. This difference may diminish as e-bikes become more widespread and drivers adapt (Petzoldt, Schleinitz, Heilmann, et al. 2017). Overall, most naturalistic studies find that e-bike riders tend to ride very much like bicyclists, even if they ride slightly faster than they would ride a bicycle (Langford, Chen, and Cherry 2015).

Because e-bikes are not consistently classified as such in police crash or hospitalization data, or because e-bike crashes are still relatively small in number, only a few studies have scrutinized e-bike crash or hospitalization data. Older riders tend to have slightly higher injury rates, but these are mostly related to loss of balance while dismounting from the heavier bike. Despite some evidence of higher hospitalization and injury rates in Europe, the average injury severity is about the same as for cyclists (Schepers et al. 2014; Weber, Scaramuzza, and Schmitt 2014). A recent article from Israel found that hospitalization rates were lower for e-bike users than for conventional bicyclists, but severity rates (Siman-Tov et al. 2018) and collisions with pedestrians were higher (DiMaggio et al. 2019).

As e-bikes continue to grow in number, their contribution to the injury burden will continue to grow. By most standards, their use and crash patterns mimic those of bicyclists. E-bike riders travel faster on some types of roadways, and the implication is that some conflicts are amplified and subsequent crash outcomes are more severe. Moreover, the weight of e-bikes can make them difficult to maneuver for some populations that could result in increases of certain crash types (e.g., dismount crashes of the elderly).

## A Path Forward for E-bikes

There is much to learn about how e-bikes affect the transport system. We know that they increase mobility—people use them to travel more often, and farther distances, than other conventional bicycle riders and themselves before they owned an e-bike. There could be some self-selection bias in some of the studies: people who travel farther buy e-bikes. However, some studies that query an individual's recalled travel history, through a travel diary, confirm higher trip frequencies on e-bikes. Also, when e-bikes are mixed in bikeshare systems, higher trip distances are reported for e-bikes. It is not a mystery why; e-bikes require less metabolic energy and result in less fatigue than with conventional bicycles.

Women with e-bikes ride more than those with conventional bikes. Part of this might stem from safety perceptions. The travel patterns of women also tend to be more complex, and women have been early adopters of cargo bikes (see Garrard, chapter 11, this volume). Female respondents to surveys report that e-bikes enable them to overcome terrain barriers while carrying goods and often children.

E-bikes can be a bridge to retain existing bicycle riders as they age (see Garrard et al., chapter 13, this volume). E-bike early adopters tended to be older riders who adopted either because of a desire to ride more and maintain physical activity or because ailments limited their ability to continue bicycling comfortably. Like conventional bicycles, e-bikes are used for both utilitarian purposes and recreation. Younger e-bike riders tend to rely on e-bikes for utilitarian purposes, while older riders and those with physical limitations use them more often for recreational purposes.

E-bikes also overcome several additional barriers to other modes of transport, such as expensive parking. E-bikes, though expensive themselves, tend to cost less than cars, and some survey respondents report delaying the purchase of or replacing personal cars because of e-bikes. E-bikes benefit from convenient bicycle infrastructure and parking; they can even be marginally cheaper than mass transit. In cities with high transport costs, e-bikes can be among the most cost-competitive modes for most trips.

While there have not been many studies on the environmental sustainability of e-bikes, their light weight, low speeds, and efficient electric drive systems make them very energy efficient. Most e-bikes consume less than 15 watt-hours of electricity per kilometer, for a cost of less than \$0.01 per kilometer. During vehicle use, the greenhouse gas and local emissions, even in electrical grids heavily reliant on fossil fuels, are just a fraction of those for operating any other motorized mode of travel. Little work has explored the environmental impact of e-bikes over one's life cycle. Most studies have found that about half of e-bike trips in the United States replaced trips formerly made by car. This is important since net benefits of this car substitution likely outweigh any added negative externalities, such as marginal increases in injury burden (most often to e-bike riders themselves) or diminished physical activity as a result of shifting away from more active conventional bicycling.

There has been some criticism that the bicycling community steers e-bike policy toward bicycle-like designs and performance that potentially limits innovation. However, in general, the bicycling community has been successful in supporting e-bike policy that is compatible with conventional cycling (see Pucher et al., chapter 20, this volume).

Much of the confusion associated with e-bikes comes from their shape rather than their performance. The challenge with other low-speed, powered mobility devices confounds (or perhaps clarifies) many of the debates about how different vehicle classes are defined by form rather than through their performance envelope. Many of these vehicles have various forms and rider postures but very compatible performance. Many emerging thoughts on infrastructure access are related mostly to performance (i.e., speed and size). To the extent that these emerging classes of vehicles begin to proliferate, there will continue to be questions about how conventional bicycles, e-bikes, and other vehicles can operate on shared infrastructure.

Overall, these different vehicle classes could increase the mobility of a larger set of users, providing critical demand for low-speed infrastructure. This could make lightweight and low-speed mobility more mainstream, particularly in cities. As shared-mobility companies begin to innovate with more diverse product offerings, more bikeshare, e-bikeshare, e-scooter share, and other technologies will evolve. It is likely that e-bikes will begin to fall into a much larger class of low-speed vehicles, and the details of their form will matter less than their performance. They will be among a host of nonmotorized, semimotorized, and fully motorized vehicles that are gaining access to appropriately low-speed and human-scaled infrastructure, formerly called bicycle lanes.

## References

Bourne, Jessica E., Sarah Sauchelli, Rachel Perry, Angie Page, Sam Leary, Clare England, and Ashley R. Cooper. 2018. Health Benefits of Electrically-Assisted Cycling: A Systematic Review. *International Journal of Behavioral Nutrition and Physical Activity* 15(1): 116. https://doi.org/10.1186/s12966-018-0751-8.

Castro, Alberto, Mailin Gaupp-Berghausen, Evi Dons, Arnout Standaert, Michelle Laeremans, Anna Clark, Esther Anaya-Boig, Tom Cole-Hunter, Ione Avila-Palencia, David Rojas-Rueda, Mark Nieuwenhuijsen, Regine Gerike, Luc Int Panis, Audrey De Nazelle, Christian Brand, Elisabeth Raser, Sonja Kahlmeier, and Thomas Götschi. 2019. Physical Activity of Electric Bicycle Users Compared to Conventional Bicycle Users and Non-cyclists: Insights Based on Health and Transport Data from an Online Survey in Seven European Cities. *Transportation Research Interdisciplinary Perspectives* 1. https://doi.org/10.1016/j.trip.2019.100017.

Cherry, C., and J. Macarthur. 2017. Are E-bikes Unsafe? A Review of European and North American Studies. Podium presentation at International Cycling Safety Conference, Davis, CA.

DiMaggio, Charles J., Marko Bukur, Stephen P. Wall, Spiros G. Frangos, and Andy Y. Wen. 2019. Injuries Associated with Electric-Powered Bikes and Scooters: Analysis of US Consumer Product Data. *Injury Prevention.* Published online first, November 11. https://doi.org/10.1136/injuryprev-2019-043418.

Dozza, Marco, Giulio Francesco Bianchi Piccinini, and Julia Werneke. 2016. Using Naturalistic Data to Assess E-cyclist Behavior. *Transportation Research Part F: Traffic Psychology and Behaviour* 41B: 217–226. https://doi.org/10.1016/j.trf.2015.04.003.

Fishman, Elliot, and Christopher Cherry. 2016. E-bikes in the Mainstream: Reviewing a Decade of Research. *Transport Reviews* 36(1): 72–91. https://doi.org/10.1080/01441647.2015.1069907.

Fyhri, Aslak, and Nils Fearnley. 2015. Effects of E-bikes on Bicycle Use and Mode Share. *Transportation Research Part D: Transport and Environment* 36: 45–52. https://doi.org/10.1016/j.trd.2015.02.005.

Fyhri, Aslak, Eva Heinen, Nils Fearnley, and Hanne Beate Sundfør. 2017. A Push to Cycling—Exploring the E-bike's Role in Overcoming Barriers to Bicycle Use with a Survey and an Intervention Study. *International Journal of Sustainable Transportation* 11(9): 681–695. https://doi.org/10.1080/15568318.2017.1302526.

Haustein, Sonja, and Mette Møller. 2016. E-bike Safety: Individual-Level Factors and Incident Characteristics. *Journal of Transport and Health* 3(3): 386–394.

Jamerson, F. E., and E. Benjamin. 2016. Electric Bikes Worldwide Report—Light Electric Vehicles / EV Technology with 2016 Update. *Electric Bike World Report.* www.ebwr.com.

Jones, Tim, Lucas Harms, and Eva Heinen. 2016. Motives, Perceptions and Experiences of Electric Bicycle Owners and Implications for Health, Wellbeing and Mobility. *Journal of Transport Geography* 53: 41–49. https://doi.org/10.1016/j.jtrangeo.2016 .04.006.

Langford, Brian Casey, Jiaoli Chen, and Christopher R. Cherry. 2015. Risky Riding: Naturalistic Methods Comparing Safety Behavior from Conventional Bicycle Riders and Electric Bike Riders. *Accident Analysis and Prevention* 82: 220–226. https://doi.org /10.1016/j.aap.2015.05.016.

Leger, Samantha J., Jennifer L. Dean, Sara Edge, and Jeffrey M. Casello. 2018. "If I Had a Regular Bicycle, I Wouldn't Be Out Riding Anymore": Perspectives on the Potential of E-bikes to Support Active Living and Independent Mobility among Older Adults in Waterloo, Canada. *Transportation Research Part A: Policy and Practice* 123: 240–254. https://doi.org/https://doi.org/10.1016/j.tra.2018.10.009.

Macarthur, J., M. Harpool, D. Scheppke, and C. Cherry. 2018. *A North American Survey of Electric Bicycle Owners*. Portland, OR: National Institute of Transportation and Communities.

People for Bikes. 2015. New E-bike Law Passes in California. People for Bikes, October 7. https://peopleforbikes.org/blog/new-e-bike-law-passes-in-california/.

Petzoldt, T., K. Schleinitz, J. F. Krems, and T. Gehlert. 2017. Drivers' Gap Acceptance in Front of Approaching Bicycles—Effects of Bicycle Speed and Bicycle Type. *Safety Science* 92: 283–289. https://doi.org/10.1016/j.ssci.2015.07.021.

Petzoldt, Tibor, Katja Schleinitz, Sarah Heilmann, and Tina Gehlert. 2017. Traffic Conflicts and Their Contextual Factors When Riding Conventional vs. Electric Bicycles. *Transportation Research Part F: Traffic Psychology and Behaviour* 46: 477–490. https://doi.org/10.1016/j.trf.2016.06.010.

Plazier, Paul A., Gerd Weitkamp, and Agnes E. van den Berg. 2017. "Cycling Was Never So Easy!" An Analysis of E-bike Commuters' Motives, Travel Behaviour and Experiences Using GPS-Tracking and Interviews. *Journal of Transport Geography* 65: 25–34. https://doi.org/10.1016/j.jtrangeo.2017.09.017.

Popovich, Natalie, Elizabeth Gordon, Zhenying Shao, Yan Xing, Yunshi Wang, and Susan Handy. 2014. Experiences of Electric Bicycle Users in the Sacramento, California Area. *Travel Behaviour and Society* 1(2): 37–44. https://doi.org/10.1016/j.tbs.2013.10.006.

Schepers, J. Paul, Elliot Fishman, P. den Hertog, Karin Klein Wolt, and Arend L. Schwab. 2014. The Safety of Electrically Assisted Bicycles Compared to Classic Bicycles. *Accident Analysis and Prevention* 73: 174–180. https://doi.org/10.1016/j.aap .2014.09.010.

Siman-Tov, Maya, Irina Radomislensky, Kobi Peleg, H. Bahouth, A. Becker, I. Jeroukhimov, I. Karawani, B. Kessel, Y. Klein, G. Lin, O. Merin, M. Bala, Y. Mnouskin,

A. Rivkind, G. Shaked, G. Sivak, D. Soffer, M. Stein, and M. Weiss. 2018. A Look at Electric Bike Casualties: Do They Differ from the Mechanical Bicycle? *Journal of Transport and Health* 11: 176–182. https://doi.org/10.1016/j.jth.2018.10.013.

Weber, T., G. Scaramuzza, and K. U. Schmitt. 2014. Evaluation of E-bike Accidents in Switzerland. *Accident Analysis and Prevention* 73: 47–52. https://doi.org/10.1016/j.aap.2014.07.020.

# 10 Bikesharing's Ongoing Evolution and Expansion

Elliot Fishman and Susan Shaheen

Bikesharing has experienced an explosion of activity in the past decade, particularly since 2015, coinciding with the emergence of dockless bikesharing. This chapter begins by reviewing the four generations of bikesharing, then illustrates bikesharing's remarkable growth, and then follows this with a brief exploration of some of the business models underpinning bikesharing. Key benefits of bikesharing and the technological innovations that have continued to widen its appeal are then described. We look at what the research says about the demographics of bikesharing users and the lessons that can be drawn from failed approaches. Finally, we draw on the outcomes of this exploration to outline how cities can get the most out of bikesharing by designing a user-oriented system.

One of the most important themes in this chapter is that people do not make transport decisions in isolation. The decision to use bikesharing will usually be a result of comparing the pros and cons of the different available modes. When bikesharing feels safe, is relatively time competitive with other modes, and involves low cost, there is a heightened possibility that it will be favored. The concept of "thinking like a rider" in designing a bikesharing program to provide a compelling value proposition to potential users is perhaps the most important implication for cities considering the introduction or renewal of a bikesharing program.

## Four Generations of Bikesharing

Some researchers have categorized the evolution of bikesharing systems into four "generations" (Parkes et al. 2013; Shaheen, Guzman, and Zhang 2012). White Bikes in Amsterdam was the first known example of bicycles being made publicly available for shared use, launching in 1965. Consisting

**Figure 10.1**
Generations of bikesharing. *Source*: Elliot Fishman.

of little more than 5 to 10 bicycles, hand painted white and left on the street for people to use freely (Luud Schimmelpennink, Institute for Sensible Transport, interview with Elliot Fishman, 2014), the program quickly disintegrated because of theft and vandalism. Second-generation programs involved a coin deposit system (similar to shopping carts or trolleys at a supermarket or airport). The first large-scale second-generation program was launched in Copenhagen in 1995, but the anonymity exposed the system to theft and vandalism (DeMaio 2009). Third-generation systems are characterized by technology enabling users to unlock a bike from a docking station by using a credit card. Fourth-generation systems include other forms of technology, such as GPS and dockless systems (Shaheen, Guzman, and Zhang 2012). Figure 10.1 captures the types of bikes that characterize each of the four bikesharing generations.

### The Growth of Bikesharing: A Global Perspective

In the past decade, the number of cities operating bikesharing programs has increased from just 13 in 2004 to over 1,000 as of 2018 (including third

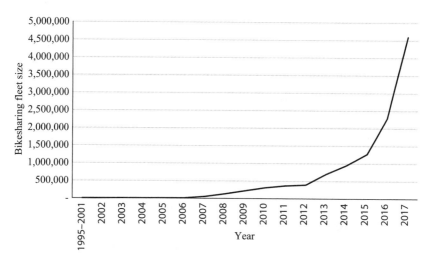

**Figure 10.2**
Estimated global bikesharing fleet size. *Source*: Russell Meddin.

generation and above). As shown in figure 10.2, the global growth in bike-sharing began around 2006 to 2007.

The rise of dockless bikesharing has accounted for much of the growth in bikesharing since 2015 (Meddin and DeMaio 2016). It is now increasingly difficult to determine the number of bicycles within bikesharing systems, as there are multiple commercial suppliers placing bicycles in the same city, competing with one another, often without any formal government involvement.

## Bikesharing Business Models

The rapid increase in the number of cities with bikesharing programs has been matched by the growth of the industry itself. For over a decade, modern-day bikesharing systems have disrupted mobility across the globe, being commonly deployed in a network within a metropolitan region, city, neighborhood, employment center, and/or university campus. Bikesharing users access traditional or electric bicycles on an as-needed basis for one-way (point-to-point) or roundtrip travel using one of the bikesharing models outlined earlier.

The majority of the world's bikesharing systems are publicly accessible, for a nominal fee, with a credit or debit card on file. Companies specialize in a wide range of products and services to support the growth of bikesharing programs. This includes specialist bicycle suppliers, the technology that supports their use, and the hardware used in docking stations and payment kiosks. While some of these suppliers also operate bikesharing services (repair, maintenance, bike redistribution, and call centers), there are also firms that focus solely on the operational responsibilities associated with bikesharing. A small number of firms also work to find sponsorship from large corporations to support bikesharing costs so that the services can be delivered at the lowest direct cost to the taxpayer.

The emergence of dockless bikesharing providers offers an additional model, in which commercial providers offer the bikesharing service, typically without any direct financial subsidy from the government. Under a typical dockless bikesharing model, the commercial operator or supplier interacts directly with the end user (rider), with minimal, if any, interaction with other parties. For dockless bikesharing, the operator is generally also the funder and the hardware and technology supplier and then deals directly with the end user. The involvement of government is minimal or nonexistent, although some jurisdictions (e.g., Seattle and San Francisco) have now established a permit application process that dockless bikesharing providers must complete in order to operate.

Some researchers have drawn comparisons to Uber, the app-based transport company (also known as ridesourcing and ridehailing), which, at least initially, entered markets without any formal authorization. The link with Uber became particularly interesting in early April 2018, when it was announced that one of the pioneers of dockless bikesharing (JUMP, formerly Social Bicycles) had been acquired by Uber. Sources suggest that Uber paid around US$200 million for JUMP, enabling the global transport platform to offer electric bikesharing within its suite of future mobility options (Techweek 2018). In spring 2020, Uber invested in Lime and turned JUMP operations over to Lime. A more detailed discussion of electric bicycles is provided by Cherry and Fishman (chapter 9, this volume).

It is important to note that many dockless bikesharing services are moving into electric standing-scooter sharing models. However, the viability of the dockless model is now being questioned, with some operators pulling out of cities only a year or less from their initial launch (e.g., Mobike in Newcastle, United Kingdom, and OBike in Melbourne and Sydney, Australia).

## The Benefits of Bikesharing: Understanding the Impacts

City governments have introduced bikesharing because of the benefits they associate with increased urban cycling: improved population health, reduced air and noise pollution, and as a means to combat climate change (see in this volume Buehler and Pucher, chapter 2; Garrard et al., chapter 3). Added to these benefits are the potential travel time savings, increased transport mode flexibility, and enhanced possibilities for integration with public transport (Fishman 2016a; Shaheen, Guzman, and Zhang 2010). There may also be some benefit when bikesharing replaces peak-hour public transport journeys, as one seat freed up on a heavily loaded train is in essence a new seat made available for someone else (Fishman 2016a; Shaheen, Guzman, and Zhang 2012). Furthermore, bikesharing is seen as a first-mile/last-mile strategy, helping to improve access to and from public transport (Parkes et al. 2013; Shaheen, Guzman, and Zhang 2010). The integration of cycling and public transport is thought to make up for the weakness of both (Pucher and Buehler 2012).

This section examines the degree to which the purported benefits of bikesharing are being realized and concludes with some of the additional measures required to improve the ability of city agencies and researchers to evaluate bikesharing's benefits.

The impacts of bikesharing are directly related to the degree to which it is used (Shaheen et al. 2014; Fishman 2016a). Figure 10.3 illustrates the number of trips per day, on a per bike basis, for a selection of cities. It shows that bike use varies widely, from almost six trips per day per bike (Lyon) to as low as 0.3 trips per day per bike (Brisbane). The reasons for the difference in ridership levels are explored comprehensively in Fishman (2019).

Many bikesharing trips replace public transport and walking (Fishman, Washington, and Haworth 2014). For bikesharing programs to optimize their impact, it is necessary to implement measures focused on encouraging those currently making trips by car to use bikesharing. Results of a survey from Brisbane of those who do not use bikesharing (see Fishman et al. 2014) suggest that getting them to do so may be best achieved via policy changes that increase the competitive advantage of bikesharing over the convenience of car use and by improving rider safety perceptions by developing a network of protected bicycle lanes and paths. Creating the infrastructure that makes cycling more attractive is discussed further by Furth (chapter 5,

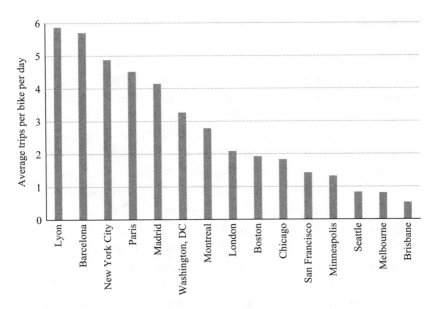

**Figure 10.3**
Bikesharing ridership for selected cities in 2015. *Source*: Fishman (2019).

this volume) and is a central theme in the success of cycling documented by Koglin, te Brömmelstroet, and van Wee (chapter 18, this volume) for Copenhagen and Amsterdam and by Geller and Marqués (chapter 19, this volume) for Portland and Seville.

### Improved Transport Choice
Enhanced mobility options and convenience are two of the most consistent themes to emerge from research on why people use bikesharing (Fishman 2016a). As implied throughout this book, social equity and inclusion are two other important benefits. Bikesharing offers a relatively affordable method of providing greater levels of accessibility to the public transport network, and for annual members it provides a very low-cost daily mobility option.

Bikesharing is popularly combined with public transport, helping to make up for the inevitable deficiencies in the coverage or frequency of a city's public transport system (Martin and Shaheen 2014). Bikesharing is able to increase the catchment area of a train station by a factor of 15 (Hudson 1982). It also allows people to make one-way bicycle trips more easily: they can cycle the component of the journey that makes sense to cycle, using other transport options when it does not (unencumbered by a private bike).

In addition, it has also been found that bikesharing helps to normalize the image of cycling (Goodman, Green, and Woodcock 2013). This is important in cities with very low bike modal share (less than ~4% of trips) because cycling is seen by many to be a marginalized activity that is restricted to so-called MAMILs (middle-aged men in Lycra). Goodman, Green, and Woodcock (2013) found that bikesharing riders in London were more likely to ride wearing regular clothes, with a gender balance closer to the average for the population. Bikesharing may therefore normalize cycling as an everyday transport option rather than a sporting activity only available to a narrow segment of the population.

## Bikesharing and Safety

Bikesharing safety research straddles the sustainable transport and population health and safety fields. The introduction of bikesharing in various cities has been associated with a marked reduction in hospital-recorded cycling injuries (Fishman and Schepers 2016). This counterintuitive finding is especially significant given the overall amount of cycling increases after the introduction of a bikesharing program. This is consistent with the "safety in numbers" phenomenon (e.g., see Elvik 2009), in which a rise in the amount of cycling does *not* lead to a proportional rise in the number of injuries. This relationship is explored in more detail by Elvik (chapter 4, this volume).

There may also be some risk reduction to private cyclists as well if bikesharing helps to make drivers more aware of the possibility of encountering cyclists (reducing the "looked but did not see" phenomenon). Moreover, if bikesharing is introduced with a host of other supportive measures, particularly protected bicycle infrastructure, to improve a city's bicycle friendliness (as was the case in cities such as Paris and New York; see Pucher, Parkin, and de Lanversin, chapter 17, this volume), it is definitely feasible to increase the safety of all people choosing to cycle (bikesharing and private).

A better understanding of the reasons for the apparent lower injury and fatality rates for bikesharing could help to maintain or improve the safety record of bikesharing and perhaps aid in the safety of bicycling more broadly. Theories for further consideration include:

- Bikesharing bicycles are more visible and recognizable. Their design puts the rider in a more upright position, helping to improve the rider's field of vision and make riders more conspicuous to other road users. Many bikesharing bikes also have full-time safety lights (Fishman and Schepers 2016).

- Bikesharing cyclists may be inherently more cautious when riding. Demographics could also play a role. Surveys of docked bikesharing users consistently suggest that they do not reflect the general population but rather are younger and more educated, among other characteristics (see Shaheen, Guzman, and Zhang 2012; Shaheen et al. 2014).
- The design of bikesharing bicycles causes slower and more stable riding, mitigating risk factors that often lead to bicycle collisions.
- Bikesharing systems often occupy the central city areas, in which bicycle infrastructure may be more mature than in the outer areas of cities (Fishman and Schepers 2016).

### Demographics of Bikesharing Users

The demographics of bikesharing users have become a common focus of attention for bikesharing operators and researchers. Social equity concerns have been raised, especially in the United States, as multiple studies have found that the demographics of docked bikesharing users do not reflect the demography of the city as a whole. For instance, only 3% of Capital Bikesharing members are African American, compared to 8% for bicycle riders in general in the Washington, DC, area (Buck et al. 2013), despite African Americans making up some 50% of the city's population (U S Census Bureau 2013). Martens, Golub, and Hamre (chapter 14, this volume) discuss social equity considerations related to cycling.

Bikesharing users tend to be better educated, wealthier, and disproportionately male. This possibly reflects the initial positioning of early systems within downtown areas with high levels of white-collar employment. It also may reflect characteristics of early adopters as well as access to credit. A credit card is typically needed to access bikesharing, although there are a number of US cities that have developed programs to assist the "unbanked" in using bikesharing. Naturally, expansion into the urban periphery may also increase the diversity of system users. Garrard (chapter 11, this volume) provides more detailed information on gender issues associated with cycling.

### Technological Trends Shaping the Industry

Technology has been central to the development of modern bikesharing programs (Fishman 2016b; Shaheen, Guzman, and Zhang 2010). As identified earlier, technology was crucial to overcoming the challenge of misuse that led to the demise of the first-generation bikesharing program (Amsterdam's White Bikes). Since that time, the technology underpinning bikesharing programs has grown increasingly sophisticated, which has enhanced both the user experience and the capacity of operators to manage the bicycle fleet.

The introduction of the credit card as part of the bikesharing enrollment process acted as a form of security (in case of damage or theft) and payment for usage. Without the widespread adoption of universal credit cards, it is difficult to see how bikesharing's growth would have eventuated.

### Smart Dock, Dumb Bike

Third-generation bikesharing systems, which required kiosks and credit card enrollment, were the breakthrough technology that spurred much of bikesharing's growth between 2006 and 2014. These systems have been characterized by some as "smart dock, dumb bike." The "brains" of the system are in the docking stations themselves, leaving the bicycle largely free of digital technology. The general trend is for cities that have adopted this style of bikesharing since 2010 or 2011 to opt for modular designs with solar power to avoid additional installation costs. The bottom left-hand corner of figure 10.1 offers an example of the "smart dock, dumb bike" system.

### Dumb Dock, Smart Bike

In the past three to five years, a new bikesharing system has emerged that generally uses global positioning system (GPS) and other technology on the bike itself, with technologically intensive docking stations being replaced by relatively ordinary bike parking hoops or simply space on the pavement. In many of these systems, the bicycle's location is monitored constantly through the use of an onboard solar-powered GPS. This provides both the operator and user with real-time bicycle tracking.

### Dockless Systems

Affordable GPS, combined with the ubiquity of the smartphone, has made it possible to operate bikesharing systems without the need for any fixed points at which to lock the bicycle: users may access (unlock) a bicycle and park it at any location within a predefined geographic region. These "floating systems" in which users locate bicycles through their smartphone are becoming increasingly popular.

## Hybrid Schemes

The technological developments just described, coupled with some of the vandalism issues associated with dockless systems, have opened up the possibility of a hybrid system that blends a combination of characteristics from both the smart dock and smart bike concepts. Users can check out a bicycle from a docking station and end their trip by returning it to either a station or a nonstation location. Alternatively, they can pick up a dockless bicycle and return it to either a docking station or a nonstation location within a designated geographic (or geofenced) area (Shaheen and Cohen 2018). In essence, a hybrid system enables the user to lock the bicycle to a fixed structure, but it also allows them to lock the freestanding bicycle and park it without a docking station (as in dockless systems) when the bicycle is outside the geocatchment area of the system or if all the docking slots are full.

## E-bikesharing

The growth of bikesharing has coincided with a similarly rapid growth in e-bike performance, affordability, and usage (see Cherry and Fishman, chapter 9, this volume). In recent years, a number of cities have launched bikesharing programs that offer electric-assist bikes, known as pedelecs (called e-bikesharing). E-bikes enable the rider to maintain speed with less effort, especially in topographically challenging cities, resulting in a strong consumer preference for e-bikesharing. Birmingham, Alabama, launched a system that uses a combination of conventional and electric-assist bikesharing bikes. The electric-assist bikes are ridden three to four times as much as the conventional bicycles (Marc Delesclefs, Bewegen Technologies, Inc., personal communication with Elliot Fishman, 2016).

## Other Technologies

A number of other technologies are now available within bikesharing systems. These include digital wayfinding, in which the GPS function is used to provide guidance to the rider via turn-by-turn directions. This is especially important to those who are new to a city (e.g., tourists). Moreover, some systems allow riders to elect their preference for a longer, safer route or a direct, less-safe route. Near-field communication is another technology that is making it easier for riders to pay for their bikesharing use. A rider's smartphone or credit card can be used directly on the bicycle itself.

## Lessons Learned from Failed Approaches

Two Australian bikesharing programs launched in 2010, in Brisbane and Melbourne, both have had ridership levels among the lowest of any city globally (Fishman et al. 2014). The purpose of this section is to provide a brief analysis of the lessons that can be drawn from the Australian experience to equip cities to make better decisions, resulting in higher usage levels.

Australia is in a somewhat unique position as one of only a handful of countries with mandatory helmet laws across all age groups and all riding contexts (Haworth et al. 2010). This makes bikesharing more challenging to implement, as it reduces the spontaneity with which people are able to use it (Fishman et al. 2014). While mandatory helmet requirements are often the first factor people point to when looking at the low usage levels of Australian bikesharing programs, it is unlikely that they are the only contributing factor (Fishman 2012). Fishman has detailed the relationship between bikesharing and helmet issues (see Fishman 2017a; Fishman 2017b; Fishman 2014). The city of Seattle also has a mandatory helmet law, and its docked bikesharing program, Pronto (now defunct), faced challenges in providing helmets to users. It is important to note that the hilly terrain of Seattle likely contributed to this program's closure. As of June 2020, Seattle's bikesharing system, operated by JUMP, is undergoing revision under the new management of Lime.

Melbourne, despite having a population twice that of Brisbane (5 million vs. 2.3 million), rolled out a bikesharing program of only 550 bicycles. Brisbane's scheme included around 1,800 bicycles. Melbourne's system was state funded, whereas much of the funding for the Brisbane scheme was provided via an outdoor advertising contract reached between the Brisbane City Council and JC Decaux. The Brisbane funding relationship is important because it distorted the incentive that many operators, such as JC Decaux, have from ridership. Because JC Decaux's revenue primarily came from street advertising, they were not as reliant on strong ridership levels. This led to some policy decisions that discouraged ridership, including the introduction of a closing time (10 p.m.) and then a reopening again (at 5 a.m.), a very rare policy for an IT-based, third-generation bikesharing program. Moreover, for the first couple of years, prospective users were unable to swipe their credit cards to gain access but rather had to call a phone number during office hours or enroll online and wait days for the delivery of a "fob," a plastic key that gives riders access to the system of bikes. Prospective users who called the hotline first had to listen to 24 minutes of a legal disclaimer before being granted access to the system.

One distinctive factor in the development of the Melbourne and Brisbane bikesharing programs was the lack of consultation and planning. Unlike most North American bikesharing programs, there were no community outreach or stakeholder engagement sessions with the transport sector or wider community regarding the possible design features to maximize the bikesharing program's success. No consultation was offered to refine the types of bicycles offered, the placement of the docking stations, the enrollment procedure, or helmet use. The following section elaborates on each of these issues to assist other cities seeking to establish a bikesharing program.

The studies that have examined the usage levels of the Brisbane and Melbourne bikesharing programs have found that infrastructure in both cities fails to encourage cycling, and this acts as a barrier to usage (Fishman et al. 2014). Those surveyed who had not used bikesharing (in either city) were found to be more sensitive to the riding environment and required protected bicycle infrastructure rather than cycling in mixed traffic or in a painted bike lane. This supports more recent work on the importance of creating separate facilities (CDM Research and ASDF Research 2017). Infrastructure considerations for cycling are discussed by Furth (chapter 5, this volume).

The size of a bikesharing catchment area is also critical to its success. Melbourne's bikesharing program, in particular, is at the global extreme (on the small side) in terms of the size of its bikesharing catchment area relative to the size of the city—both in geographical and population terms. Former Chicago transport commissioner Gabe Klein, explaining bikesharing, said: "I knew that any sort of nodal business was only as effective as the number of nodes you have" (Vanderbilt 2013). The New York City Department of Transportation, in a lengthy investigation of the applicability of bikesharing for that city, reached a conclusion similar to Klein's, noting that "evidence from bikesharing programs around the world suggests that small programs do not provide meaningful transport, health, or economic development gains nor do they provide a significant basis from which the city could evaluate the effectiveness of the program" (New York City Department of

City Planning 2009, 100). The level of utility (or satisfaction or usefulness) depends to some degree on the size of the docking station catchment area, as this largely governs where one is able to travel. Melbourne's bikesharing catchment area is among the world's smallest, on a people per bike basis, with less than 600 bicycles and five million people. A system of this scale fails to offer significant convenience or mobility enhancement, particularly given that it is competing with one of the densest, most well connected, and frequent public transport services in Australia, including the center of the world's largest tram/streetcar network.

Melbourne's bikesharing program was launched at the beginning of winter. There is evidence that winter weather reduces private bicycle usage levels, even in Australia (Ahmed et al. 2010), and it is difficult to imagine how bikesharing would be immune to this effect. The winter start date of the Melbourne program meant bikes sat underused for approximately the first 100 days of the program (i.e., each bike being used approximately once every three days), until the weather became more conducive to riding. This served to reinforce a perception that the bikes are unpopular, which as shown in focus groups (see Fishman, Washington, and Haworth 2012) can create a self-reinforcing downward spiral ("I don't see anyone using them, so I won't use them"). The public's attitude toward bikesharing was formed in these early critical months and has persisted even in the face of change (e.g., easier access to helmets, warmer weather). Citing a lack of usage, the Victorian government, which had been funding the Melbourne bikesharing program, decided to discontinue it in late 2019.

### Some Lessons from Dockless Bikesharing

As this chapter demonstrates, while the massive scale of dockless systems has made some progress toward improving transport sustainability (for example, the bike modal share in Beijing rose from 5% to 12% after the introduction of dockless bikesharing, according to operator Mobike), these systems have also proven to be a challenge for governmental agencies responsible for managing public rights-of-way, particularly in relation to parking issues. There are several ways that dockless bikesharing impacts cities in ways that docked systems do not. Members of the public can (and often do) park dockless bikes in ways that can cause a public nuisance, and there is greater potential for major system imbalances (too many bikes in one area and/or not enough in another). The potential spatial imbalance of docked systems is limited to the number of available docking points within a given vicinity (Shaheen and Cohen 2018).

Without bikesharing stations colocated next to mobility hubs or public transport nodes, additional policies may be needed to encourage multimodality (Shaheen and Cohen 2018). Without the ability to physically colocate dockless bikesharing with other transport services, fare payment integration is critical to maximize ridership (Shaheen and Cohen 2018). For example, digital integration can include leveraging application programming interfaces (APIs) to integrate dockless bikesharing with public transport apps and multimodal trip planners.

With the growth of bikesharing fleets, cities are increasingly confronting questions about curb space management: how to prevent dockless bicycles (and scooters) from parking in inconvenient or dangerous areas that impede the rights-of-way of pedestrians, cyclists, and vehicles. Seattle has developed a policy for curbside management and to guide where dockless bicycles should be parked in urban areas. Seattle's policy defines three key zones: (1) a landscape/furniture zone, (2) a pedestrian zone, and (3) a frontage zone. Seattle requires that dockless bicycles be parked in the landscape/furniture zone and has painted labels on several curbs to highlight appropriate parking places. Additionally, Seattle prohibits bicycles from being parked on corners, in driveways, or on curb ramps, or being parked in a way that blocks access to buildings, parking meters, benches, bus stops, or fire hydrants.

In addition to curbside and bicycle parking management, a number of cities also employ "geofencing," the process of designating a certain region of a city or metropolitan area as off limits, to prevent bicycles from being parked in distant, less urban environments. For example, dockless operators in San Diego use geofencing to prohibit cyclists from parking and leaving their bicycles on Coronado Island. Similarly, JUMP geofenced Union Square in San Francisco to discourage bicycle parking in the busy pedestrian plaza. In the event that dockless bicycles end up in prohibited locations, a number of public agencies have developed fees and impounding policies to address these situations. For example, Seattle requires dockless bikesharing companies to move improperly parked bicycles and to correct parking violations within two hours of a problem being reported during normal business hours. In Washington, DC, the National Park Service prohibits parking dockless bicycles on the National Mall and impounds illegally parked bicycles.

### Getting the Most out of Bikesharing: Designing a User-Oriented System

A number of best-practice principles can be drawn from the literature covered in this chapter. Safety, convenience, and spontaneity are the key principles on which current and prospective bikesharing planners should focus their efforts to increase usage and overall system effectiveness. These principles and their subcomponents are identified in figure 10.4.

Dockless bikesharing programs can adapt the framework identified in figure 10.4. All the safety components still apply, as do the convenience elements, with the exception of docking stations, of course. For high-usage areas, however (e.g., around public transit hubs), it may be necessary to have a bike parking corral that offers designated space to park dockless bikes. The spontaneity component applies to dockless bikes as it does docked bikes, although a dockless bike very often provides a high degree of spontaneity to the user, primarily through smartphone scanning of a QR code to unlock the dockless bike.

The best-practice principles identified in figure 10.4 were developed by analyzing examples of bikesharing success as well as those cities that have

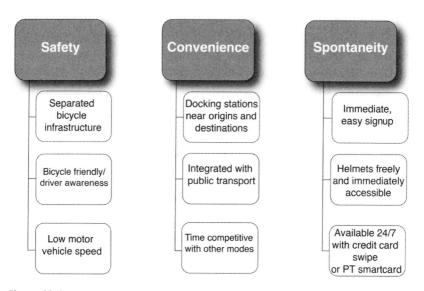

**Figure 10.4**
Synthesis of best-practice principles. The helmet requirement only pertains to jurisdictions in which their use is legislated. *Source*: Fishman (2014).

hosted underperforming bikesharing programs. One overarching consideration that may be helpful for cities beginning the process of examining a future bikesharing program is the value proposition offered. Just as public transport planners have been encouraged to "think like a passenger," it is helpful for bikesharing planners to "think like a rider." What compelling value proposition do people have to use bikesharing? Is it faster than competing travel options? Is it more pleasant to use? Is it cheaper than other forms of transport? Are docking stations or dockless bikes located close to where people live, work, and shop? Does it feel safe to use? Is sign-up easy and fast? Can users easily integrate bikesharing with public transport? Designing a system with these questions in mind is crucial to the success of a future bikesharing program and is discussed below in more detail.

**Safety**

As highlighted earlier, safety concerns are the primary reason people choose not to cycle in many countries across the globe, and people who do not currently cycle are even more sensitive than current cyclists to levels of protection from motorized transport (Fishman 2014). In terms of best practices, London, Paris, and New York City all embarked on bicycle infrastructure campaigns in the years before they established a bikesharing system (see Pucher, Parkin, and de Lanversin, chapter 17, this volume). This infrastructure included a large number of separated bicycle facilities, and these helped to provide prospective users with the confidence necessary to start using the bikesharing program. Furth (chapter 5, this volume) provides a good set of examples of the type of protected bicycle infrastructure that is helping to make streets safer for all road users and boosting the attractiveness of this mode.

In Paris, the infrastructure also included many targeted "contraflow" lanes, which allowed two-way travel for cyclists but only one-way travel for motorists. This helps to increase the value proposition for bikesharing relative to motorized travel modes. In London, a large number of separated bicycle lanes, which extend for many kilometers, have now been added to the network, helping to provide safe conditions for cycling. By contrast, very little bicycle infrastructure was developed when planning the bikesharing programs in Melbourne and Brisbane, and both cities have such a level of hostility toward prospective cyclists that it has detracted from usability (Fishman, Washington, and Haworth 2012). The overwhelming theme emerging from the international assessment of bikesharing program success is that substantial separated, connected bicycle lanes and paths need to be constructed, and vehicle speed limits need to be lowered, before bikesharing is likely to succeed. In a survey of the general public in Brisbane, when nonbikesharing members were asked "What factors would encourage you to become a CityCycle member?," "More bike lanes and paths" received the highest mean score (Fishman 2014). Moreover, in addition to helping to bolster the bikesharing program itself, these lanes and paths will also help to increase road safety for all road users, including those cycling on private bikes.

## Convenience

One of the most consistent findings in the international literature on bikesharing is that people choose to use bikesharing when it meets their convenience criteria. In practice, this means that bikes (whether in docking stations or dockless) need to be close to their home, workplace, and other destinations they frequently visit (the threshold distance appears to be about 250 m). Because of their lower unit costs, a set budget will always buy more dockless bikes than docked bikes, meaning the system can be larger, resulting in better average coverage across a city. The current practice in many US cities, however, is to limit the number of bicycles dockless bikesharing companies are able to deploy.

Proximity to shared bicycles might also help explain why dockless bikesharing has been so popular in many cities. Seattle's dockless bikesharing providers (which largely increased the total number of bicycles available) have had many more trips taken in six months than the earlier, docked system (Pronto, now defunct), partly because people are able to drop bikes much closer to their final destination. A frustration with popular docked stations is the inevitable imbalance between bicycle supply and demand, sometimes resulting in empty docks when people are looking for a bike and full docks when looking to return a bicycle (Transport for London 2014). Rebalancing can consume a large proportion of an operator's budget and result in additional truck use in dense city areas (Fishman, Washington, and Haworth 2014). The pricing scheme of some programs (e.g., visitor pricing and free use up to 20 minutes) has also been found to discourage usage (Shaheen and Christensen 2014).

Cities can support dockless bikesharing and minimize disruption by proactively developing policies to guide (1) identifying locations where

bicycles can be parked; (2) development of agreements with private operators that indemnify the public agency from liability for any loss or injury
that could result from a dockless bicycle being operated or parked in the
public rights-of-way; (3) enumeration of the enforcement procedures for
illegally parked bicycles, such as fines or impoundment; (4) development
of a process for requesting access to the use of the public rights-of-way (e.g.,
curb space); (5) identifying fees that should be charged or permits issued
for dockless bikesharing to operate within a municipality; (6) establishing
standards for dockless parking signage and/or markings to identify proper
parking areas; and (7) development of data-sharing requirements and/or
impact studies as a condition for allowing dockless bicycles to be parked in
public rights-of-way (Shaheen and Cohen 2018).

Best-practice bikesharing cities, in addition to providing relatively large
systems, have also worked to integrate them into the existing public transport
service. Data on user attitudes and behavior show a very strong desire to make
multimodal journeys that involve segments of walking, bikesharing, and public transport (Shaheen, Martin, and Cohen 2013). Although by no means an
industry standard yet, integrating bikesharing access into public transport
smartcard ticketing taps into a very strong user preference to have access to
a pass that works for bikesharing and other forms of public transport (Shaheen, Cohen, and Martin 2017). For instance, the revamped Velib system in
Paris is integrated with the Navigo smartcard public transport pass. Emerging
technological developments allow bikesharing users to be able to access both
bikesharing and public transport through the use of a single smartphone app.
As part of the US Federal Transit Administration's Mobility on Demand Sand-
box, the Chicago Transit Authority (CTA) is currently partnering with Divvy
bikes, a station-based bikesharing operator, to integrate bikesharing into their
Ventra app and allow customers to pay for their bikesharing use with their
Ventra card (fare payment). Ventra cards store public transit credit for use
on the CTA and Pace, the suburban bus and regional paratransit division of
the Regional Transportation Authority in the Chicago metropolitan area. The
aim is for Chicago public transit riders to be able to open the Ventra app, add
transit value to their account, pay for a bikesharing pass, go to a bikesharing
station, and start cycling (Shaheen and Cohen 2018).

Ultimately, much research and user surveys suggest that bikesharing
needs to be time competitive with competing modes of transport. This is

especially the case for commuting trips (in which users are more time sensitive). Convenience is all about the value proposition. What value proposition does bikesharing offer a potential user in terms of time and money? Central to the success of bikesharing is the degree to which it competes with car use for short to medium trips. If car use is faster and cheaper, door to door, it will be difficult to attract people to bikesharing. Thinking about bikesharing services in relation to competing modes of transport is critical to their ability to provide the level of convenience and cost effectiveness necessary to attract ridership.

### Spontaneity

The average bikesharing trip lasts between 12 and 16 minutes (Fishman 2016a). These short trips are very often not planned well in advance. Programs that require a user to enroll days in advance have lost large numbers of potential users (Fishman 2014). Generally, existing best practice is to allow credit card swipe/tap/QR scan sign-up, which offers users immediate access, 24/7.

As identified earlier, in Brisbane, especially in the early phase, users were asked to listen to over 20 minutes of a legal disclaimer and were not permitted to use the system between 10 p.m. and 5 a.m. These factors severely limited the ability of new users to sign up as fast as they would have liked, with many forgoing the opportunity. Moreover, mandatory helmet use hampered the spontaneity with which people could use the bikesharing systems in Australia and the handful of other jurisdictions in which a blanket helmet law applies. No city with a well-enforced mandatory helmet law has run a bikesharing system averaging more than one trip per day per bike.

Finally, the strongest marketing tool for a new bikesharing program is seeing someone else using the system. For this reason, it is vital that everything be done from a program's launch to incentivize early use. Having bikesharing bikes sitting idle must be minimized. This can be achieved by having a "brand champion" (such as a local sports hero or other celebrity), trial days, ambassadors, and pricing to incentivize early use (e.g., half-price memberships for people who sign up with a friend). Seeing others using the system creates the social norm allowing others to see the possibility of using the system themselves. The ability to sign up spontaneously is central to maximizing this possibility.

## Conclusion

Ultimately, bikesharing's ability to deliver benefits will be related to the degree to which it acts as a replacement for car use. Cities need to view bikesharing in terms of what it offers in meeting their wider strategic objectives. What does your city want to be like 50 years from now? When most people are asked this question, they come up with a similar picture in their mind: a sustainable city, a city that has clean air and healthy citizens, a city that is accessible without the need to get in a car. Bikesharing has a role to play in the city of the future. City politicians and planners need to think strategically about what they can do now to help transform their cities into the one they would like to leave for their children and grandchildren 50 years from now. There has never been more opportunity or more reason to start planning better cities, and bikesharing is a notable strategy to help cities become more attractive places to live, work, and visit.

## Acknowledgment

The global bikesharing community mourns the loss of Russell Meddin, who passed away in April 2020. Anyone fortunate enough to have known Russell will miss his infectious enthusiasm for bikesharing and active transport, along with the vibrancy they add to urban life. For bikesharing researchers, Russell's work was essential. He diligently monitored bikesharing developments (programs and size) across the globe for about a decade, and he generously shared these data and insights with the world. Whenever either of this chapter's authors (and this book's co-editors) had bikesharing questions, Russell was typically among the first to come to mind, and more often than not, he had a credible answer. He was generous with his knowledge, and he did not hesitate to criticize the ideas he opposed. In addition, he was a tireless supporter of social equity in bikesharing. His work helped to advance a critical discussion and action in shared mobility that has and will continue to have lasting effects. The micromobility world and the urban transport community have lost a true friend and champion.

## References

Ahmed, Farhana, Geoffrey Rose, and Christian Jakob. 2010. Impact of Weather on Commuter Cyclist Behaviour and Implications for Climate Change Adaptation. Presented at Australasian Transport Research Forum, Canberra.

Buck, Darren, Ralph Buehler, Patricia Happ, Bradley Rawls, Payton Chung, and Natalie Borecki. 2013. Are Bikesharing Users Different from Regular Cyclists? *Transportation Research Record* 2387: 112–119.

CDM Research and ASDF Research. 2017. Near-Market Research. Melbourne: City of Melbourne.

DeMaio, Paul. 2009. Bike-Sharing: History, Impacts, Models of Provision, and Future. *Journal of Public Transportation* 12: 41–56.

Elvik, Rune. 2009. The Non-linearity of Risk and the Promotion of Environmentally Sustainable Transport. *Accident Analysis and Prevention* 41(4): 849–855.

Fishman, Elliot. 2012. Fixing Australian Bike Share Goes Beyond Helmet Laws. *The Conversation.* https://theconversation.edu.au/fixing-australian-bike-share-goes-beyond -helmet-laws-10229.

Fishman, Elliot. 2014. Bikesharing: Barriers, Facilitators and Impacts on Car Use. PhD thesis, Queensland University of Technology.

Fishman, Elliot. 2016a. Bikesharing: A Review of Recent Literature. *Transport Reviews* 36(1): 92–113.

Fishman, Elliot. 2016b. Cycling as Transport. *Transport Reviews* 36(1): 1–8.

Fishman, Elliot. 2017a. *Bike Share Parking Infrastructure Guidelines.* Melbourne: Prepared by the Institute for Sensible Transport for VicRoads.

Fishman, Elliot. 2017b. *Canberra Bike Share: Initial Considerations.* Melbourne: Institute for Sensible Transport, commissioned by Transport Canberra and City Services Directorate.

Fishman, Elliot. 2019. *Bike Share.* New York: Routledge.

Fishman, Elliot, and Paul Schepers. 2016. Global Bike Share: What the Data Tells Us about Road Safety. *Journal of Safety Research* 56: 41–45.

Fishman, Elliot, Simon Washington, and Narelle Haworth. 2012. Barriers and Facilitators to Public Bicycle Scheme Use: A Qualitative Approach. *Transportation Research Part F: Traffic Psychology and Behaviour* 15(6): 686–698.

Fishman, Elliot, Simon Washington, and Narelle Haworth. 2014. Bike Share's Impact on Car Use: Evidence from the United States, Great Britain, and Australia. *Transportation Research Part D: Transport and Environment* 31: 13–20.

Fishman, Elliot, Simon Washington, Narelle Haworth, and Armando Mazzei. 2014. Barriers to Bikesharing: An Analysis from Melbourne and Brisbane. *Journal of Transport Geography* 41: 325–337.

Fishman, Elliot, Simon Washington, Narelle Haworth, and Angela Watson. 2015. Factors Influencing Bike Share Membership: An Analysis of Melbourne and Brisbane. *Transportation Research Part A: Policy and Practice* 71: 17–30.

Goodman, Anna, Judith Green, and James Woodcock. 2013. The Role of Bicycle Sharing Systems in Normalising the Image of Cycling: An Observational Study of London Cyclists. *Journal of Transport and Health* 1: 5–8.

Haworth, Narelle, Amy Schramm, Mark King, and Dale Steinhardt. 2010. *Bicycle Helmet Research*. Brisbane: Centre for Accident Research and Road Safety—Queensland.

Hudson, Mike. 1982. *Bicycle Planning: Policy and Practice*. Oxford: Architectural Press.

Martin, Elliot, and Susan Shaheen. 2014. Evaluating Public Transit Modal Shift Dynamics in Response to Bikesharing: A Tale of Two U.S. Cities. *Journal of Transport Geography* 41: 315–424.

Meddin, Russell, and Paul DeMaio. 2016. The Bike-Sharing World the Last Week of December 2015. http://bike-sharing.blogspot.com.au/2015/12/the-bike-sharing -world-last-week-of.html.

New York City Department of City Planning. 2009. *Bike-Share: Opportunities in New York City*. New York: New York City Department of City Planning.

Parkes, Stephen, Greg Marsden, Susan Shaheen, and Adam Cohen. 2013. Understanding the Diffusion of Public Bikesharing Systems: Evidence from Europe and North America. *Journal of Transport Geography* 31: 94–103.

Pucher, John, and Ralph Buehler, eds. 2012. *City Cycling*. Cambridge, MA: MIT Press.

Shaheen, Susan, and Matthew Christensen. 2014. Bikesharing Pricing Could Slow Trend's Rapid Expansion. *The Conversation*. https://theconversation.com/bikesharing -pricing-could-slow-trends-rapid-expansion-31888.

Shaheen, Susan, and Adam Cohen. 2018. How Dockless Bikesharing Is Transforming Cities. *Move Forward Blog*. https://www.move-forward.com/how-dockless-bikesharing -is-transforming-cities-seven-policy-recommendations-to-minimize-disruption/.

Shaheen, Susan, Adam Cohen, and Elliot Martin. 2017. *Smartphone App Evolution and Early Understanding from a Multimodal App User Survey*. UC Berkeley, Transportation Sustainability Research Center. http://dx.doi.org/10.7922/G2CZ35CH.

Shaheen, Susan, Stacey Guzman, and Hua Zhang. 2010. Bikesharing in Europe, the Americas, and Asia. *Transportation Research Record* 2143: 159–167.

Shaheen, Susan, Stacey Guzman, and Hua Zhang. 2012. Bikesharing across the Globe. In *City Cycling*, edited by John Pucher and Ralph Buehler, 183–210. Cambridge, MA: MIT Press.

Shaheen, Susan, Elliot Martin, Nelson Chan, Nelson Cohen, and Michael Pogodzinski. 2014. *Public Bikesharing in North America during a Period of Rapid Expansion: Understanding Business Models, Industry Trends and User Impacts*. San Jose, CA: Mineta Transportation Institute.

Shaheen, Susan, Elliot Martin, and Adam Cohen. 2013. Public Bikesharing and Modal Shift Behavior: A Comparative Study of Early Bikesharing Systems in North America. *International Journal of Transportation* 1: 35–53.

Shaheen, Susan, Elliot Martin, Adam Cohen, and Rachel Finson. 2012. *Public Bikesharing in North America: Early Operator and User Understanding.* San Jose, CA: Mineta Transportation Institute.

Techweek. 2018. How This Non-profit Turned into a Mini-unicorn That Uber Scooped Up. *Techweek*, May 29. https://techweek.com/new-york-jump-bikes-uber-acquisition/.

Transport for London. 2014. *Barclays Cycle Hire Customer Satisfaction and Usage Survey: Members Only.* London: Transport for London.

US Census Bureau. 2013. *State and Country QuickFacts.* Washington, DC: US Department of Commerce. http://quickfacts.census.gov/qfd/states/11000.html.

Vanderbilt, Tom. 2013. The Best Bike-Sharing Program in the United States. *Slate .com.* https://slate.com/human-interest/2013/01/capital-bikeshare-how-paul-demaio -gabe-klein-adrian-fenty-and-other-dc-leaders-launched-the-best-bike-sharking -program-in-the-united-states.html.

# 11  Women and Cycling: Addressing the Gender Gap

Jan Garrard

Cycling as a daily means of urban travel varies markedly across countries and cities (see Buehler and Pucher, chapter 2, this volume, figures 2.1, 2.2, and 2.3). Not only does cycling vary geographically, but there are also some striking variations in the demographic characteristics of people who cycle. In bicycle-friendly cities and countries, cycling is an inclusive, population-wide activity that includes large numbers of children, older adults, and women (see in this volume Buehler and Pucher, chapter 2, figures 2.4 and 2.5; McDonald, Kontou, and Handy, chapter 12; Garrard et al., chapter 13; Koglin, te Brömmelstroet, and van Wee, chapter 18). In these cities and countries, the "cycling population" reflects the overall population, with roughly equal proportions of women and men using a bicycle as an every-day means of transport.

In contrast, in car-oriented countries with low levels of cycling, the majority of cyclists are young to middle-aged men. In Australia, the United States, and the United Kingdom, where overall levels of cycling are low, women constitute about one-third of recreational cyclists and one-quarter of commuter cyclists.

The strong positive relationship between overall bicycle mode share of travel and female participation in cycling is now well established and is evident at national, state, city, and local levels (Garrard, Handy, and Dill 2012). This consistent relationship in diverse locations has led to frequent references to women being an "indicator species" for a cycling-friendly environment (Baker 2009). Research interest has now focused on gaining a better understanding of this relationship and consequently what can be done to improve both bicycle mode share and gender equity in cycling in those countries and cities with low levels of cycling for transport and low proportions of women who cycle. A key question is whether the focus should be on increasing the bicycle mode share of travel in general, in the hope that female representation will automatically improve, or whether gender-specific measures are needed. Moreover, if gender-specific measures are required, what are they?

This chapter presents the most recent data and research addressing these issues. It documents the benefits of cycling for women, describes differing patterns of female participation in cycling, explores reasons for the highly variable rates of female cycling, and discusses strategies for establishing cycling as a viable transport option for women.

## Benefits of Cycling for Women

The health benefits of cycling described by Garrard et al. (chapter 3, this volume) are similar for men and women; however, women stand to gain more from improved conditions that support cycling because in most countries they are less likely to be adequately active than men (World Health Organization 2018). Many women do not enjoy or do not have the time or resources to participate in structured sport and exercise activities. However, when physical activity is an incidental, "nonathletic," habitual part of everyday life that does not require additional exercise time, women are more likely to achieve physical activity parity with men. In the Netherlands, where men and women cycle for transport at comparable levels, on average both men and women exceed the minimum recommended level of physical activity through active transport alone (mainly cycling) (Fishman, Bocker, and Helbich 2015). In Germany, about half of adults (48%) achieve the recommended 150+ minutes of physical activity through active transport alone (about one-third of which is cycling), with women more likely than men to achieve recommended levels of physical activity through active transport (Buehler et al. 2019).

In addition to the health benefits of utility cycling, women may benefit more than men from the establishment of accessible, safe cycling conditions that can lead to reduced escort trips for children and older adults. Women's travel patterns differ from those of men, with women more likely to undertake more but shorter trips and more "household responsibility" and child-escort trips, including when both parents are employed (Motte-Baumvol, Bonin, and Belton-Chevallier 2014; Tremblay-Breault, Vandersmissen, and Thériault 2014; Department for Transport 2015). When conditions preclude children and older adults traveling independently, including by bicycle, it is more often women who serve the travel needs of other family members. Consequently, when cycling-friendly conditions support independent bicycle trips by children and other dependents, women are the principal beneficiaries of a reduction in escort trips.

## Differing Patterns of Women's Participation in Cycling

Improving female participation in cycling requires an understanding of contrasting patterns of cycling in countries that have high and low levels of cycling in general and the associated high and low proportions of females cycling. Table 11.1 compares patterns of cycling in a low-cycling country (Australia) and a high-cycling country (the Netherlands). Cycling in Australia is typical of low-cycling countries such as the United States, United Kingdom, Canada, and New Zealand, while the patterns and trends for the Netherlands are similar to those of other high-cycling developed countries such as Denmark, Japan, and Germany (see in this volume Buehler and Pucher, chapter 2; Koglin, te Brömmelstroet, and van Wee, chapter 18).

## Female Participation and Type of Cycling

Not all types of cycling in the Netherlands follow the high-participation, gender-inclusive patterns described in table 11.1. Cycling for sport in the Netherlands has relatively low levels of participation, together with a gender gap in cycling similar to that seen in countries such as the United States, United Kingdom, and Australia. Data on sports participation from Statistics Netherlands (CBS) indicate that 16% of males, compared with only 9% of females, participate in "cycle racing/mountain biking" (CBS 2010).

These differing patterns of "everyday" and "sports" cycling in the Netherlands suggest that the generic terms "cyclist" and "cycling" actually encompass two very different cycling styles, one of which is gender neutral and the other more male oriented. Indeed, in the Dutch language, a "cyclist" refers to a sports cyclist (a *wielrenner*, literally "wheel runner") riding a *wielrenfiets* (a road or racing bike). On the other hand, "riding a bicycle" generally refers to ordinary people in everyday clothes riding a *fiets* (meaning a sit-up bicycle), who would not be referred to as "cyclists" but simply people on bicycles (Bicycle Dutch 2012). Interestingly, a video of "sport cyclists" in

**Table 11.1**
Cycling patterns and trends in Australia and the Netherlands

| Australia | The Netherlands |
| --- | --- |
| Low rates of cycling, most of which is occasional rather than regular: 34% of Australians cycled in the past year, compared with only 15.5% in the past week. | High rates of cycling, most of which is regular rather than occasional: 24% of the Dutch population cycle every day. |
| Large gender differences in cycling: only 26% of trips to work by bicycle are by females. | Small or no gender differences in cycling participation. |
| Cycling declines substantially for females during the transition from childhood to adolescence and from middle age to older age. Reductions in cycling also occur for adult males but at a later age and to a lesser extent (see figure 11.1). | High levels of participation in cycling across the life course, including for adolescent girls, women in the child-rearing years, and older women. |
| Most cycling is for recreation rather than transport. Only 1.2% of trips to work in the 2016 Australian census were by bicycle. | Most cycling is for transport rather than recreation. |
| Pockets of increased cycling in some locations (mainly in some inner-city areas), but at the overall population level declining rates of cycling over time, indicating that cycling for transport is largely an inner-city phenomenon. | While large cities such as Amsterdam have the highest mode share of cycling in the Netherlands, cycling rates are also high in many suburban, regional, and rural areas, thereby contributing to a high bicycle mode share of travel nationally (27%). |
| Declining rates of cycling nationally (2011 to 2019), with an increase in gender inequality in cycling as overall cycling rates decline. | No decline in bicycle mode share over time. |
| Relatively high rates of cyclist fatalities and serious injuries (compared with other road users and compared with developed countries with high cycling rates). | Relatively low rates of cyclist fatalities and serious injuries. |

*Sources*: Munro (2019); Garrard, Greaves, and Ellison (2010); Australian Bureau of Statistics (2016); Harms and Kansen (2018); Ministry of Transport, Public Works, and Water Management (2009); Hembrow (2011).

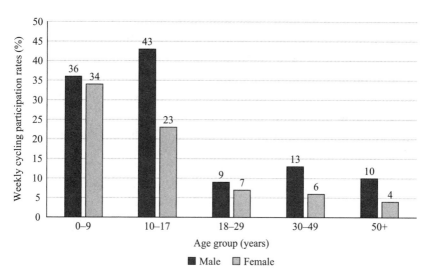

**Figure 11.1**
Weekly cycling participation rates by gender and age in Australia in 2019. *Source*:
Munro (2019).

the Netherlands captures this minority form of cycling in that country. The
video "Spandex in the streets of the Netherlands" features cyclists wearing
bicycle helmets and Lycra (spandex), riding on road/racing bicycles, and
traveling at relatively high speed. As far as can be ascertained, all the cyclists
captured in the four-minute video are male (Bicycle Dutch 2012).

These minority, predominantly male sports cyclists in the Netherlands
comprise a much higher proportion of cyclists in low-cycling countries,
such as Australia, the United States, and the United Kingdom. Following
World War II, rising car ownership and use sharply reduced cycling levels in the United Kingdom from 15% before the war to only 2% by 1970;
they have remained at that low level over the past five decades (Pucher
and Buehler 2008). In Australia and the United States, the bicycle share of
trips has never risen above 1%. In all three countries, cycling is primarily
for sports, recreation, and exercise. Current efforts to increase utilitarian
cycling may have limited impacts on cycling rates and gender equity in
cycling if the focus is on encouraging a style of cycling that is more suited
to cycling for fitness and exercise. Thus, it is essential to implement the
kinds of cycling infrastructure and programs that are necessary to establish a utilitarian cycling culture where people cycle to get to places as part

of daily life. While there is some overlap between the two, there are also several differences, many of which are highlighted by the marked gender differences in the two cycling cultures described in table 11.1. What differentiates the Netherlands from the low-cycling English-speaking countries such as Australia and the United States is not its cycling culture per se but rather its *utilitarian* cycling culture.

As described elsewhere in this volume, many countries, states/provinces, cities, and towns are now attempting to increase rates of utilitarian cycling, with varying degrees of success (Buehler and Pucher, chapter 2; Pucher, Parkin, and de Lanversin, chapter 17; Koglin, te Brömmelstroet, and van Wee, chapter 18; Pucher et al., chapter 20). To understand and encourage female participation in cycling, it is useful to explore how female participation in cycling changes as the overall mode share of cycling changes over time in these emerging cycling areas.

### What Happens to Women's Share of Bicycle Trips When Overall Bicycle Mode Share of Transport Changes?

As described elsewhere in this volume, trends in cycling participation over time vary by country, city, and local area, covering the spectrum from increased rates, to stable rates, to declining rates of cycling (Buehler and Pucher, chapter 2; Pucher et al., chapter 15; Pardo and Rodriguez, chapter 16; Pucher, Parkin, and de Lanversin, chapter 17; Koglin, te Brömmelstroet, and van Wee, chapter 18; Geller and Marqués, chapter 19). In view of the association between cycling rates and female representation in cycling, it is useful to explore trends in female participation in cycling for these three scenarios. Longitudinal data add a level of explanatory power not available from comparative cross-sectional data alone.

There is consistent evidence that *declining* rates of cycling in countries, cities, or local areas are associated with reduced proportions of female cyclists, with evidence of this trend in Australia (Munro 2019), England and Wales (Aldred, Woodcock, and Goodman 2016), and Ireland (Geddes 2009). In the high-cycling countries where cycling rates are steady over time, roughly equal representation of women has been continuing (Cycling Embassy of Denmark n.d.; Harms and Kansen 2018). This leaves the question of what is happening to female participation in cycling in areas where cycling rates have recently increased from a low base. Area-specific longitudinal data are useful for exploring this question, but currently these data do not show a consistent pattern.

Recently, data from the English and Welsh Census 2001 and 2011 were used to investigate the relationship between changes in rates of bicycle travel to work and changes in gender representation for these trips for 356 local authority areas in England and Wales (Aldred, Woodcock, and Goodman 2016). The study found that, while a decrease in bicycle mode share was associated with a decrease in gender equity, an increase in bicycle mode share for the trip to work was not associated with an increase in gender equity in cycling to work.

A separate analysis was conducted for 13 Inner London local authorities that had experienced an overall increase in bicycle mode share of work trips from 3.8% in 2001 to 7.2% in 2011. There was little change overall in female representation of cyclists, with eight showing an increase and five showing a decrease (Aldred, Woodcock, and Goodman 2016), a finding consistent with the national data described earlier.

Similar findings have been reported in Vancouver, Canada, where longitudinal count data indicate that while cycling volume at observation sites increased by 15.5% from 2012 to 2016, there was no change over time in the proportion who were women (Winters and Zanotto 2017).

In contrast, Portland, Oregon, has experienced an increase in both bicycle mode share of travel (from 1.1% in the early 1990s to 7.2% in 2014) and the proportion of bicycle trips undertaken by women (from 21% in the early 1990s to 32% in 2014), based on two-hour peak-hour counts at 218 locations citywide (Portland Bureau of Transportation n.d.). These changes accompanied a large increase in the total length of the bikeway network, from 66 mi (106 km) in 1990 to 274 mi (441 km) in 2007, with a more recent emphasis on improving the quality of bikeways. The female share of cyclists varied across locations, with the highest proportions in neighborhoods closest to the city center and those where the quality of the bikeway network and traffic conditions are better (Roger Geller, personal communication, 2018; see also Geller and Marqués, chapter 19, this volume).

Similar increases in both bicycle mode share and women's share of bicycle trips have occurred in Seville, Spain, where a 4.6-fold increase in bicycle trips (from 3.1 million in 2006 to 14.4 million in 2017) was accompanied by an increase in women's share of bicycle trips from 25% in 2006 to 36% in 2017. In the same period, protected bicycle paths increased from 12 km to 175 km, and the number of cyclists killed or seriously injured declined from 0.98 per million bicycle trips to 0.37 per million bicycle trips, highlighting improvements in two key concerns for women—protection from motor vehicle traffic and safety (see Geller and Marqués, chapter 19, this volume). Two magazine articles have also commented about the cycling renaissance in Seville. Walker (2015) stated that "Segregation [i.e., protected bicycle paths] … makes cycling accessible to people of all ages, allowing them to trundle along at slow speeds in everyday clothes. This is in contrast to the scene in most UK cities, where mainly young, generally male riders speed alongside motor traffic dressed in helmets and luminous high-vis jackets. The ordinariness comes in their approach. The overall sense is of cycling not as a pursuit, or a sport but, in the Dutch style, a deeply everyday activity, little more than a more efficient means of walking." Marqués, García-Cebrían, and Calvo-Salazar (2018) also commented that "almost none of them wear Lycra garments nor other special clothing."

The data described here suggest that an increase in bicycle mode share of travel from a low base may be associated with increased female representation but that an increase in female participation does not occur automatically. It is therefore important to explore the circumstances and conditions under which female representation does improve. Area-specific measures of many of these factors are not currently available, but more general data and research provide some insights into factors that appear to be important.

## What Factors Are Associated with Increased Female Participation in Cycling?

Many factors have been proposed to help explain the large gender difference in utility cycling in low-cycling countries. Although direct evidence of their impact is slim, a considerable body of indirect evidence points to a number of likely factors. The following discussion is structured around the social-ecological model of health-related behaviors shown in figure 11.2 (Sallis et al. 2006). This model describes four mutually interactive domains of influence on the behavior of interest; in this case, women cycling.

### Intrapersonal Factors
Intrapersonal factors include demographic characteristics as well as personal attributes such as beliefs, attitudes, skills, and preferences. A key intrapersonal

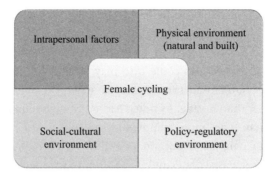

**Figure 11.2**
Social-ecological model of the determinants of female cycling. *Source*: Based on Sallis et al. (2006).

factor that is likely to influence women's participation in cycling lies in their preferences for and participation in different types of physical activity. Compared to men, women are more likely to prefer moderate-intensity rather than vigorous-intensity physical activity, activities that can be incorporated into a busy lifestyle and that do not require set time commitments, activities that have minimal impact on family commitments, and activities that do not require specialized and expensive equipment or club membership (van Uffelen, Khan, and Burton 2017; VicHealth 2015).

Adolescent girls (who have very low rates of cycling in low-cycling countries—see figure 11.1) tend to prefer physical activities that are informal and unstructured in nature and that are perceived to be appropriately "feminine" based on social, cultural, and psychological influences (Whitehead and Biddle 2008). Participation in more vigorous forms of sports and exercise can be seen by some young women as undermining their femininity and requiring them to "put their bodies on show" in ways that might facilitate unfavorable comparisons with social ideals of the "body beautiful" (Whitehead and Biddle 2008).

While it is important to challenge these notions of female identity, appearance, and appropriate activities, they may help to explain why the "ordinariness" of cycling for transport at a moderate pace on a sit-up bicycle in everyday clothes could be appealing to young women, and particularly young women who do not see themselves as being "sporty" and are not attracted to fast-paced on-road cycling. These factors may help to explain the sharp decline in cycling among adolescent girls in low-cycling countries such as Australia (see figure 11.1), which does not occur in high-cycling countries such as the Netherlands.

In addition to factors related to participation in physical activity, transport-related factors also come into play. Women's patterns of travel behavior support utilitarian trips by bicycle when cycling conditions are supportive rather than hostile. As noted earlier, women tend to make more trips per day than men, more complex linked trips, for shorter distances, within their local neighborhood, and for a wider variety of purposes, including escort trips such as accompanying children to school (Motte-Baumvol, Bonin, and Belton-Chevallier 2014; Department for Transport 2015). The ability to combine exercise time with travel time is also likely to be more appealing to women. Lack of time is a major constraint on women's participation in physical activity, with women spending more time than men on child care and household activities, regardless of employment status (Horne et al. 2018; Wilkins and Lass 2018).

These gender differences in physical activity preferences, time constraints, and travel patterns suggest that cycling at a moderate pace on bicycles that enable cycling with children and carrying shopping items and other goods is likely to be appealing to women. This style of cycling is common among women in high-cycling countries that have high rates of female cycling.

A crucial element of female-friendly cycling conditions is cycling safety. While actual risks of cycling (in terms of deaths and injuries) are generally lower for women than for men (Department for Transport 2015), women's *concerns* about traffic risks are consistently greater than those of men (Chataway et al. 2014) and more likely to act as a barrier to cycling for women than for men (Heesch, Sahlqvist, and Garrard 2012).

These findings serve as a reminder that perceptions of risk and sensitivity to risk may be more important than actual risks as a barrier to cycling, particularly for women. Recent research also indicates that traffic risks extend beyond risk of fatality or serious injury to include risk of near misses and harassment, with women reporting a higher rate of near-miss incidents than men (see Elvik, chapter 4, this volume; Aldred and Crosweller 2015; Aldred and Goodman 2018; Poulos et al. 2017). Media reports that disproportionately focus on fatalities associated with cycling also play a role in contributing to the fear of cycling (Macmillan et al. 2016).

Consequently, the concept of traffic risk as a barrier to cycling for women might best be viewed as a "risk iceberg," with a small number of fatalities at the tip, followed by broadening layers of serious injuries, minor injuries, near misses, and harassment. In terms of risk assessment, which is commonly conceptualized as the product of potential harm and probability of occurrence (Kirch 2008), a large number of low-harm incidents can also contribute to substantially heightened risk concerns.

In summary, women's physical activity preferences and transport needs are potential supports for utilitarian cycling, but safety concerns are a major barrier. The establishment of supportive built, policy-regulatory, and social-cultural environments will assist in building on these supports while also meeting women's needs for safe cycling conditions.

## The Built Environment

Extensive research over the past two decades has shown that the provision of safe, convenient, widespread, well-designed cycling infrastructure is essential for increasing utilitarian cycling (see in this volume Buehler and Pucher, chapter 2; Furth, chapter 5). Good cycling infrastructure is common in developed countries with high levels of cycling by both women and men and is important for increasing female cycling in low-cycling countries and cities.

However, the mixed success that emerging cycling cities such as London and Vancouver have had in achieving more equal gender representation suggests that women and men may have different infrastructure requirements. On-road bicycle lanes are a common form of cycling infrastructure in emerging cycling cities, but these have less appeal to women than to men, with a number of studies demonstrating that women prefer more protected bicycle paths and lanes (Aldred et al. 2017; Winters and Zanotto 2017). The provision of good-quality, separated bicycle paths in Portland and Seville, as described earlier, is likely to have contributed to increases in bicycle trips and in women's share of trips in these cities.

In addition to the type of cycling infrastructure provided, the location of cycling infrastructure is also important. In order to provide for multiple, relatively short trips within local neighborhoods, a greater focus on bicycle networks within suburban neighborhoods is needed. In some emerging cycling cities, the focus to date has been on radial cycling routes into the city center from suburban areas. This enables women who live in inner suburban areas to commute short distances by bicycle to the city center but will attract relatively few women to commute longer distances by bicycle from the middle and outer suburbs or to make local trips within suburbs with poor cycling infrastructure. Women are consistently found to cycle shorter distances than men (Dolati 2014).

These findings regarding different infrastructure requirements for men and women (in terms of both type and location) help to explain why cycling improvements in some countries, states/provinces, cities, and towns may lead to increased bicycle mode share of travel without necessarily improving gender equity in cycling (Aldred, Woodcock, and Goodman 2016; Winters and Zanotto 2017).

In summary, cycling infrastructure is important for increasing cycling, but an increase in the number of kilometers of bicycle facilities in emerging cycling cities (an oft-cited measure) may be insufficient to improve gender equity in cycling. Infrastructure type and location are also important. Addressing the gender gap in cycling will require an increased focus on improved cycling infrastructure, including for all utilitarian cycling trips, rather than the current focus on bicycle commuting to work in large city centers. It is likely that the five main requirements for cycle-friendly infrastructure developed in the Netherlands but now internationally recognized provide a useful framework by which to assess female-friendly cycling infrastructure. According to that standard, cycling infrastructure should be safe, direct, cohesive, attractive, and comfortable (Dufour 2010).

It is essential that safe cycling infrastructure for women also includes consideration of social safety (i.e., personal safety from possible attacks or harassment). A recent study suggests that commonly used cycling level of service (CLoS) tools fail to account for gender differences in perceptions of social safety (Xie and Spinney 2018). As a consequence, traditional CLoS ratings may mistakenly assess a cycling route as rideable when women consider the route socially unsafe.

A key characteristic of different types of cycling infrastructure is their different levels of interaction between cyclists and motor vehicles. For example, the on-road bicycle lanes that are common in emerging cycling cities often lack continuity. They are frequently blocked by parked cars and trucks and usually end at intersections or when roads narrow, requiring cyclists to merge with motorized traffic. Physically separated bicycle lanes (with barriers protecting them from motor vehicles) also require interactions with drivers, mainly at intersections. These interactions with motorists are therefore important. They are shaped by both the *policy-regulatory environment* (e.g., speed limits, driver licensing, cyclist education, and road rules and their enforcement) and the *social-cultural environment* (e.g., drivers' attitudes toward "sharing the road" with cyclists), as discussed in the following sections.

### The Policy-Regulatory Environment

When high-cycling and low-cycling countries and cities are compared, the provision (or lack) of well-designed cycling infrastructure is usually identified as a key determining factor in explaining their different cycling rates. While good cycling infrastructure is essential, it is also important to bear in mind that noninfrastructure cycling conditions are also more supportive of cycling in high-cycling countries and cities. As documented by Pucher, Dill, and Handy (2010), a comprehensive, integrated package of mutually reinforcing measures is required to increase cycling overall and improve gender equity in cycling. These measures include not only a well-designed, integrated network of cycling facilities but also many complementary measures: traffic calming of residential neighborhoods (e.g., 30 km/h speed limits); presumed motorist liability for collisions between motorists and cyclists; restrictions on where motor vehicles can be operated and parked in urban areas; cycling education for all children; and strict driver training and licensing requirements that include a substantial cycling component emphasizing motorists' responsibility to avoid endangering cyclists.

Failure to establish road rules that protect cyclists, to enforce the limited rules that do exist, and to impose appropriate penalties for traffic violations that threaten or injure cyclists also contributes to hostile cycling conditions that constrain female cycling (Aldred and Goodman 2018; Bailey and Woolley 2017; Voelcker 2007). The development of a more balanced, less car-focused policy-regulatory environment that provides greater protection for cyclists is likely to improve both safety and feelings of safety for the high-risk and high-fear interactions between cyclists and drivers that characterize "sharing the road" in countries with limited separated cycling infrastructure.

### The Social-Cultural Environment

The social-cultural environment has many components, together with multiple interactions with intrapersonal factors and factors within the built environment and the policy-regulatory environment. Social-cultural influences are often subtle, indirect, unrecognized, and difficult to identify and quantify. They are mostly neglected in research and evaluation related to female participation in cycling. Nevertheless, qualitative, observational, and case study analyses point to a number of social influences on women's participation in cycling.

A key social influence is the status of cycling as a mode of transport. In countries where the mode share of cycling is high, cycling is considered a legitimate form of transport, and the rights of cyclists to use the transport network are respected. In contrast, in countries where motor vehicle travel is the norm, cyclists can be viewed as a minority "out group" of "deviant" road users (Basford et al. 2002), who are viewed negatively and subjected to considerable harassment, abuse, and discrimination (Heesch, Sahlqvist, and Garrard 2011).

It has been proposed that, in car-oriented countries, cyclists represent one of the last minority groups to experience socially accepted discrimination, a perspective that is supported and reinforced through negative portrayals of cyclists and cycling in the media (Walker 2017; Rimano et al. 2015; Rissel et al. 2010). Not surprisingly, being a member of an "out group" of disparaged road users may have little appeal to women, as women have been shown to be more concerned about, and therefore more constrained by, harassment and abuse from drivers (Garrard, Crawford, and Hakman 2006).

Discrimination against cyclists in car-centric countries and cities has been described as a form of social vulnerability—the social equivalent of the physical vulnerability associated with being at greater risk of physical injury than more highly protected motor vehicle occupants (Aldred and Jungnickel 2014; Aldred 2013). As is the case for physical vulnerability, social vulnerability associated with cycling in countries and cities where car travel is the norm appears to be a greater constraint on cycling for women than for men.

Another social constraint on female cycling is many women's and adolescent girls' reluctance to "put their bodies on show" by conducting vigorous exercise in public while wearing figure-hugging sporting gear (Whitehead and Biddle 2008). This particular barrier to female physical activity reflects social norms and expectations regarding appropriate and inappropriate female appearance and behavior. As noted, everyday cycling in ordinary clothes using sit-up bicycles helps to mitigate these concerns. However, when the dominant cycling culture is male, fast, and Lycra-clad, the utilitarian cycling style common in high-cycling countries may be subjected to social disapproval by the more prevalent sports and fitness cyclists. In an article in the *Sydney Morning Herald* titled "Male Cyclists of Melbourne, What the Hell Is Your Problem?," the female author, recently arrived from the Netherlands and cycling Dutch style in Melbourne, described harassment from male cyclists for being too slow, taking up too much space, not riding the way she should, and sometimes, apparently, just for being female. The author commented that, "I think it is often overlooked that what makes women feel unsafe is not only the poor bike infrastructure but also the harassment they endure. Cyclists as a whole are disrespected as road users, so add being a woman to the mix and you'll be disrespected even more" (Polkamp 2017).

An overarching social-cultural influence on female cycling is what de Madariaga (2013) has described as the "mobility of care"; that is, the mobility requirements of (predominantly) women undertaking care- and home-related tasks on a daily basis, in contrast to the "employment-related mobility" that historically has focused on the transport needs of men. Gender equity in cycling will improve when cycling equipment, infrastructure, and conditions support the complex patterns of mobility required for the fulfillment of caregivers' daily tasks.

## What Can Be Done to Improve Gender Equity in Cycling?

As discussed earlier in this chapter, elements of the built, policy-regulatory, and social-cultural environments all contribute to cycling participation and to gender equity in cycling. Emerging cycling countries and cities have largely focused on improving cycling infrastructure as a means of increasing utilitarian cycling. However, in some locations, this approach has been more effective for men than for women, possibly because the infrastructure provided does not meet women's preferences for more protected bicycle paths and lanes.

While improved cycling infrastructure and cycling safety are important for both men and women, they are much more important for women. Women prefer bicycle paths and lanes that provide maximum separation from motor vehicles in order to minimize potentially harmful interactions with motorists (Garrard, Handy, and Dill 2012; Winters and Zanotto 2017; Aldred et al. 2017). When interactions with drivers and other cyclists are necessary, these need to be respectful and harmonious rather than hostile and stressful. These interactions can be improved by implementing a range of "soft infrastructure" measures such as reduced speed limits and traffic volumes; better education of drivers through licensing processes that emphasize safe, courteous interactions with cyclists; increasing the duty of care on drivers to avoid collisions with cyclists through presumed driver liability or related measures; road rules that move from optimizing motor vehicle flow and motor vehicle occupant safety to optimizing cyclist safety and improving bicycle flow; and increased enforcement and penalties for motorists who break road rules that endanger cyclists.

In addition to higher-order infrastructure and safety improvements, a number of female-specific measures fall within what can be broadly termed a change in cycling culture to one that embraces everyday utility cycling. While there is some overlap between sport/fitness cycling and utilitarian cycling, there are sufficient differences to warrant a targeted shift in focus from the former to the latter. Gentler-paced recreational cycling on off-road, multiuse greenways and trails is also likely to be appealing to women.

A more "everyday cycling" culture will support diverse population groups (including women, children, adolescents, and older adults) riding sit-up bicycles (that can also be used to carry children and goods) in ordinary clothes, at relatively low speed, in local neighborhoods, and to local destinations on a network of separated bike paths and traffic-calmed streets.

The establishment of a permeable network of cycling facilities and traffic-calmed streets *within local neighborhoods* differs somewhat from the "principal bicycle routes" approach that supports higher-speed, longer-distance bicycle commuting to city centers, which is more appealing to men. Utilitarian cyclists, both men and women, prefer direct routes, and this applies to trips in the local neighborhood as well as longer bicycle commutes to work.

Bicycle design will also influence how comfortable women feel cycling. If women need to carry infants and young children, groceries, or other items with them while traveling, they will require bicycles with appropriate carrying capacity (Lovejoy and Handy 2012). These bicycles also enable women to cycle in everyday clothing, including cycling to work. The increasing popularity of e-bikes might also contribute to more women cycling, as e-bikes are more comfortable, require less effort, and enable longer bicycle trips than ordinary bicycles (see Cherry and Fishman, chapter 9, this volume). A female-friendly cycling culture will therefore be supported by awareness, promotion, availability, and use of bicycles designed for everyday cycling; that is, a *fiets* (sit-up bicycle) rather than a *wielrenner* (road or racing bicycle).

## Conclusions

This chapter commenced by posing the question, "Is an increase in bicycle mode share of travel in a low-cycling country or city sufficient to improve gender equity in cycling?" Research in this area is recent and limited, but, based on available data, the answer is that it depends on several factors. A small number of cities and local areas have increased both cycling rates and gender equity, but others have increased cycling rates without any change in gender equity. A number of possible reasons for these different outcomes have been proposed, and further research is required to more systematically examine these tentative explanations. In particular, indicators of an "everyday cycling" culture should be assessed, including both hard and soft infrastructure measures, social norms and discourses on cycling, types of bicycles used, clothing worn, cycling speed, and trip distance, purpose, and location.

The pattern of evidence described in this chapter suggests that if emerging cycling countries, cities, and local areas are to achieve the gender equity of the high-cycling countries, they will need to plan for cycling improvements with women, utility cycling, and moderately paced recreational cycling on off-road greenways and trails in mind. Doing more of the same will probably lead to more of the same: small (if any) countrywide or citywide improvements in overall bicycling mode share without diminishing the current gender gap.

### References

Aldred, Rachel. 2013. Incompetent or Too Competent? Negotiating Everyday Cycling Identities in a Motor Dominated Society. *Mobilities* 8(2): 252–271. https://doi.org/10.1080/17450101.2012.696342.

Aldred, Rachel, and Sian Crosweller. 2015. Investigating the Rates and Impacts of Near Misses and Related Incidents among UK Cyclists. *Journal of Transport and Health* 2(3): 379–393. https://doi.org/10.1016/j.jth.2015.05.006.

Aldred, Rachel, Bridget Elliott, James Woodcock, and Anna Goodman. 2017. Cycling Provision Separated from Motor Traffic: A Systematic Review Exploring Whether Stated Preferences Vary by Gender and Age. *Transport Reviews* 37(1): 29–55. https://doi.org/10.1080/01441647.2016.1200156.

Aldred, Rachel, and Anna Goodman. 2018. Predictors of the Frequency and Subjective Experience of Cycling Near Misses: Findings from the First Two Years of the UK Near Miss Project. *Accident Analysis and Prevention* 110: 161–170. https://doi.org/10.1016/j.aap.2017.09.015.

Aldred, Rachel, and Katrina Jungnickel. 2014. Why Culture Matters for Transport Policy: The Case of Cycling in the UK. *Journal of Transport Geography* 34: 78–87. http://dx.doi.org/10.1016/j.jtrangeo.2013.11.004.

Aldred, Rachel, James Woodcock, and Anna Goodman. 2016. Does More Cycling Mean More Diversity in Cycling? *Transport Reviews* 36(1): 28–44.

Australian Bureau of Statistics. 2016. *2016 Census*. Canberra: Australian Bureau of Statistics.

Bailey, Trevor, and Jeremy Woolley. 2017. Vulnerable Road Users in a Safe System. *Journal of the Australasian College of Road Safety* 28(1): 40–49.

Baker, Linda. 2009. How to Get More Bicyclists on the Road: To Boost Urban Cycling, Figure Out What Women Want. *Scientific American* 301(4): 28–29.

Basford, L., S. Reid, T. Lester, J. Thomson, and A. Tolmie. 2002. *Drivers' Perceptions of Cyclists*. TRL Report TRL549. Glasgow: Department of Transport, University of Strathclyde.

Bicycle Dutch. 2012. Lycra in the Streets of the Netherlands. https://bicycledutch.wordpress.com/2012/05/28/lycra-on-the-streets-of-the-netherlands/.

Buehler, Ralph, Tobias Kuhnimhof, Adrian Bauman, and Christine Eisenmann. 2019. Active Travel as Stable Source of Physical Activity for One Third of German Adults: Evidence from Longitudinal Data. *Transportation Research Part A: Policy and Practice* 123: 105–118. https://doi.org/10.1016/j.tra.2018.09.022.

Chataway, Elijah S., Sigal Kaplan, Thomas A. S. Nielsen, and Carlo G. Prato. 2014. Safety Perceptions and Reported Behavior Related to Cycling in Mixed Traffic: A Comparison between Brisbane and Copenhagen. *Transportation Research Part F: Traffic Psychology and Behaviour* 23: 32–43.

Cycling Embassy of Denmark. n.d. Facts about Cycling in Denmark. Copenhagen: Cycling Embassy of Denmark. http://www.cycling-embassy.dk/facts-about-cycling-in-denmark/statistics/.

de Madariaga, Inés Sánchez. 2013. From Women in Transport to Gender in Transport: Challenging Conceptual Frameworks for Improved Policymaking. *Journal of International Affairs* 67(1): 43–65.

Department for Transport. 2015. *National Travel Survey: England 2014*. London: Department for Transport.

Dolati, Haleh. 2014. *Biking Distance: Exploring Gender, Race, and Climate*. Thesis, Master of City and Regional Planning, Ohio State University, Columbus.

Dufour, Dirk. 2010. *PRESTO Cycling Policy Guide: Cycling Infrastructure*. A project of the EU's Intelligent Energy—Europe Programme. Brussels: Ligtermoet and Partners.

Fishman, Elliot, Lars Bocker, and Marco Helbich. 2015. Adult Active Transport in the Netherlands: An Analysis of Its Contribution to Physical Activity Requirements. *PLoS One* 10(4). https://doi.org/10.1371/journal.pone.0121871.

Garrard, Jan, Sharinne Crawford, and Natalie Hakman. 2006. *Revolutions for Women: Increasing Women's Participation in Cycling for Recreation and Transport*. Melbourne: Deakin University.

Garrard, Jan, Stephen P. Greaves, and Adrian B. Ellison. 2010. Cycling Injuries in Australia: Road Safety's Blind Spot? *Journal of the Australasian College of Road Safety* 21(3): 37–43.

Garrard, Jan, Susan Handy, and Jennifer Dill. 2012. Women and Cycling. In *City Cycling*, edited by John Pucher and Ralph Buehler, 211–234. Cambridge, MA: MIT Press.

Geddes, Mark. 2009. Gender and Cycling Use in Dublin: Lessons for Promoting Cycling. MS thesis, Spatial Planning. Dublin: Dublin Institute of Technology.

Harms, Lucas, and Maarten Kansen. 2018. *Cycling Facts*. The Hague: Ministry of Infrastructure and Water Management, Netherlands Institute for Transport Policy Analysis.

Heesch, Kristiann C., Shannon Sahlqvist, and Jan Garrard. 2011. Cyclists' Experiences of Harassment from Motorists: Findings from a Survey of Cyclists in Queensland, Australia. *Preventive Medicine* 53(6): 417–420.

Heesch, Kristiann C., Shannon Sahlqvist, and Jan Garrard. 2012. Gender Differences in Recreational and Transport Cycling: A Cross-Sectional Mixed-Methods Comparison of Cycling Patterns, Motivators, and Constraints. *International Journal of Behavioral Nutrition and Physical Activity* 9. https://doi.org/10.1186/1479-5868-9-106.

Hembrow, David. 2011. Who Cycles in the Netherlands? Everyone Cycles in the Netherlands! *A View from the Cycle Path*. http://www.aviewfromthecyclepath.com/2011/02/who-cycles-in-netherlands.html.

Horne, Rebecca M., Matthew D. Johnson, Nancy L. Galambos, and Harvey J. Krahn. 2018. Time, Money, or Gender? Predictors of the Division of Household Labour across Life Stages. *Sex Roles* 78(11): 731–743. https://doi.org/10.1007/s11199-017-0832-1.

Kirch, Wilhelm. 2008. Risk Assessment. In *Encyclopedia of Public Health*, edited by Wilhelm Kirch, 1261–1264. New York: Springer.

Lovejoy, Kristin, and Susan Handy. 2012. Developments in Bicycle Equipment and Its Role in Promoting Cycling as a Travel Mode. In *City Cycling*, edited by John Pucher and Ralph Buehler, 75–104. Cambridge, MA: MIT Press.

Macmillan, Alex, Alex Roberts, James Woodcock, Rachel Aldred, and Anna Goodman. 2016. Trends in Local Newspaper Reporting of London Cyclist Fatalities 1992–2012:

The Role of the Media in Shaping the Systems Dynamics of Cycling. *Accident Analysis and Prevention* 86: 137–145. https://doi.org/10.1016/j.aap.2015.10.016.

Marqués, R., J. A. García-Cebrían, and M. Calvo-Salazar. 2018. Seville: How a Small Spanish City Became a Cycling Hub for All. *Euronews*, October 12, 2018. https://www.euronews.com/2018/10/12/seville-how-a-small-spanish-city-became-a-cycling-hub-for-all-view.

Ministry of Transport, Public Works, and Water Management. 2009. *Cycling in the Netherlands*. The Hague: Ministry of Transport, Public Works, and Water Management.

Motte-Baumvol, Benjamin, Olivier Bonin, and Leslie Belton-Chevallier. 2014. Gender Differences in Escorting Children among Dual-Earner Families in the Paris Region. Paper presented at Bridging the Gap: The 5th International Conference on Women's Issues in Transportation, Paris, April 14–16.

Munro, Cameron. 2019. *Australian Cycling Participation 2019*. Sydney: Austroads.

Polkamp, Merie. 2017. Male Cyclists of Melbourne, What the Hell Is Your Problem? *The Sydney Morning Herald*, October 27. https://www.smh.com.au/opinion/male-cyclists-of-melbourne-what-the-hell-is-your-problem-20171017-gz2igy.html.

Portland Bureau of Transportation. n.d. *Portland Bicycle Count Report 2013–2015*. Portland: Portland Bureau of Transportation.

Poulos, Roslyn G., Julie Hatfield, Chris Rissel, Lloyd K. Flack, L. Shaw, Raphael Grzebieta, and Andrew S. McIntosh. 2017. Near Miss Experiences of Transport and Recreational Cyclists in New South Wales, Australia: Findings from a Prospective Cohort Study. *Accident Analysis and Prevention* 101: 143–153. https://doi.org/10.1016/j.aap.2017.01.020.

Pucher, John, and Ralph Buehler. 2008. Making Cycling Irresistible: Lessons from the Netherlands, Denmark, and Germany. *Transport Reviews* 28(4): 495–528.

Pucher, John, Jennifer Dill, and Susan Handy. 2010. Infrastructure, Programs and Policies to Increase Bicycling: An International Review. *Preventive Medicine* 50(Suppl 1): S106–S125.

Rimano, Alessandra, Maria Paola Piccini, Paola Passafaro, Renata Metastasio, Claudia Chiarolanza, Aurora Boison, and Franco Costa. 2015. The Bicycle and the Dream of a Sustainable City: An Explorative Comparison of the Image of Bicycles in the Mass-Media and the General Public. *Transportation Research Part F: Traffic Psychology and Behaviour* 30: 30–44. http://dx.doi.org/10.1016/j.trf.2015.01.008.

Rissel, Chris, Catriona Bonfiglioli, Adrian Emilsen, and Ben J. Smith. 2010. Representations of Cycling in Metropolitan Newspapers—Changes Over Time and Differences between Sydney and Melbourne, Australia. *BMC Public Health* 10. https://doi.org/10.1186/1471-2458-10-371.

Sallis, James F., Robert B. Cervero, William Ascher, Karla A. Henderson, M. Katherine Kraft, and Jacqueline Kerr. 2006. An Ecological Approach to Creating Active Living

Communities. *Annual Review of Public Health* 27: 297–322. https://doi.org/10.1146 /annurev.publhealth.27.021405.102100.

Statistics Netherlands (CBS). 2010. *Men and Women Plump for Different Sports.* The Hague: Statistics Netherlands.

Tremblay-Breault, Martin, Marie-Hélène Vandersmissen, and Marius Thériault. 2014. Gender and Daily Mobility: Did the Gender Gap Change between 1996 and 2006 in the Quebec Urban Area? Paper presented at Bridging the Gap: The 5th International Conference on Women's Issues in Transportation, Paris, April 14–16.

van Uffelen, Jannique G. Z., Asaduzzaman Khan, and Nicola W. Burton. 2017. Gender Differences in Physical Activity Motivators and Context Preferences: A Population-Based Study in People in Their Sixties. *BMC Public Health* 17: 624. https://doi.org/10 .1186/s12889-017-4540-0.

VicHealth. 2015. *Female Participation in Sport and Physical Activity: A Snapshot of the Evidence.* Carlton: Victorian Health Promotion Foundation.

Voelcker, Jake. 2007. A Critical Review of the Legal Penalties for Drivers Who Kill Cyclists or Pedestrians. Unpublished MA dissertation, School for Policy Studies, University of Bristol, Bristol, UK.

Walker, Peter. 2015. How Seville Transformed Itself into the Cycling Capital of Southern Europe. *The Guardian*, January 29. https://www.theguardian.com/cities /2015/jan/28/seville-cycling-capital-southern-europe-bike-lanes.

Walker, Peter. 2017. Why Does the BBC Feel It's OK to Demonise Cyclists? *The Guardian*, November 17. https://www.theguardian.com/environment/bike-blog/2017/nov /17/why-are-cyclists-the-one-minority-group-the-bbc-feels-its-ok-to-demonise.

Whitehead, Sarah, and Stuart Biddle. 2008. Adolescent Girls' Perceptions of Physical Activity: A Focus Group Study. *European Physical Education Review* 14(2): 243–262.

Wilkins, Roger, and Inga Lass. 2018. *The Household, Income and Labour Dynamics in Australia Survey: Selected Findings from Waves 1 to 16.* Melbourne: Melbourne Institute of Applied Economic and Social Research.

Winters, Meghan, and Moreno Zanotto. 2017. Gender Trends in Cycling over Time: An Observational Study in Vancouver, British Columbia. *Journal of Transport and Health* 5: S37–S38. https://doi.org/10.1016/j.jth.2017.05.324.

World Health Organization. 2018. *Fact Sheets: Physical Activity.* Geneva: World Health Organization.

Xie, Linjun, and Justin Spinney. 2018. "I Won't Cycle on a Route Like This; I Don't Think I Fully Understood What Isolation Meant": A Critical Evaluation of the Safety Principles in Cycling Level of Service (CLoS) Tools from a Gender Perspective. *Travel Behaviour and Society* 13: 197–213.

# 12  Children and Cycling

Noreen McDonald, Eleftheria Kontou, and Susan Handy

Learning to ride a bicycle is an important milestone for children. The ability to control a bicycle provides evidence of the child's physical and cognitive development. For a child, riding a bicycle brings newfound independence and the ability to travel faster and farther, bringing destinations that previously were far out of reach within grasp.

Beyond the excitement and sense of achievement that being able to ride a bike may bring to the child and the family, there are larger societal trends at play. Increasing levels of childhood obesity have been associated with declines in everyday physical activities such as cycling or walking (Sallis et al. 2012). Biking around the neighborhood or to destinations such as school and parks is a healthful activity that is increasingly rare. While no known studies establish a causal link between children's cycling and reduced levels of obesity, Garrard et al. (chapter 3, this volume) and other researchers have shown that being active is good for children's health and that being active as a child may create lifelong habits (Steinbeck 2001).

While we know that bicycling is good, it turns out that how much children bicycle varies widely from country to country and, within countries, from family to family. Getting more children on bicycles will require a comprehensive effort to build communities where children can safely ride bikes. This chapter explores the infrastructure, community design, education, and family practices that we know are critical to making children's cycling more common and safer in the future.

## Who Is Biking?

Tracking children's bicycling is difficult. We focus on the trip to school because it is a daily trip for children and is the type of trip that is most often included in travel surveys from many nations. Table 12.1 summarizes how

**Table 12.1**

Children's school travel mode

| | Survey year | Respondent ages | Bike (%) | Walk (%) | Automobile (%) | Other (%) | Categories | Data source |
|---|---|---|---|---|---|---|---|---|
| **Countries** | | | | | | | | |
| United States | 2017 | 5–17 | 1.1 | 9.6 | 50.2 | 39.1 | | Kontou et al. (2020) |
| Canada | 2010 | 5–12 | 25 | | 62 | | age based | Gray et al. (2014) |
| Canada | 2010 | 13–17 | 23 | | 65 | | age based | Gray et al. (2014) |
| United Kingdom | 2017 | 5–10 | 2 | 51 | 41 | 6 | age based | Department for Transport (2017) |
| United Kingdom | 2017 | 11–16 | 3 | 39 | 26 | 31 | age based | Department for Transport (2017) |
| Netherlands | 2003 | 0–11 | 29 | 29 | 40 | 2 | age based | Wegman and Aarts (2008) |
| Netherlands | 2003 | 12–17 | 52 | 18 | 17 | 13 | age based | Wegman and Aarts (2008) |
| Australia | 2002 | 5–12 | 1 | 21.5 | 58.7 | 18.8 | urban | Merom et al. (2006) |
| Australia | 2002 | 5–12 | 2.5 | 11.4 | 38.7 | 47.4 | rural | Merom et al. (2006) |
| **Regions/Cities** | | | | | | | | |
| Dresden, Germany | 2008 | 10–19 | 8 | 21 | 10 | 61 | winter term | Müller, Tscharaktschiew, and Haase (2008) |
| Dresden, Germany | 2009 | 10–19 | 24 | 18 | 6 | 52 | summer term | Müller, Tscharaktschiew, and Haase (2008) |
| Flanders, Belgium | 2008 | 6–12 | 21 | 8 | 70 | 1 | boys | Nevelsteen et al. (2012) |
| Flanders, Belgium | 2008 | 6–12 | 16 | 10 | 73 | 1 | girls | Nevelsteen et al. (2012) |
| Beijing, China | 2014 | 7–18 | 30 | 29 | 15 | 26 | n.a. | Zhang, Yao, and Liu (2017) |
| Tehran, Iran | 2011 | 12–17 | 0.6 | 15.5 | 7.3 | 76.6 | boys | Ermagun and Levinson (2015) |
| Tehran, Iran | 2012 | 12–17 | 0 | 26.8 | 10.2 | 63 | girls | Ermagun and Levinson (2015) |
| Okinawa, Japan | 2008 | 15–18 | 11 | 18 | 51 | 20 | n.a. | Alemu and Tsutsumi (2011) |
| Louisiana, US | 2014 | 5–14 | 0.3–4.5 | 2.4–17.4 | n.a. | n.a. | n.a. | Gustat et al. (2015) |
| Toronto, Canada | 2006 | 11 | 0.5 | 56.9 | 25.5 | 17.1 | n.a. | Mitra, Papaioannou, and Nurul Habib (2015) |

*Note:* n.a. = not applicable.

children are reaching school at the national and city levels. Rates of bicycling to school are low (<3%) in the United States, United Kingdom, and Australia but significantly higher in northern Europe, China, and Japan. In the Netherlands, 29% of elementary school children get to school by cycling (vs. 29% by walking and 40% driven by car); 52% of middle and high school children get to school by cycling (vs. 18% by walking and 17% driven by car). In Canada, about a fourth of youths biked or walked to school in 2010, while almost two-thirds were driven by car.

Rates of walking and bicycling to school are decreasing in North American countries (Buliung, Mitra, and Faulkner 2009; McDonald et al. 2011). Walking has declined more than bicycling, but the declines in bicycling are still dramatic and troubling. For example, 2% of US high school students biked to school in 1977, but the share dropped to 0.8% in 2017 (Kontou et al. 2020). Results are similar for the Toronto region, where 1.7% of students at age 11 were bicycling to school in 1986 but the share had dropped to 0.7% in 2006 (Buliung, Mitra, and Faulkner 2009).

Many studies find that bicycling is more common for boys than for girls. For example, in the Flanders region in Belgium, among students 6 to 12 years old, the share of those bicycling to school was 21% for boys in 2008 but only 16% for girls. That is consistent with survey evidence suggesting that girls bicycle to school less frequently because of perceptions of safety and social inhibitions (Frater and Kingham 2018; McDonald 2010). Similarly, Hispanic and Latino youths bicycle to school less than non-Hispanic whites (McDonald et al. 2011). Differences in cycling levels are exacerbated by inequitable policies that do not offer equal opportunities to all youths, such as training and participation in educational programs, as discussed later (Goodman, van Sluijs, and Ogilvie 2015; Goodman, van Sluijs, and Ogilvie 2016). Inequality gaps across gender, ethnicity, and income are increased by the absence of equity objectives in providing youth access to infrastructure that encourages physical activity (Gilbert et al. 2018; Hämäläinen et al. 2016).

The limited data we have on children's recreational biking suggest that it has been declining, at least in the United States (Buehler, Pucher, and Bauman 2020; Schoeppe et al. 2013). Explanations are complex, but the move to organized children's activities and changing community design may mean that children have to travel farther to participate in activities.

## How Safe Is Biking?

Road safety is an important issue for children and adolescents. In the United States, motor vehicle fatalities are the leading cause of death for teenagers, accounting for one-third of all deaths (Miniño 2010). Because bicyclists often have to share the road with other modes of transport that are heavier and move faster, they are especially vulnerable to injuries caused by traffic crashes. In the United Kingdom, 10% of the 2016 bicyclists who were killed or injured were children (Royal Society for the Prevention of Accidents 2017). The European Commission published a report in 2015 noting that 4% of bicyclist fatalities are children under 15 years old and 2% are high schoolers (European Commission 2015).

Comparing bicyclist injury and fatality rates across countries is complicated by data collection protocols and differences in exposure. According to the National Highway Traffic Safety Administration (NHTSA 2018), bicyclists' fatality rate in 2016 in the United States was 0.97 per million population for children less than 14 years old and 1.89 per million population for middle schoolers. In the Netherlands, the fatality rate of adolescent cyclists in 2017 was 2.40 per million population, but the average Dutch adolescent cycled 6.37 km per day—significantly more than in the United States (Buehler, Pucher, and Bauman 2020; CBS 2017). Thus, the fatality rate per kilometer for adolescents is much lower in the Netherlands than in the United States. That is consistent with figure 2.7 (see Buehler and Pucher, chapter 2, this volume), which reports a national cyclist fatality rate per kilometer (all ages in aggregate) that is seven times higher in the United States than in the Netherlands (5.6 vs. 0.8).

### Safety Education and Regulations

Many countries promote bicycling safety through education interventions and by regulating helmet use. Some countries, such as the Netherlands and Germany, integrate cycling education into school curricula and have a formal process for testing children on their bicycling competence (Pucher and Dijkstra 2003). Many other countries rely on voluntary bicycle training programs that may or may not be integrated into school settings. Documented components of these educational programs include skills and knowledge tests, hands-on training, and helmet fitting and instruction on how to wear helmets correctly (Hamann and Conrad 2018). Programs such as Safe

Routes to School (SRTS) in the United States, Cycle Safe in New Zealand, CAN-BIKE in Canada, and Let's Ride in Australia offer online education, onsite courses, and school and community activities (Christchurch City Council 2018; Cycling Australia 2018; Cycling Canada 2018; National Center for Safe Routes to School 2018). The success of such interventions has been reported in the literature.

Laws mandating helmet use for children have been in effect in several countries (Bonander, Nilson, and Andersson 2014; Insurance Institute for Highway Safety and Highway Loss Data Institute 2018). Data from a study in Valencia, Spain, found that helmet regulations improve children's helmet use without having a negative effect on the frequency of bicycling to school (Molina-Garcia and Queralt 2016). A study focusing on the efficacy of US helmet laws showed that helmet regulations increased helmet use among youths, with some evidence of reducing bicycling levels, suggesting that such regulations need to be accompanied by active encouragement of bicycling (Kraemer 2016). In a similar manner, a study on youth injury reduction resulting from the Swedish helmet regulation showed that head injuries to male youth cyclists decreased by 7.8% (Bonander, Nilson, and Andersson 2014).

**Neighborhood Design and Bicycle Infrastructure**

Getting more children to bike requires communities with safe cycling routes and destinations within bikeable distances. The spatial distribution of educational facilities and homes, as well as other recreational facilities and the infrastructure connecting them, is crucial when considering alternative mode choices, particularly for youth travel (e.g., Kaplan, Nielsen, and Prato 2016). When youth trip destinations are closer and safe infrastructure is in place, bicycling becomes a viable travel option. Such facility location decisions and infrastructure that support bicycling are greatly affected by land-use planning and urban development decisions. Characteristics of the natural environment, such as steep hills or weather conditions, may also affect travel mode choices (Wegman, Zhang, and Dijkstra 2012). Shorter distances encourage youth bicycling. For example, based on data from the 2017 *US National Household Travel Survey*, 82.8% of those 5 to 17 years old who bicycled to school lived less than 2 mi away (Kontou et al. 2020).

## Bicycle Infrastructure

The infrastructure supporting safe bicycling between destinations is critical for enabling youths' bicycle use and shaping guardians' comfort with letting them ride (Mandic et al. 2017). Several studies link bicycling activity to the presence of bike paths (Helbich et al. 2016; Larouche et al. 2018; Sallis et al. 2013; Wang et al. 2016). The Dutch Sustainable Safety approach highlights the importance of separating modes when vehicles are traveling at higher speeds because of the risks to pedestrians and bicyclists (Wegman, Zhang, and Dijkstra 2012). Bicycling paths separated from motor vehicle traffic also help improve parents' perceptions of safety and convenience of infrastructure for their children's cycling (Ghekiere et al. 2018; Winters et al. 2018).

While bicycle paths and separated cycling facilities would benefit all residents, some important interventions are specifically for children. Secondary schools in some Dutch cities are investing in guarded bicycle parking to encourage students to ride to school by alleviating concerns about bicycle theft. A small secondary school in The Hague had a tenfold increase in the number of students bicycling after installing a guarded parking facility (Netherlands Ministry of Transport 2009). Another Dutch strategy for increasing safe mobility is Delft's effort to create *kindlint*, or "child ribbons" (Netherlands Ministry of Transport 2009). These ribbons connect destinations important to children, such as schools and playgrounds, and provide a safe route with special markings and traffic calming to connect the destinations. This type of intervention can be used in new communities or to retrofit existing developments.

## Siting of Schools and School Bicycling Policies

Recently, more attention has been paid to how decisions about school location affect how students travel. The most recent American school planning guidelines were developed for rapidly growing suburban areas and emphasized acquiring large school sites with sufficient room for recreation facilities and future expansion (McDonald 2010). These guidelines encourage the building of large campuses, generally 15 or more acres for an elementary school and 30 acres or more for a high school. Finding such large sites is often impossible in developed areas; consequently, many districts have built schools far from residential neighborhoods (Gurwitt 2004). The result is that, over recent decades, US students live farther and farther from their school. American elementary school students traveled an average of 5.8

km (3.6 mi) to school in 2009 compared with 2.4 km (1.5 mi) for British 5–10-year-olds and Swiss 6–14-year-olds (Department for Transport 2010; Grize et al. 2010).

Schools also influence bicycling through their policies. There are examples of schools actively encouraging bicycling to school but also others discouraging it. Liability and safety concerns have caused some schools to prohibit or make it difficult to bike to school. No large-scale studies have cataloged the extent of this anti-bicycling policy, but an informal survey by a Chicago nonprofit found bicycle bans in place in several suburban school districts. For example, Oak Park, Illinois, a Chicago suburb, had bicycle bans in place at 5 of 11 schools, and a nearby school district banned bicycles at 2 of 27 schools (Glowacz 2004). In the United Kingdom, schools cannot ban bicycling, but they can prohibit students from bringing bicycles onto school property. One British secondary school took this approach (Schlesinger 2009). The result was several stolen bicycles and concerns that fewer students would bike to school.

In contrast, many schools are working to encourage students to walk and bike to school. Communities employ multifaceted approaches featuring aspects of education, encouragement, engineering, and enforcement. Evaluations show these efforts have been effective at increasing walking and bicycling to school, although the impacts vary greatly (Buttazzoni et al. 2018; McDonald et al. 2014; Villa-González et al. 2018). Funding for these interventions often comes from specialized government programs focused on improving safe access to school. In the United States, the most prominent program has been the Safe Routes to School program funded by the US Department of Transportation (National Center for Safe Routes to School 2018).

### Personal Factors, Parents, and Peers

Good infrastructure is necessary to increase rates of bicycling among children and adolescents, but its effectiveness also depends on the degree to which they are able and willing to bicycle. Who is willing to ride a bike depends on a host of social and psychological factors—the mentality of youths themselves but also the mentality of their parents and their peers. Indeed, one study of high school students found that attitudes are more important than infrastructure in predicting which students bicycled to school (Fitch, Rhemtulla, and Handy 2019). Although research on these factors is more limited than research on infrastructure, programs that promote a pro-bicycling way of thinking are clearly essential to efforts to increase bicycling among children and adolescents. Such programs play an important role in Safe Routes to School and related efforts.

### Personal Factors

A child's own views about bicycling are essential to getting children to ride a bike. One of the most important predictors of bicycling among children (as well as adults) is their personal confidence in and comfort with bicycling. In a survey of high school students in Davis, California, the one-third of students who regularly bicycled to school reported a significantly higher agreement than other students that they were confident in their bicycling ability and that they felt comfortable bicycling on a busy street with a bike lane (Emond and Handy 2012). Similarly, in a New Zealand study, high school students who rated their cycling capability more highly reported a higher intention to bicycle to school than other students (Frater et al. 2017). But, in both studies, other attitudes also mattered. Davis students who strongly agreed that they like bicycling were far more likely to bicycle to school, and the New Zealand students who expressed positive views about the experience and usefulness of bicycling were more likely to bicycle to school. Studies show that many children of all ages cycle for fun (Thigpen and Handy 2018).

Training programs can help to increase bicycling confidence and comfort as well as pro-bicycling attitudes more generally. As discussed earlier, some countries integrate training into their school curricula, ensuring that all students receive training. In the United Kingdom, nearly half of children complete bicycle training, in part because the government-promoted Bikeability training program offered free of charge at over half of primary schools (Goodman, van Sluijs, and Ogilvie 2015). Studies of such programs show that they consistently increase confidence, knowledge, and skills, though they do not always produce an increase in overall bicycling or bicycling to school (Sersli et al. 2018). The Belgian study that found that a cycle training course for fourth-grade children increased bicycling skills also found that it did not increase bicycle use (Ducheyne et al. 2014). In New Zealand, a skills training program improved bicycling confidence in adolescent girls but did not change their bicycling habits (Mandic et al. 2018). Getting adolescents into training programs may itself be challenging: the students who already had positive attitudes about bicycling also had the most positive perceptions of the benefits of such programs (Mandic et al. 2016). Nevertheless, there is strong evidence to support continued investment in such programs as a strategy for increasing bicycling among children and adolescents.

Most children learn how to bicycle in the first place with help from their parents. In the United States, this training tends to focus on bicycling in parks or on neighborhood streets as a recreational activity for the family, and many adults remember with fondness both their own learning experience and the experience of teaching their children (Handy and Lee 2018). Dutch and Danish parents, among other Europeans, are known for riding with their children from a young age as a daily means of travel. The parents coach their children as they ride about proper positioning on the street, navigating intersections, and negotiating around other users of the street (Mcilvenny 2012). The dissemination of strategies for teaching children to cycle is no longer dependent on word-of-mouth: at the time of this writing, a Google search produced nearly 12 million hits on the phrase "how to teach a child to ride a bike."

## Parents

Parents play a key role in their children's bicycling in many ways. In general, parents have control over how children travel, with essentially complete control for younger children who are not able to travel independently. Parents are especially influential when it comes to active travel (McDonald and Aalborg 2009; McMillan 2007). Evidence suggests that parents may be suppressing active travel, particularly bicycling. A study in Toronto found that while only 2%–3% of children cycled to or from school, 40% preferred to do so (Larouche et al. 2016).

One way that parents influence the bicycling of their children is by determining whether bicycling is possible. Most obviously, parents who buy a bicycle for their children and teach them how to bicycle have taken the first step toward making bicycling possible. Many school-age children in the United States have access to a bicycle and have learned how to ride one. However, the possibility of bicycling is influenced by other factors that depend on parental choices, and these factors help to explain the low share of bicycling to school in the United States. The choice of where to live, for example, determines the quality of the bicycling infrastructure and traffic conditions. Parents' choice of a school for their children determines the distance to the school, a factor that significantly influences bicycle commuting. Convincing parents that the possibility of bicycling should factor into such choices could thus be one strategy for increasing bicycling among children or at least increase the possibility.

Even when parents have made choices that enable their children to bicycle, they may not encourage it or even allow it. Before children are old enough to travel on their own, their travel is dependent on their parents' choices about their own travel. Parents who need to or prefer to drive to work may not give their young children the option of bicycling simply as a matter of time and convenience. When children are older, parents may allow them to travel independently but set rules about where, when, and how they travel, including whether they are allowed to bicycle. Such rules often depend on parental perceptions about the safety of bicycling both in general and at specific locations (Hopkins and Mandic 2017). Some evidence suggests that parents who bicycle are less concerned about safety than parents who do not (Hopkins and Mandic 2017) and that they put more restrictions on girls than on boys (Driller and Handy 2013). These restrictions influence the amount that children bicycle. Whether parents allowed their children to bicycle alone on main roads, for example, was a significant predictor of bicycling to school for Australian schoolchildren (Carver, Timperio, and Crawford 2015). The age at which parents give children the freedom to bicycle on their own varies from country to country and from community to community depending on the bicycling environment as well as cultural and other factors.

Beyond simply allowing bicycling, parents can actively encourage it. In the study of high school students in Davis, the children of parents who encouraged them to bicycle had twice the odds of bicycling as those whose parents did not (Emond and Handy 2012). The influence of parental encouragement was also evident in a later study that included two additional California high schools (Fitch, Rhemtulla, and Handy 2019). A New Zealand study found similar results: high school students who reported that their parents thought they should ride a bicycle and would approve of their riding a bicycle to school had greater intention to bicycle (Frater et al. 2017). The corollary to encouragement is whether parents make other options possible for their children. In the Davis study, high school students who had a driver's license and access to a car, conditions that usually require some help from parents, were far less likely to bicycle than those who didn't (Emond and Handy 2012);

students who said that they could rely on their parents to drive them to school and other activities were also less likely to bicycle. Encouragement of driving generally overrides encouragement of bicycling.

Whether parents enable and encourage bicycling depends on their own experiences with bicycling, which have shaped their attitudes toward bicycling over time. Promotional programs such as those discussed by Heinen and Handy (chapter 7, this volume) can help to develop more positive attitudes toward bicycling among parents. In turn, whether parents enable and encourage bicycling influences their children's own confidence and other attitudes toward bicycling (Driller and Handy 2013). Pro-bicycling attitudes are thus self-reinforcing within families, with bicycling rates highest among families in which both parents and children regularly bicycle (Tal and Handy 2008). Therefore, the most important way that parents can encourage bicycling is to do it themselves (Thigpen and Handy 2018). Of course, as most parents well know, encouragement can sometimes backfire: children in Davis whose parents required them to bicycle tended to have negative rather than positive feelings about bicycling (Driller and Handy 2013).

### Peers

Another important influence on children's bicycling is peers, including friends as well as the broader student population. Prior to high school, friends tend to have a positive effect on bicycling. For children of elementary school age, bicycling is seen as a fun and exciting activity with friends, at least as remembered by adults in the United States (Underwood et al. 2014). In middle school, an age at which many children are able to travel independently, bicycling can be a way to socialize with friends on the way to school or a means of meeting up with friends outside school, at least in a community like Davis, where bicycling is relatively safe (Thigpen and Handy 2018).

For high school students, the role of peers becomes more ambiguous. In Davis, students who bicycled to school were more likely to report that their friends bicycled to school. Neither bicyclists nor nonbicyclists admitted that they worried what their peers would think of them if they bicycled (Emond and Handy 2012). A New Zealand study concluded that perceived social pressure by friends was the most significant influence—in a positive way—on intention to bicycle to high school (Frater, Kuijer, and Kingham 2017). On the other hand, interviews with adults in the United States suggest that bicycling had become distinctly "uncool" by the time they reached high school, as they remembered it, with girls being especially sensitive to negative images of bicycling (Underwood et al. 2014). Even so, not all high school students are deterred from bicycling. Some manage to maintain positive attitudes toward bicycling despite the negative attitudes of their peers, and these teenagers are more likely to bicycle as adults (Underwood et al. 2014). Some high school students in a New Zealand study valued the alone time that bicycling gave them (Hopkins and Mandic 2017).

Evidence showing the importance of friends and other peers suggests the need for strategies to enhance pro-bicycling attitudes among children and especially adolescents. What strategies might prove effective at changing attitudes among this population is less clear. In countries such as the Netherlands and Denmark, bicycling is just what children do; children and parents alike view it unquestioningly as normal. In places such as the United States and New Zealand, promotional programs targeted at teenagers may prove helpful, though such programs are rare and evidence on their effectiveness is not available. Student-led efforts may ultimately do the most to create a social movement toward increased bicycling.

## Conclusion

Bicycling is a skill, and childhood is a critical time for people to learn how to ride. Getting children on bicycles provides them with independence and an opportunity to explore their neighborhoods, visit friends, and get to school. Researchers have shown that it can have long-term health benefits (as documented in Garrard et al., chapter 3, this volume). Yet, across the globe, we see highly varying rates of bicycling among children and adolescents. Children's bicycling requires supportive infrastructure, including separate facilities that protect them from high-speed traffic; courses or training sessions teaching safe bicycling techniques; and communities designed so that trip origins and destinations are not far apart, and thus within cycling distance. Encouraging children to ride bicycles also requires children and parents to view cycling positively and want to participate. This chapter highlights the most recent research on these topics and identifies positive actions, programs, and approaches that can potentially increase children's bicycling.

## References

Alemu, Dubbale Daniel, and Jun-ichiro Giorgos Tsutsumi. 2011. Determinants and Spatial Variability of After-School Travel by Teenagers. *Journal of Transport Geography* 19(4): 876–881.

Bonander, Carl, Finn Nilson, and Ragnar Andersson. 2014. The Effect of the Swedish Bicycle Helmet Law for Children: An Interrupted Time Series Study. *Journal of Safety Research* 51: 15–22.

Buehler, Ralph, John Pucher, and Adrian Bauman. 2020. Physical Activity from Walking and Cycling for Daily Travel in the United States, 2001–2017: Demographic, Socioeconomic, and Geographic Variation. *Journal of Transport and Health* 16. https://doi.org/10.1016/j.jth.2019.100811.

Buliung, Ron N., Raktim Mitra, and Guy Faulkner. 2009. Active School Transportation in the Greater Toronto Area, Canada: An Exploration of Trends in Space and Time (1986–2006). *Preventive Medicine* 48(6): 507–512.

Buttazzoni, Adrian N., Emily S. Van Kesteren, Tayyab I. Shah, and Jason A. Gilliland. 2018. Active School Travel Intervention Methodologies in North America: A Systematic Review. *American Journal of Preventive Medicine* 55(1): 115–124.

Carver, Alison, Anna F. Timperio, and David A. Crawford. 2015. Bicycles Gathering Dust Rather than Raising Dust—Prevalence and Predictors of Cycling among Australian Schoolchildren. *Journal of Science and Medicine in Sport* 18(5): 540–544.

Centraal Bureau voor de Statistiek (CBS). 2017. *More Deaths among Cyclists Than Car Occupants in 2017*. Centraal Bureau voor de Statistiek. https://www.cbs.nl/en-gb/news/2018/17/more-deaths-among-cyclists-than-car-occupants-in-2017.

Christchurch City Council. 2018. Cycle Safe Programme. https://ccc.govt.nz/transport/getting-to-school/resources-for-schools/cyclesafe/.

Cycling Australia. 2018. Let's Ride. https://letsride.com.au.

Cycling Canada. 2018. CAN-BIKE Courses for Children. http://canbikecanada.ca.

Department for Transport. 2010. *Table NTS0614. Trips to School by Main Mode, Trip Length and Age: Great Britain, 2008/09*. London: Department for Transport.

Department for Transport. 2017. *Transport Statistics Great Britain 2017*. London: Department for Transport.

Driller, Brigitte, and Susan Handy. 2013. Exploring the Impact of Parent-Child Relationships on Children's Bicycling in Davis, CA. Paper presented at Annual Meeting of the Transportation Research Board, Washington, DC, January.

Ducheyne, Fabian, Ilse De Bourdeaudhuij, Matthieu Lenoir, and Greet Cardon. 2014. Effects of a Cycle Training Course on Children's Cycling Skills and Levels of Cycling to School. *Accident Analysis and Prevention* 67: 49–60.

Emond, Catherine R., and Susan L. Handy. 2012. Factors Associated with Bicycling to High School: Insights from Davis, CA. *Journal of Transport Geography* 20(1): 71–79.

Ermagun, Alireza, and David Levinson. 2015. *Physical Activity in School Travel: A Cross-Nested Logit Approach*. Minneapolis: University of Minnesota. http://nexus.umn.edu.

European Commission. 2015. *Road Safety in the European Union—Trends, Statistics and Main Challenges*. Brussels: European Commission.

Fitch, Dillon T., Mijke Rhemtulla, and Susan L. Handy. 2019. The Relation of the Road Environment and Bicycling Attitudes to Usual Travel Mode to School in Teenagers. *Transportation Research Part A: Policy and Practice* 123: 35–53.

Frater, Jillian, and Simon Kingham. 2018. Gender Equity in Health and the Influence of Intrapersonal Factors on Adolescent Girls' Decisions to Bicycle to School. *Journal of Transport Geography* 71: 130–138.

Frater, Jillian, Roeline Kuijer, and Simon Kingham. 2017. Why Adolescents Don't Bicycle to School: Does the Prototype/Willingness Model Augment the Theory of Planned Behaviour to Explain Intentions? *Transportation Research Part F: Traffic Psychology and Behaviour* 46: 250–259.

Frater, Jillian, John Williams, Debbie Hopkins, Charlotte Flaherty, Antoni Moore, Simon Kingham, Roeline Kuijer, and Sandra Mandic. 2017. A Tale of Two New Zealand Cities: Cycling to School among Adolescents in Christchurch and Dunedin. *Transportation Research Part F: Traffic Psychology and Behaviour* 49: 205–214.

Ghekiere, Ariane, Benedicte Deforche, Ilse De Bourdeaudhuij, Peter Clarys, Lieze Mertens, Greet Cardon, Bas de Geus, Jack Nasar, and Jelle Van Cauwenberg. 2018. An Experimental Study Using Manipulated Photographs to Examine Interactions between Micro-scale Environmental Factors for Children's Cycling for Transport. *Journal of Transport Geography* 66: 30–34.

Gilbert, Hulya, Carolyn Whitzman, Johannes Pieters, and Andrew Allan. 2018. Children's Everyday Freedoms: Local Government Policies on Children and Sustainable Mobility in Two Australian States. *Journal of Transport Geography* 71: 116–129.

Glowacz, Dave. 2004. Safe Routes to Suits: Cracking the Liability Lies in Walking and Biking to School. *Mr. Bike.com*. http://www.mrbike.com/rants.php.

Goodman, Anna, Esther M. F. van Sluijs, and David Ogilvie. 2015. Cycle Training for Children: Which Schools Offer It and Who Takes Part? *Journal of Transport and Health* 2(4): 512–521.

Goodman, Anna, Esther M. F. van Sluijs, and David Ogilvie. 2016. Impact of Offering Cycle Training in Schools upon Cycling Behaviour: A Natural Experimental Study. *International Journal of Behavioral Nutrition and Physical Activity* 13(1): 1–12.

Gray, Casey E., Richard Larouche, Joel D. Barnes, Rachel C. Colley, Jennifer Cowie Bonne, Mike Arthur, Christine Cameron, Jean-Philippe Chaput, Guy Faulkner, Ian

Janssen, Angela M. Kolen, Stephen R. Manske, Art Salmon, John C. Spence, Brian W. Timmons, and Mark S. Tremblay. 2014. Are We Driving Our Kids to Unhealthy Habits? Results of the Active Healthy Kids Canada 2013 Report Card on Physical Activity for Children and Youth. *International Journal of Environmental Research and Public Health* 11(6): 6009–6020.

Grize, Leticia, Bettina Bringolf-Isler, Eva Martin, and Charlotte Braun-Fahrländer. 2010. Trend in Active Transportation to School among Swiss School Children and Its Associated Factors: Three Cross-Sectional Surveys, 1994, 2000 and 2005. *International Journal of Behavioral Nutrition and Physical Activity* 7: 1–8.

Gurwitt, Rob. 2004. Governing The States and Localities: Edge-Ucation, What Compels Communities to Build Schools in the Middle of Nowhere? *Governing.* http://www.governing.com/topics/education/Edge-Ucation.html.

Gustat, Jeanette, Katherine Richards, Janet Rice, Lori Andersen, Kathryn Parker-Karst, and Shalanda Cole. 2015. Youth Walking and Biking Rates Vary by Environments Around 5 Louisiana Schools. *Journal of School Health* 85(1): 36–42.

Hämäläinen, Riitta Maija, Petru Sandu, Ahmed M. Syed, and Mette W. Jakobsen. 2016. An Evaluation of Equity and Equality in Physical Activity Policies in Four European Countries. *International Journal for Equity in Health* 15(1): 1–13.

Hamann, Cara J., and Alyssa Conrad. 2018. *Inventory of Child Bicycle Education Programs.* Transportation Research Board. https://trid.trb.org/view/1501943.

Handy, Susan, and Amy Lee. 2018. What Is It about Bicycling? Evidence from Davis, California. Paper 18–01485. *Proceedings of TRB Annual Meeting Online 2018.*

Helbich, Marco, Maarten J. Zeylmans van Emmichoven, Martin J. Dijst, Mei-Po Kwan, Frank H. Pierik, and Sanne I. de Vries. 2016. Natural and Built Environmental Exposures on Children's Active School Travel: A Dutch Global Positioning System-Based Cross-Sectional Study. *Health and Place* 39: 101–109.

Hopkins, Debbie, and Sandra Mandic. 2017. Perceptions of Cycling among High School Students and Their Parents. *International Journal of Sustainable Transportation* 11(5): 342–356.

Insurance Institute for Highway Safety, and Highway Loss Data Institute. 2018. *Pedestrians and Bicyclists.* Insurance Institute for Highway Safety. https://www.iihs.org/iihs/topics/laws/bicycle-laws.

Kaplan, Sigal, Thomas Alexander Sick Nielsen, and Carlo Giacomo Prato. 2016. Walking, Cycling and the Urban Form: A Heckman Selection Model of Active Travel Mode and Distance by Young Adolescents. *Transportation Research Part D: Transport and Environment* 44: 55–65.

Kontou, Eleftheria, Noreen C. McDonald, Kristen Brookshire, Nancy Pullen-Seufert, and Seth LaJeunesse. 2020. U.S. Active School Travel in 2017: Prevalence and

Correlates. *Preventive Medicine Reports* 17. https://doi.org/10.1016/j.pmedr.2019 .101024.

Kraemer, John D. 2016. Helmet Laws, Helmet Use, and Bicycle Ridership. *Journal of Adolescent Health* 59(3): 338–344.

Larouche, Richard, George Mammen, David A. Rowe, and Guy Faulkner. 2018. Effectiveness of Active School Transport Interventions: A Systematic Review and Update. *BMC Public Health* 18(1): 1–18.

Larouche, Richard, Michelle Stone, Ron N. Buliung, and Guy Faulkner. 2016. "I'd Rather Bike to School!": Profiling Children Who Would Prefer to Cycle to School. *Journal of Transport and Health* 3(3): 377–385.

Mandic, Sandra, Charlotte Flaherty, Christina Ergler, Chiew Ching Kek, Tessa Pocock, Dana Lawrie, Palma Chillón, and Enrique García Bengoechea. 2018. Effects of Cycle Skills Training on Cycling-Related Knowledge, Confidence and Behaviour in Adolescent Girls. *Journal of Transport and Health* 9: 253–263. https://doi.org/10 .1016/j.jth.2018.01.015.

Mandic, Sandra, Charlotte Flaherty, Tessa Pocock, Alex Mintoft-Jones, Jillian Frater, Palma Chillón, and Enrique García Bengoechea. 2016. Attitudes towards Cycle Skills Training in New Zealand Adolescents. *Transportation Research Part F: Traffic Psychology and Behaviour* 42: 217–226.

Mandic, Sandra, Debbie Hopkins, Enrique García Bengoechea, Charlotte Flaherty, John Williams, Leiana Sloane, Antoni Moore, and John C. Spence. 2017. Adolescents' Perceptions of Cycling versus Walking to School: Understanding the New Zealand Context. *Journal of Transport and Health* 4: 294–304.

McDonald, Noreen C. 2010. School Siting. *Journal of the American Planning Association* 76(2): 184–198.

McDonald, Noreen C., and Annette E. Aalborg. 2009. Why Parents Drive Children to School. *Journal of the American Planning Association* 75(3): 331–342.

McDonald, Noreen C., Austin L. Brown, Lauren M. Marchetti, and Margo S. Pedroso. 2011. U.S. School Travel, 2009: An Assessment of Trends. *American Journal of Preventive Medicine* 41(2): 146–151.

McDonald, Noreen C., Ruth Steiner, Chanam Lee, Tori Rhoulac Smith, Xuemei Zhu, and Yizhao Yang. 2014. Impact of the Safe Routes to School Program on Walking and Bicycling. *Journal of the American Planning Association 80(2): 153–167.*

Mcilvenny, Paul Bruce. 2012. Learning to Be Vélomobile. Abstract from Cycling and Society Annual Symposium, London, United Kingdom. http://vbn.aau.dk/files /74646539/C_S_Provisional_Schedule.pdf.

McMillan, Tracy E. 2007. The Relative Influence of Urban Form on a Child's Travel Mode to School. *Transportation Research Part A: Policy and Practice* 41(1): 69–79.

Merom, Dafna, Catrine Tudor-Locke, Adrian Bauman, and Chris Rissel. 2006. Active Commuting to School among NSW Primary School Children: Implications for Public Health. *Health and Place* 12(4): 678–687.

Miniño, Arialdi M. 2010. Mortality among Teenagers Aged 12–19 Years: United States, 1999–2006. *NCHS Data Brief* 37: 1–8. http://www.ncbi.nlm.nih.gov/pubmed /20450538.

Mitra, Raktim, Elli M. Papaioannou, and Khandker M. Nurul Habib. 2015. Past and Present of Active School Transportation: An Exploration of the Built Environment Effects in Toronto, Canada from 1986 to 2006. *Journal of Transport and Land Use* 9(2): 25–41.

Molina-Garcia, Javier, and Ana Queralt. 2016. The Impact of Mandatory Helmet-Use Legislation on the Frequency of Cycling to School and Helmet Use among Adolescents. *Journal of Physical Activity and Health* 13(6): 649–653.

Müller, Sven, Stefan Tscharaktschiew, and Knut Haase. 2008. Travel-to-School Mode Choice Modelling and Patterns of School Choice in Urban Areas. *Journal of Transport Geography* 16(5): 342–357.

National Center for Safe Routes to School. 2018. Safe Routes. http://www.saferoutesinfo .org.

National Highway Traffic Safety Administration (NHTSA). 2018. *Traffic Safety Facts 2016*. https://crashstats.nhtsa.dot.gov/Api/Public/ViewPublication/812554.

Netherlands Ministry of Transport. 2009. *Cycling in the Netherlands*. http://www .fietsberaad.nl/library/repository/bestanden/CyclingintheNetherlands2009.pdf.

Nevelsteen, Kristof, Thérèse Steenberghen, Anton Van Rompaey, and Liesbeth Uyttersprot. 2012. Controlling Factors of the Parental Safety Perception on Children's Travel Mode Choice. *Accident Analysis and Prevention* 45: 39–49.

Pucher, John, and Lewis Dijkstra. 2003. Promoting Safe Walking and Cycling to Improve Public Health: Lessons from the Netherlands and Germany. *American Journal of Public Health* 98(9): 1509–1516.

The Royal Society for the Prevention of Accidents. 2017. *Road Safety Factsheet: Cycling Accidents*. November. The Royal Society for the Prevention of Accidents. www.rospa .com.

Sallis, James F., Terry L. Conway, Lianne I. Dillon, Lawrence D. Frank, Marc A. Adams, Kelli L. Cain, and Brian E. Saelens. 2013. Environmental and Demographic Correlates of Bicycling. *Preventive Medicine* 57(5): 456–460.

Sallis, James F., Myron F. Floyd, Daniel A. Rodriguez, and Brian E. Saelens. 2012. The Role of Built Environments in Physical Activity, Obesity, and CVD. *Circulation* 125(5): 729–737.

Schlesinger, Fay. 2009. Cotton-Wool Culture Blamed as School Bans All 1,000 Pupils from Bringing Their Bikes to School. *The Daily Mail, November 17.*

Schoeppe, Stephanie, Mitch J. Duncan, Hannah Badland, Melody Oliver, and Carey Curtis. 2013. Associations of Children's Independent Mobility and Active Travel with Physical Activity, Sedentary Behaviour and Weight Status: A Systematic Review. *Journal of Science and Medicine in Sport* 16(4): 312–319.

Sersli, Stephanie, Danielle DeVries, Maya Gislason, Nicholas Scott, and Meghan Winters. 2018. Changes in Bicycling Frequency in Children and Adults after Bicycle Skills Training: A Scoping Review. *Transportation Research Part A: Policy and Practice 123: 170–187.*

Steinbeck, K. S. 2001. The Importance of Physical Activity in the Prevention of Overweight and Obesity in Childhood: A Review and an Opinion. *Obesity Reviews* 2(2): 117–130.

Tal, Gil, and Susan Handy. 2008. Children' s Biking for Non-school Purposes: Getting to Soccer Games in Davis. *Transportation Research Record* 2074: 40–45.

Thigpen, Calvin, and Susan Handy. 2018. Effects of a Building Stock of Bicycling Experience in Youth. *Transportation Research Record 2672(36): 12–23.*

Underwood, Sarah K., Susan L. Handy, Debora A. Paterniti, and Amy E. Lee. 2014. Why Do Teens Abandon Bicycling? A Retrospective Look at Attitudes and Behaviors. *Journal of Transport and Health* 1(1): 17–24.

Villa-González, Emilio, Yaira Barranco-Ruiz, Kelly R. Evenson, and Palma Chillón. 2018. Systematic Review of Interventions for Promoting Active School Transport. *Preventive Medicine* 111: 115–134.

Wang, Yu, Chi Kwan Chau, Jackie Ng, and Tzeming Leung. 2016. A Review on the Effects of Physical Built Environment Attributes on Enhancing Walking and Cycling Activity Levels within Residential Neighborhoods. *Cities* 50: 1–15.

Wegman, Fred, and Letty Aarts. 2008. Advancing Sustainable Safety: National Road Safety Outlook for The Netherlands for 2005–2020. *Safety Science* 46(2): 323–343.

Wegman, Fred, Fan Zhang, and Atze Dijkstra. 2012. How to Make More Cycling Good for Road Safety? *Accident Analysis and Prevention* 44(1): 19–29.

Winters, Meghan, Michael Branion-Calles, Suzanne Therrien, Daniel Fuller, Lise Gauvin, David G. T. Whitehurst, and Trisalyn Nelson. 2018. Impacts of Bicycle Infrastructure in Mid-sized Cities (IBIMS): Protocol for a Natural Experiment Study in Three Canadian Cities. *BMJ Open* 8(1): 1–11.

Zhang, Rui, Enjian Yao, and Zhili Liu. 2017. School Travel Mode Choice in Beijing, China. *Journal of Transport Geography* 62: 98–110.

# 13   Older Adults and Cycling

Jan Garrard, Jennifer Conroy, Meghan Winters, John Pucher, and Chris Rissel

Older adults make up a steadily increasing proportion of the population of most developed countries, with the proportion of people aged 60 years and older in high-income countries rising from 11.8% in 1950 to 23.6% in 2017 and projected to increase to nearly one-third (32.9%) of the population in 2050 (United Nations Department of Economic and Social Affairs Population Division 2017).

Older age is associated with increases in chronic diseases such as cardiovascular disease, type 2 diabetes, cancer, and dementia. Physical activity reduces the risk for these conditions (Weggemans et al. 2018), with benefits occurring at all ages (Sherman et al. 1999; Bauman et al. 2017). Consequently, it is recommended that people who are already active should maintain a physically active lifestyle into older age, and previously inactive older adults will achieve a health benefit if they commence physical activity (British Heart Foundation 2012; Bauman et al. 2017).

Physical activity is crucial to the health of older adults, but physical activity levels decline markedly with age in many developed countries. Data from the 2016 US *National Health Interview Survey* show that while 60% of those aged 18 to 24 years met physical activity guidelines for aerobic activity through leisure-time aerobic activity (150 minutes per week), only 44% of those aged 65 to 74 years, and 29% of those aged 75 years or older met the guidelines (Clarke, Norris, and Schiller 2017).

Where cycling conditions are favorable, however, as in countries such as the Netherlands, Germany, Denmark, and Japan, high rates of cycling contribute to the maintenance of adequate levels of physical activity into older age, as noted by Buehler and Pucher (chapter 2, this volume, figure 2.5). In the Netherlands, for example, older adults are *more* likely than younger adults to meet physical activity guidelines (71% of people aged 55 years and older meet guidelines compared to 51% of those aged 18 to 54 years), in part because older Dutch people spend more time cycling than younger adults (European Commission and World Health Organization Regional Office for Europe 2015).

Cycling is a popular type of physical activity, a convenient transport mode for short to medium trips, and sufficiently intensive to achieve health benefits (Ainsworth et al. 2011). As a low-impact form of physical activity, cycling can appeal to older adults seeking an alternative to high-impact activities. Consequently, cycling for recreation as well as daily travel for utilitarian purposes can help older adults meet physical activity guidelines and achieve the health benefits of remaining active as they grow older (see Garrard et al., chapter 3, this volume).

This chapter provides an overview of the health benefits of cycling for older adults, current patterns and trends in cycling among older adults, factors that influence cycling among older adults, and strategies and policies for increasing cycling among older adults.

## Health Benefits of Cycling for Older Adults

For people of all ages, including older adults, the health benefits of cycling include improved physical, psychological, and social health and well-being, with many of these benefits flowing from cycling as a form of physical activity (see Garrard et al., chapter 3, this volume). Cycling provides especially important benefits for older people. It is minimally weight bearing and puts less stress on the joints than walking, jogging, and running (Base 2012). Cycling can contribute to enhancing social networks and building empowerment (Hansson et al. 2011) and can be incorporated easily into a daily routine. The advent of electric bikes (e-bikes) provides additional opportunities for older adults to cycle more frequently and for greater distances, especially for groups at risk for physical inactivity, such as women and individuals who are overweight or cannot easily walk because of arthritis, for example (Van Cauwenberg, de Geus, and Deforche 2018; see also Cherry and Fishman, chapter 9, this volume).

For older adults, cycling has been shown to reduce the risk of cardiovascular disease (Koolhaas et al. 2016), enhance physical and mental function (Bouaziz et al. 2017; Varela et al. 2018), improve quality of life, and contribute to maintaining physical performance and confidence in maintaining balance among older adults, which are important components of falls prevention, mobility, and independence (Harvey, Rissel, and Pijnappels 2018). Cycling also reduces joint pain and stiffness associated with osteoarthritis and improves functional capacity and quality of life for older adults with osteoarthritis (Alkatan et al. 2016). In a study investigating types of physical activity (walking, cycling, domestic work, sports, and gardening) and health-related quality of life in older adults, cycling was found to make the greatest contribution to improved health-related quality of life (Koolhaas et al. 2018).

Older adults value their mobility, autonomy, and social connection, all of which have a positive impact on health (Tsunoda et al. 2015). Social connections through cycling in recreational groups build interdependence and camaraderie, extend social networks, and increase mobility for older adults. Older adult cyclists have larger life spaces (a measure that assesses the range, independence, and frequency of movement) than their peers who do not cycle (Van Cauwenberg et al. 2019). Older people who cycle regularly are more likely to be in regular contact with friends and family, and more able to participate fully in desired activities (Ryan et al. 2016). Riding with friends, husbands, partners, children, and grandchildren is a consistent theme in the cycling histories of older women (Rowe et al. 2016). Based on interviews with 12 older men in a cycling club in Ontario, Canada, Minello and Nixon (2017) reported on the passion and intensity that bonded the cyclists, who described the bicycle as an extension of the self, while also recalling pleasure, struggle, immersion, ritual, and freedom experienced through cycling.

Cycling cessation can be a source of distress for older cyclists, perceived as a loss of independence, social connection, and well-being. In a study of 456 older adults living in Malmo, Sweden, over two-thirds said they traveled by bike for most of their lives and that they intended to continue for years to come (Ryan et al. 2016). Many described cycling cessation because of injury or illness as an unhappy time, and they were impatient to start again. For example, one said, "I couldn't cycle for nearly two months and I started to get crazy" (Ryan et al. 2016). Faced with the idea of no longer riding their bicycles, participants expressed resilience and determination to overcome fears and obstacles.

In view of the physical, psychological, and social health benefits of cycling for older adults, cycling is well placed to facilitate active aging, a process that has been described as enabling older adults to continue participating in "social, economic, cultural, spiritual and civic affairs" and maintain a good quality of life (World Health Organization 2002).

## Cycling Injuries among Older Adults

The perception of cycling as dangerous can be a barrier to cycling for older people. Older people are more vulnerable than younger people to impact-related injury. However, there is inconsistent evidence on whether older cyclists experience higher crash and injury rates than younger age groups. Most evidence points to increased injury *severity* among older adults in the event of a crash, although there is little evidence that older adults are more likely to be involved in a bicycle crash (Schepers et al. 2015; Dubbeldam et al. 2017; Schepers 2012). Indeed, a recent study of exposure-adjusted crash and injury rates in Australia found that cyclists 60 years of age or older were less likely than younger cyclists to experience a crash. Older cyclists tended to be more likely to experience a crash requiring medical attention, but the difference was not statistically significant (Poulos et al. 2015). Poulos et al. (2015) suggested that the higher injury rates among older cyclists that have been reported elsewhere may result from age-related factors (such as reduced physiological reserves, reduced muscle strength, osteopenia, or osteoporosis) rather than from an elevated crash risk.

A review of cyclist injury data from mainly Western countries found that between 60% and 95% of cyclists admitted to hospitals or treated at hospital emergency departments are victims of single-bicycle crashes (i.e., not involving a collision with a motor vehicle or another bicycle) (Schepers et al. 2015). Thus, the prevention of cycling injuries, including for older adults, needs to include attention to the factors that contribute to single-bicycle crashes in addition to collisions involving motor vehicles. While bicycle crashes involving motor vehicles are more likely to result in fatal or serious injuries, these are substantially smaller in number than injuries resulting from single-bicycle crashes (Schepers et al. 2015).

Factors that contribute to single-bicycle crashes and consequent injuries include a number of individual and environmental factors related to both the risk of a bicycle crash occurring, and the risk of injury in the event of a crash. Older age can affect control of the bicycle (including when mounting, dismounting, or traveling at low speed), maneuvering around obstacles, interacting with other road or path users, and anticipation of potential hazards. As noted above, the risk of injury in the event of a crash may also be related to age. However, other factors may reduce the risk of injury. For example, older adults may take greater care while cycling, including their choice of when and where to cycle, the type of bicycle used, cycling style, and cycling purpose. Some physical and cognitive limitations that may put older adults at risk of a bicycle injury (e.g., muscle strength, balance, and cognitive abilities) are also attributes that are improved by cycling (Harvey, Rissel, and Pijnappels 2018; Leyland et al. 2019), further complicating the pattern of benefits and risks associated with cycling for older adults. Overall, the benefits of cycling for older adults appear to outweigh the risks (Fishman, Schepers, and Kamphuis 2015), which is consistent with the positive benefit-risk ratios that many studies have estimated for adult cycling in general (see Garrard et al., chapter 3, this volume).

## Patterns and Trends in Cycling among Older Adults

When cycling conditions are favorable, people will continue to cycle for recreation and transport well into older age, a time when many other forms of physical activity decline. For example, in the Netherlands, 17% of people over 65 years of age cycle every day, a level somewhat lower than for the population as a whole (24%) but nevertheless a substantial rate of daily cycling (Bicycle Dutch 2018). Similarly, a Eurobarometer survey of over 27,000 respondents from the 28 European Union countries found that there was little change with age in the proportion of people reporting their most frequently used mode of transport was a bicycle: 11% of those aged 15–24, 7% of those aged 25–39, 7% of those aged 40–54, and 8% of people 55 and older (European Union 2014).

National travel surveys confirm that cycling levels remain high among older adults in Germany, Denmark, the Netherlands, and Japan (see Buehler and Pucher, chapter 2, this volume, figure 2.5). Indeed, cycling rates increase slightly between middle-age groups and the 65–69 age group and fall only slightly for the 70 and older age group. The oldest group (70 years and older) makes 10% of their daily trips by bicycle in Denmark, 12% in Germany and Japan, and 23% in the Netherlands. These patterns show that older age need not be a barrier to cycling, which remains physically possible for many older people, provided conditions are safe and convenient. In stark contrast, those aged 70 and older make only 0.4% of daily trips by bicycle in the United States and 0.9% in the United Kingdom, two countries where cycling rates decline markedly with age, as shown by Buehler and Pucher (chapter 2, this volume, figure 2.5). Cycling facilities in the Netherlands, Denmark, and Germany are far more extensive and better designed than those in the United States and the United Kingdom (Pucher and Buehler 2008), which may explain these differences. Safe, convenient, and comfortable cycling infrastructure is key to encouraging cycling by older adults (see in this volume Buehler and Pucher, chapter 2; Furth, chapter 5).

The contribution that cycling can make to achieving adequate levels of physical activity for older adults is illustrated by data from the Netherlands. Analysis of data from the *Netherlands National Travel Survey* found that among persons aged 20–90 years, those aged 65–69 had the highest cycling levels (94 minutes per week, on average), possibly linked to retirement and increased discretionary time (Fishman, Bocker, and Helbich 2015). Among those aged 60–90 years, the average time spent cycling was 79 minutes per week, which is over half (52.7%) the recommended 150 minutes per week of moderate to vigorous physical activity. A meta-analysis of the effect of cycling on all-cause mortality reported a 10% reduction in all-cause mortality for 100 minutes of cycling per week (Kelly et al. 2014). Based on that 10% estimate, the level of cycling among older adults in the Netherlands suggests a 7.9% mortality risk reduction, controlling for other physical activity and health-related behaviors such as smoking.

These data illustrate that when cycling conditions are favorable, people will continue to cycle for daily, utilitarian travel and recreation into older age, thereby making a substantial contribution to achieving adequate levels of physical activity and its associated health benefits. Thus, cycling has the potential to attenuate or even reverse the decline in physical activity with age that occurs in many developed countries such as the United States (Chodzko-Zajko et al. 2009).

## Factors that Influence Cycling among Older Adults

The social-ecological model of health behavior (Sallis et al. 2006) is useful for understanding and modifying the multiple wide-ranging factors that influence cycling among older adults (see Garrard, chapter 11, this volume). The model describes a set of four interacting levels of individual and environmental influences on cycling. Individual factors include sociodemographic characteristics, together with knowledge, skills, attitudes, beliefs, physical functioning, and health status. The three domains of environmental influences comprise the built environment (e.g., cycling infrastructure), the policy-regulatory environment (e.g., transport planning and road safety regulations and enforcement), and the social-cultural environment (e.g., social norms and expectations related to older adults cycling for daily travel and recreation). These are discussed in the following sections.

## Intrapersonal Factors

There is considerable heterogeneity among older adults, perhaps more than among younger people, with regard to health status and functional capacity (Novak, Campbell, and Northcott 2014). Diversity among older populations is reflected in their health, mobility, and cycling histories and preferences. Importantly, sensory changes can be gradual and compounding, thus making it harder for an individual to recognize subtle changes in themselves, compared to more obvious declining muscle strength. Changes in vision, cognition, and balance can narrow a cyclist's perceptual or attentional field of vision, slowing response times to traffic scenarios. However, these age-related changes in vision, hearing, and muscle strength do not occur for everyone at the same time or rate or in the same way (Novak, Campbell, and Northcott 2014). Indeed, some studies have found that active older adults can have a functional capacity superior to that of younger adults. A recent study found that older cyclists had immune systems and senses that compared favorably with those of inactive peers and younger adults (Duggal et al. 2018). These cyclists, some in their eighties, rode bicycles for most of their adult lives and had maintained physiological functions that otherwise tend to decline with age.

In recent decades, older adults have been making more diverse choices in their social roles, work, time use, and travel. For example, retirement is no longer a fixed transition point, with some extending their working years or continuing part-time employment. This affects levels of both commuting and recreational cycling.

Functional heterogeneity among older adults, together with the diversity of cycling experience, raises issues about the built environments required to support older adults' continued or newly adopted cycling. For example, less physically active older adults, those with less cycling experience, or those who may be experiencing physical limitations may require cycling infrastructure that differs from that of more experienced, fit, and resilient cyclists.

### The Built Environment

Safe, convenient, and comfortable cycling facilities, as discussed by Furth (chapter 5, this volume), are a crucial prerequisite for cycling by older adults. Reviews show that older adults, along with other age groups, prefer to cycle in facilities separated from other traffic, with even surfaces and highly visible markings (Van Cauwenberg et al. 2018; Ryan et al. 2016). Preferred facilities include traffic-calmed residential streets and off-road bike paths. Consistent with the crash data described in this chapter, many older cyclists prefer separation not only from vehicle traffic but also from bus stops, roundabouts, pedestrians, and faster cyclists. Older adults also express concerns about inattention and intimidation by other cyclists and pedestrians. Signage and wayfinding can be used to support older adult cyclists, not just for easier routing but as a general reminder to be aware of and considerate of vulnerable riders of any age.

Creating safe and appealing conditions for older cyclists also includes measures that influence interactions between road users on roads and bicycle networks. These conditions and interactions are shaped by the policy and regulatory environment.

### The Policy-Regulatory Environment

The policy-regulatory environment, which includes traffic laws and enforcement as well as education and training programs for all road users, can create safe and appealing conditions for older cyclists. Good cycling infrastructure must be complemented by supportive programs such as cycling training, group cycling programs for older adults, and traffic laws that favor children and older adults. Traffic regulations in many northern European countries explicitly prioritize the safety of cyclists and pedestrians, with extra protection for children and older adults. Nationally standardized driver training courses and motorist licensing tests in Europe emphasize the importance of protecting vulnerable road users. Strict enforcement of these laws has a substantial impact on driver behavior and is one of the reasons why cycling is so much safer in northern Europe than in the United States, Canada, and Australia (see in this volume Buehler and Pucher, chapter 2, and Heinen and Handy, chapter 7; Buehler et al. 2017; ECF 2017; ECF 2018).

## The Social-Cultural Environment

The social-cultural environment includes both everyday social interactions and the broader social influences on older adults' cycling behaviors. Examples include the social support (or discouragement) for physical activity that older adults receive from family, friends, and wider social networks (Barnett, Guell, and Ogilvie 2013). Older adults also report challenges within the social environment while cycling, wherein they feel less safe and confident because of pressure from faster cyclists (Winters et al. 2015; Van Cauwenberg et al. 2018). The idea of intergenerational cycling spaces is notable here; for example, the provision of spaces for faster cyclists along with slower, side-by-side, or social riding (Jones and Spencer 2016).

The social-cultural environment also includes social norms and expectations. In car-centric countries and cities, the expectation is that people will drive to most places, and cycling is considered an inferior mode of transport (Jacobsen, Racioppi, and Rutter 2009). When the bicycle mode share of travel is low, cyclists can experience negative attitudes and behaviors associated with being a member of an "out group" of road users (Basford et al. 2002); judged as being irresponsible for participating in a "dangerous" activity (Pooley et al. 2013; Jacobsen, Racioppi, and Rutter 2009); and disparaged in the mass media (Rissel et al. 2010a; Rimano et al. 2015). Not surprisingly, being a member of an "out group" of denigrated road users may have little appeal to older adults in low-cycling countries, particularly when they, like most people, have developed a habit of driving to most destinations.

Travel mode "habits" are a powerful but often unrecognized influence on utilitarian cycling (Willis, Manaugh, and El-Geneidy 2015; de Bruijn et al. 2009), with "habitual" travel behavior shaped by a range of sociocultural, individual, and environmental factors. Consequently, changing travel mode habits is facilitated by establishing good cycling infrastructure and supportive social norms of everyday cycling. This multifaceted approach to encouraging older adults to remain or become active through cycling is consistent with the World Health Organization's health promotion strategy of creating supportive environments that "make the healthy choice the easy choice" (World Health Organization 1986). Similarly, "nudge theory" from the field of behavioral economics involves "nudging" people to make healthier choices by changing their "choice architecture" (Thaler and Sunstein 2008; Thaler 2015).

High-cycling countries support cycling for all age groups by establishing supportive built, policy-regulatory, and social-cultural environments that truly make cycling an easy and habitual choice that readily transitions into older age (see in this volume Buehler and Pucher, chapter 2; Furth, chapter 5; Heinen and Handy, chapter 7; Koglin, te Brömmelstroet, and van Wee, chapter 18).

## Strategies and Policies Aimed at Increasing Cycling among Older Adults

In view of the wide-ranging influences on older adults' cycling described in this chapter, comprehensive, integrated approaches are necessary to have the greatest impact on promoting and sustaining cycling at all ages (Pucher, Dill, and Handy 2010). These approaches, which span the domains of the social-ecological framework, include improvements in cycling infrastructure, policies and regulations, and attitudes toward cycling. Social-cultural change also includes addressing the "car culture" of many developed countries that is increasingly being recognized as incompatible with healthy, vibrant, livable, sustainable cities and towns of the future (Newman and Kenworthy 2015).

Two additional strategies relevant to older adults that also form part of an integrated approach are educational initiatives and bicycle technology.

### Educational Initiatives
Educational programs can play a role in either reengaging older adults who have moved in and out of cycling over time because of changes in housing, career, health, and family status (Chatterjee, Sherwin, and Jain 2013) or in attracting new cyclists (see also Buehler, Heinen, and Nakamura, chapter 6, this volume). In many cities, group rides for older cyclists are offered through senior centers, bike shops, and cycling groups, providing social support, encouragement, and mentorship to older adults. There are also local efforts to equip older adults with skills in bike handling, riding technique, and safety.

There is little published evidence on the effectiveness of cycling education for any age group (Sersli et al. 2018) (see also Buehler, Heinen, and Nakamura, chapter 6, this volume), but the literature yields suggestions as to how educational efforts could help support cycling among older adults. Rissel et al. (2010b) found significant increases in cycling following a community-based promotion campaign that included group rides, events, and skills courses. About half the program participants were 45 years old or older, and nearly one-quarter were 60 or older. Reviewing a national adult cycling training program, Rissel and Watkins (2014) documented very positive feedback, significant improvements in skills and confidence, and significant reductions in weight and BMI among participants. Zander et al. (2013) described an age-targeted cycling course that included peer support and encouragement through ongoing mentorship. In post-program interviews, two key enablers for cycling were identified for older adults: confidence through reinforcement and repetition, and the support of peers and mentors.

While few in number, these studies suggest that educational efforts, in combination with social support, can assist adults in getting started cycling or in continuing to cycle as they age (see also Heinen and Handy, chapter 7, this volume, which discusses cycling education programs more broadly).

**Bicycle Technology**

Emerging developments in bicycle design and technology have an important role to play in supporting cycling among older adults. Adult tricycles are increasingly available, either recumbent or upright. They sometimes have cargo bins in the back to facilitate carrying things such as groceries. The Dutch SOFIE pedal bike was specifically designed for the safety and comfort of older adults. It is an electric-assist bicycle with features that help prevent older adults from falling off, such as a low step-through frame and an automatic adjustable-height saddle (Dubbeldam et al. 2017).

As discussed in detail by Cherry and Fishman (chapter 9, this volume), electric bikes (e-bikes) are particularly appealing to older adults. Compared with conventional cycling, e-biking involves lower muscle activation and reduced cardiorespiratory and metabolic effort, with users reporting less perceived exertion and more enjoyment (Sperlich et al. 2012). Notably, the intensity of e-biking is sufficient to meet guidelines for moderate-intensity physical activity, suggesting that e-bikes could provide an appealing opportunity for older adults to achieve adequate levels of physical activity through cycling.

In a study of e-biking among older adults, participants were thrilled by their experiences and felt they could more easily overcome physical limitations, steep hills, and longer distances (Jones et al. 2016). E-bike riders reported a sense of freedom and achievement, and a greater capacity to ride with others, discover new routes, and extend their cycling reach in less time (Jones, Harms, and Heinen 2016; Gojanovic et al. 2011; Simons, Van

Es, and Hendriksen 2009). E-bike use is increasing in the Netherlands, Germany, Belgium, and China (Harms, Bertolini, and te Brömmelstroet 2014), and wider options in weight, price, and battery life could make e-bikes even more accessible. Similarly, better access to transit, storage, charging, and repair services could normalize e-bike ownership and increase e-bike use among older adults (Jones et al. 2016).

While older adults report a number of benefits associated with e-bike use, the speed, weight, and stability of e-bikes can also be a concern. E-bikes can be harder to balance and steer, and pose a higher risk of falling while mounting or dismounting (Twisk, Platteel, and Lovegrove 2017). Optional power can encourage riding at speeds that are harder to control and traveling faster than other road users expect, which may cause collisions (Boele-Vos et al. 2017). In terms of safety, a Swedish study found that while rates of single-bicycle crashes by e-bike users were similar across age groups, older adults were more likely to experience a serious injury (Hertach et al. 2018). This finding is consistent with injury patterns for conventional bicycles and is probably related to increased vulnerability to injury for older adults in the event of a bicycle crash.

There is limited evidence indicating that crash and injury risks differ for e-bikes and conventional bicycles when controlling for distance traveled (Hertach et al. 2018). These assessments are complicated by several factors, such as differences in the reporting of bicycle injuries (e.g., by police, hospitals, or self-reported), different styles of cycling, and different sociodemographic attributes of cyclists.

## Conclusions

Older age need not be a barrier to cycling, which remains physically possible for many people well beyond the age of 65, provided conditions are safe and convenient. Continuing to cycle or getting started cycling as an older adult can contribute to achieving recommended levels of physical activity and the associated health benefits. International comparative data indicate that cycling for daily utilitarian travel and recreation is well placed to counter the decline in physical activity with older age that occurs in most low-cycling countries.

Cycling can contribute to reducing the high levels of chronic disease associated with older age and to enhancing the mobility, social access, independence, self-worth, mental health, and quality of life of older adults. For older adults, the health risks of cycling include heightened risk of injury in the event of a bicycle crash but not increased risk of having a crash. The improved muscle strength and aerobic fitness achieved through cycling may also contribute to a reduced risk of other injuries, such as falls. As for younger age groups, the benefits of cycling for older adults far outweigh the risks of injury.

For all age groups, an integrated package of measures is likely to have the greatest impact on cycling participation (Pucher, Dill, and Handy 2010). These measures include bicycle infrastructure that is, and is perceived to be, safe, well designed, well maintained, and well connected; road rules and road user education predicated on the protection of vulnerable road users; the establishment of social norms that support cycling as an acceptable and legitimate mode of transport for people of all ages; bicycle education and skills programs tailored for older adults; and advances in bicycle design and technology.

## References

Ainsworth, Barbara E., William L. Haskell, Stephen D. Herrmann, Nathanael Meckes, David R. Bassett Jr., Catrine Tudor-Locke, Jennifer L. Greer, Jesse Vezina, Melicia C. Whitt-Glover, and Arthur S. Leon. 2011. Compendium of Physical Activities: A Second Update of Codes and MET Values. *Medicine and Science in Sports and Exercise* 43(8): 1575–1581.

Alkatan, Mohammed, Jeffrey R. Baker, Daniel R. Machin, Wonil Park, Amanda S. Akkari, Evan P. Pasha, and Hirofumi Tanaka. 2016. Improved Function and Reduced Pain after Swimming and Cycling Training in Patients with Osteoarthritis. *Journal of Rheumatology* 43(3): 666–672. https://doi.org/10.3899/jrheum.151110.

Barnett, Inka, Cornelia Guell, and David Ogilvie. 2013. How Do Couples Influence Each Other's Physical Activity Behaviours in Retirement? An Exploratory Qualitative Study. *BMC Public Health* 13: 1197. https://doi.org/10.1186/1471-2458-13-1197.

Base, B. 2012. How to Maintain Healthy Joints. *The Pharmacist*, May 10, 2012. http://www.thepharmacist.co.uk/how-to-maintain-healthy-joints/.

Basford, L., S. Reid, T. Lester, J. Thomson, and A. Tolmie. 2002. *Drivers' Perceptions of Cyclists*. TRL Report 549. Wokingham, UK: Transport Research Laboratory.

Bauman, Adrian E., Anne C. Grunseit, Vegar Rangul, and Berit L. Heitmann. 2017. Physical Activity, Obesity and Mortality: Does Pattern of Physical Activity Have Stronger Epidemiological Associations? *BMC Public Health* 17(1): 788. https://doi.org/10.1186/s12889-017-4806-6.

Bicycle Dutch. 2018. Dutch Cycling Figures. *Bicycle Dutch*. https://bicycledutch.wordpress.com/2018/01/02/dutch-cycling-figures/.

Boele-Vos, M. J., K. Van Duijvenvoorde, M. J. A. Doumen, C. W. A. E. Duivenvoorden, W. J. R. Louwerse, and R. J. Davidse. 2017. Crashes Involving Cyclists Aged 50 and Over in the Netherlands: An In-Depth Study. *Accident Analysis and Prevention* 105: 4–10. https://doi.org/10.1016/j.aap.2016.07.016.

Bouaziz, Walid, Thomas Vogel, Elise Schmitt, Georges Kaltenbach, Bernard Geny, and Pierre O. Lang. 2017. Health Benefits of Aerobic Training Programs in Adults Aged 70 and Over: A Systematic Review. *Archives of Gerontology and Geriatrics* 69: 110–127. https://doi.org/10.1016/j.archger.2016.10.012.

British Heart Foundation. 2012. *Physical Activity for Older Adults (65+ Years): Evidence Briefing.* Loughborough: Loughborough University.

Buehler, Ralph, John Pucher, Regine Gerike, and Thomas Götschi. 2017. Reducing Car Dependency in the Heart of Europe: Lessons from Germany, Austria, and Switzerland. *Transport Reviews* 37(1): 4–28.

Chatterjee, Kiron, Henrietta Sherwin, and Juliet Jain. 2013. Triggers for Changes in Cycling: The Role of Life Events and Modifications to the External Environment. *Journal of Transport Geography* 30: 183–193. https://doi.org/10.1016/j.jtrangeo.2013.02.007.

Chodzko-Zajko, Wojtek J., David N. Proctor, Maria A. Fiatarone Singh, Christopher T. Minson, Claudio R. Nigg, George J. Salem, and James S. Skinner. 2009. American College of Sports Medicine Position Stand. Exercise and Physical Activity for Older Adults. *Medicine and Science in Sports and Exercise* 41(7): 1510–1530. https://doi.org /10.1249/MSS.0b013e3181a0c95c.

Clarke, Tainya C., Tina Norris, and Jeannine S. Schiller. 2017. *Early Release of Selected Estimates Based on Data from the 2016 National Health Interview Survey.* Atlanta: US Department of Health and Human Services, Centers for Disease Control and Prevention, National Center for Health Statistics.

de Bruijn, Gert-Jan, Stef P. J. Kremers, Amika Singh, Bas van den Putte, and Willem van Mechelen. 2009. Adult Active Transportation: Adding Habit Strength to the Theory of Planned Behaviour. *American Journal of Preventive Medicine* 36(3): 189–194.

Dubbeldam, R., C. Baten, J. H. Buurke, and J. S. Rietman. 2017. SOFIE, a Bicycle That Supports Older Cyclists? *Accident Analysis and Prevention* 105: 117–123. https://doi .org/10.1016/j.aap.2016.09.006.

Duggal, Nihariker A., Ross D. Pollock, Norman R. Lazarus, Stephen Harridge, and Janet M. Lord. 2018. Major Features of Immunesenescence, Including Reduced Thymic Output, Are Ameliorated by High Levels of Physical Activity in Adulthood. *Aging Cell* 17. https://doi.org/10.1111/acel.12750.

European Commission and World Health Organization Regional Office for Europe. 2015. *Netherlands Physical Activity Fact Sheet.* Copenhagen: European Commission and World Health Organization Regional Office for Europe.

European Cyclists' Federation (ECF). 2017. *EU Cycling Strategy*. Brussels: European Cyclists' Federation.

European Cyclists' Federation (ECF). 2018. *National Cycling Policies Overview*. Brussels: European Cyclists' Federation.

European Union. 2014. *Special Eurobarometer 422a Quality of Transport Report*. Brussels: European Union.

Fishman, Elliot, Lars Bocker, and Marco Helbich. 2015. Adult Active Transport in the Netherlands: An Analysis of Its Contribution to Physical Activity Requirements. *PLoS One* 10(4). https://doi.org/10.1371/journal.pone.0121871.

Fishman, Elliot, Paul Schepers, and Carlijn Barbara Maria Kamphuis. 2015. Dutch Cycling: Quantifying the Health and Related Economic Benefits. *American Journal of Public Health* 105(8): e13–e15. https://doi.org/10.2105/AJPH.2015.302724.

Gojanovic, Boris, Joris Welker, Katia Iglesias, Chantal Daucourt, and Gerald Gremion. 2011. Electric Bicycles as a New Active Transportation Modality to Promote Health. *Medicine and Science in Sports and Exercise* 43(11): 2204–2210. https://doi.org/10.1249/MSS.0b013e31821cbdc8.

Hansson, Erik, Kristoffer Mattisson, J. Jonas Bjork, Per-Olof Ostergren, and Kristina Jakobsson. 2011. Relationship between Commuting and Health Outcomes in a Cross-Sectional Population Survey in Southern Sweden. *BMC Public Health* 11: 834. https://doi.org/10.1186/1471-2458-11-834.

Harms, Lucas, Luca Bertolini, and Marco te Brömmelstroet. 2014. Spatial and Social Variations in Cycling Patterns in a Mature Cycling Country Exploring Differences and Trends. *Journal of Transport and Health* 1(4): 232–242. https://doi.org/10.1016/j.jth.2014.09.012.

Harvey, Stephen, Chris Rissel, and Mirjam Pijnappels. 2018. Associations between Bicycling and Reduced Fall-Related Physical Performance in Older Adults. *Journal of Aging and Physical Activity* 26(3): 514–519. https://doi.org/10.1123/japa.2017-0243.

Hertach, Patrizia, Andrea Uhr, Steffen Niemann, and Mario Cavegn. 2018. Characteristics of Single-Vehicle Crashes with E-bikes in Switzerland. *Accident Analysis and Prevention* 117: 232–238. https://doi.org/10.1016/j.aap.2018.04.021.

Jacobsen, Peter L., Francesca Racioppi, and Harry Rutter. 2009. Who Owns the Roads? How Motorised Traffic Discourages Walking and Cycling. *Injury Prevention* 15(6): 369–373.

Jones, T., K. Chatterjee, J. Spinney, E. Street, C. Van Reekum, B. Spencer, H. Jones, L. A. Leyland, C. Mann, S. Williams, and N. Beale. 2016. *Cycle BOOM. Design for Lifelong Health and Wellbeing. Summary of Key Findings and Recommendations*. Oxford: Oxford Brookes University.

Jones, Tim, Lucas Harms, and Eva Heinen. 2016. Motives, Perceptions and Experiences of Electric Bicycle Owners and Implications for Health, Wellbeing and Mobility. *Journal of Transport Geography* 53: 41–49. https://doi.org/10.1016/j.jtrangeo.2016.04.006.

Jones, Tim, and Ben Spencer. 2016. Can Urban Streets and Spaces Be Intergenerational Cycling Zones? In *Intergenerational Contact Zones: A Compendium of Applications*, edited by Matthew Kaplan, Leng Leng Thang, Mariano Sanchez, and Jaco Hoffman, 1–5. University Park, PA: Penn State Extension.

Kelly, Paul, Sonja Kahlmeier, Thomas Götschi, Nicola Orsini, Justin Richards, Nia Roberts, Peter Scarborough, and Charlie Foster. 2014. Systematic Review and Meta-analysis of Reduction in All-Cause Mortality from Walking and Cycling and Shape of Dose Response Relationship. *International Journal of Behavioral Nutrition and Physical Activity* 11(1). https://doi.org/10.1186/s12966-014-0132-x.

Koolhaas, Chantal M., Dhana Klodian, Rajna Golubic, Josje D. Schoufour, Albert Hofman, Frank J. van Rooij, and Oscar H. Franco. 2016. Physical Activity Types and Coronary Heart Disease Risk in Middle-Aged and Elderly Persons: The Rotterdam Study. *American Journal of Epidemiology* 183(8): 729–738. https://doi.org/10.1093/aje/kwv244.

Koolhaas, Chantal M., Dhana Klodian, Josje D. Schoufour, Lies Lahousse, Frank J. A. van Rooij, Arfan M. Ikram, Guy Brusselle, Henning Tiemeier, and Oscar H. Franco. 2018. Physical Activity Types and Health-Related Quality of Life among Middle-Aged and Elderly Adults: The Rotterdam Study. *Journal of Nutrition, Health and Aging* 22(2): 246–253. https://doi.org/10.1007/s12603-017-0902-7.

Leyland, Louise-Ann, Ben Spencer, Nick Beale, Tim Jones, and Carien M. Van Reekum. 2019. The Effect of Cycling on Cognitive Function and Well-being in Older Adults. *PloS One* 14(2). https://doi.org/10.1371/journal.pone.0211779.

Minello, Karla, and Deborah Nixon. 2017. "Hope I Never Stop": Older Men and Their Two-Wheeled Love Affairs. *Annals of Leisure Research* 20(1): 75–95. https://doi.org/10.1080/11745398.2016.1218290.

Newman, Peter, and Jeffrey Kenworthy. 2015. *The End of Automobile Dependence: How Cities Are Moving beyond Car-Based Planning*. Washington, DC: Island Press/Center for Resource Economics.

Novak, Mark W., Lori Anne Campbell, and Herbert C. Northcott. 2014. *Aging and Society: Canadian Perspectives*. Toronto: Nelson Education.

Pooley, Colin G., Dave Horton, Griet Scheldeman, Caroline Mullen, Tim Jones, Miles Tight, Ann Jopson, and Alison Chisholm. 2013. Policies for Promoting Walking and Cycling in England: A View from the Street. *Transport Policy* 27: 66–72. http://dx.doi.org/10.1016/j.tranpol.2013.01.003.

Poulos, R. G., J. Hatfield, C. Rissel, L. K. Flack, S. Murphy, R. Grzebieta, and A. S. McIntosh. 2015. An Exposure Based Study of Crash and Injury Rates in a Cohort of Transport and Recreational Cyclists in New South Wales, Australia. *Accident Analysis and Prevention* 78: 29–38. https://doi.org/10.1016/j.aap.2015.02.009.

Pucher, John, and Ralph Buehler. 2008. Making Cycling Irresistible: Lessons from the Netherlands, Denmark and Germany. *Transport Reviews* 28(4): 495–528.

Pucher, John, Jennifer Dill, and Susan Handy. 2010. Infrastructure, Programs and Policies to Increase Bicycling: An International Review. *Preventive Medicine* 50(Suppl 1): S106–S125.

Rimano, Alessandra, Maria Paola Piccini, Paola Passafaro, Renata Metastasio, Claudia Chiarolanza, Aurora Boison, and Franco Costa. 2015. The Bicycle and the Dream of a Sustainable City: An Explorative Comparison of the Image of Bicycles in the Mass-Media and the General Public. *Transportation Research Part F: Traffic Psychology and Behaviour* 30: 30–44. http://dx.doi.org/10.1016/j.trf.2015.01.008.

Rissel, Chris, Catriona Bonfiglioli, Adrian Emilsen, and Ben J. Smith. 2010a. Representations of Cycling in Metropolitan Newspapers—Changes Over Time and Differences between Sydney and Melbourne, Australia. *BMC Public Health* 10. https://doi.org/10.1186/1471-2458-10-371.

Rissel, Chris, Carolyn New, Li Ming Wen, Dafna Merom, Adrian Bauman, and Jan Garrard. 2010b. The Effectiveness of Community-Based Cycling Promotion: Findings from the Cycling Connecting Communities Project in Sydney. *International Journal of Behavioral Nutrition and Physical Activity* 7. https://doi.org/10.1186/1479-5868-7-8.

Rissel, Chris, and G. Watkins. 2014. Impact on Cycling Behavior and Weight Loss of a National Cycling Skills Program (AustCycle) in Australia 2010–2013. *Journal of Transport and Health* 1(2): 134–140. https://doi.org/10.1016/j.jth.2014.01.002.

Rowe, Katie, David Shilbury, Lesley Ferkins, and Erica Hinckson. 2016. Challenges for Sport Development: Women's Entry Level Cycling Participation. *Sport Management Review* 19(4): 417–430. https://doi.org/10.1016/j.smr.2015.11.001.

Ryan, J., H. Svensson, J. Rosenkvist, S. M. Schmidt, and A. Wretstrand. 2016. Cycling and Cycling Cessation in Later Life: Findings from the City of Malmö. *Journal of Transport and Health* 3(1): 38–47. https://doi.org/10.1016/j.jth.2016.01.002.

Sallis, James F., Robert B. Cervero, William Ascher, Karla A. Henderson, M. Katherine Kraft, and Jacqueline Kerr. 2006. An Ecological Approach to Creating Active Living Communities. *Annual Review of Public Health* 27: 297–322. https://doi.org/10.1146/annurev.publhealth.27.021405.102100.

Schepers, Paul. 2012. Does More Cycling Also Reduce the Risk of Single-Bicycle Crashes? *Injury Prevention* 18(4): 240–245. https://doi.org/10.1136/injuryprev-2011-040097.

Schepers, Paul, Niels Agerholm, Emmanuelle Amoros, Rob Benington, Torkel Bjorns-kau, Stijn Dhondt, Bas de Geus, Carmen Hagemeister, Becky P. Loo, and Anna Niska. 2015. An International Review of the Frequency of Single-Bicycle Crashes (SBCs) and Their Relation to Bicycle Modal Share. *Injury Prevention* 21(e1): e138–e143. https://doi.org/10.1136/injuryprev-2013-040964.

Sersli, Stephanie, Danielle DeVries, Maya Gislason, Nicholas Scott, and Meghan Winters. 2018. Changes in Bicycling Frequency in Children and Adults after Bicycle Skills Training: A Scoping Review. *Transportation Research Part A: Policy and Practice 123: 170–187.* https://doi.org/10.1016/j.tra.2018.07.012.

Sherman, S. E., R. B. D'Agostino, H. Silbershatz, and W. B. Kannel. 1999. Comparison of Past versus Recent Physical Activity in the Prevention of Premature Death and Coronary Artery Disease. *American Heart Journal* 138(5): 900–907. https://doi.org/10.1016/S0002-8703(99)70015-3.

Simons, Monique, Eline Van Es, and Ingrid Hendriksen. 2009. Electrically Assisted Cycling: A New Mode for Meeting Physical Activity Guidelines? *Medicine and Science in Sports and Exercise* 41(11): 2097–2102. https://doi.org/10.1249/MSS.0b013e3181a6aaa4.

Sperlich, Billy, Christoff Zinner, Kim Hebert-Losier, Dennis-Peter Born, and Hans-Christer Holmberg. 2012. Biomechanical, Cardiorespiratory, Metabolic and Perceived Responses to Electrically Assisted Cycling. *European Journal of Applied Physiology* 112(12): 4015–4025. https://doi.org/10.1007/s00421-012-2382-0.

Thaler, Richard. 2015. *Misbehaving: The Making of Behavioural Economics.* New York: Norton.

Thaler, Richard, and Cass Sunstein. 2008. *Nudge: Improving Decisions about Health, Wealth, and Happiness.* New Haven, CT: Yale University Press.

Tsunoda, K., N. Kitano, Y. Kai, T. Tsuji, Y. Soma, T. Jindo, J. Yoon, and T. Okura. 2015. Transportation Mode Usage and Physical, Mental and Social Functions in Older Japanese Adults. *Journal of Transport and Health* 2(1): 44–49.

Twisk, D. A. M., S. Platteel, and G. R. Lovegrove. 2017. An Experiment on Rider Stability While Mounting: Comparing Middle-Aged and Elderly Cyclists on Pedelecs and Conventional Bicycles. *Accident Analysis and Prevention* 105: 109–116. https://doi.org/10.1016/j.aap.2017.01.004.

United Nations Department of Economic and Social Affairs Population Division. 2017. *Profiles of Ageing 2017.* New York: United Nations.

Van Cauwenberg, Jelle, Peter Clarys, Ilse De Bourdeaudhuij, Ariane Ghekiere, Bas de Geus, Neville Owen, and Benedicte Deforche. 2018. Environmental Influences on Older Adults' Transportation Cycling Experiences: A Study Using Bike-Along Interviews. *Landscape and Urban Planning* 169: 37–46. https://doi.org/10.1016/j.landurbplan.2017.08.003.

Van Cauwenberg, Jelle, Bas de Geus, and Benedicte Deforche. 2018. Cycling for Transport among Older Adults: Health Benefits, Prevalence, Determinants, Injuries and the Potential of E-bikes. In *Geographies of Transport and Ageing*, edited by Angela Curl and Charles Musselwhite, 133–151. Cham, Switzerland: Springer.

Van Cauwenberg, Jelle, Paul Schepers, Benedicte Deforche, and Bas De Geus. 2019. Differences in Life Space Area between Older Non-cyclists, Conventional Cyclists and E-bikers. *Journal of Transport and Health* 14. https://doi.org/10.1016/j.jth.2019 .100605.

Varela, Silvia, Jose M. Cancela, Manuel Seijo-Martinez, and Carlos Ayan. 2018. Self-Paced Cycling Improves Cognition on Institutionalized Older Adults without Known Cognitive Impairment: A 15-Month Randomized Controlled Trial. *Journal of Aging and Physical Activity* 26(4): 614–623. https://doi.org/10.1123/japa.2017–0135.

Weggemans, Rianne M., Frank J. G. Backx, Lars Borghouts, Mai Chinapaw, Maria T. E. Hopman, Annemarie Koster, Stef Kremers, Luc J. C. van Loon, Anne May, Arend Mosterd, Hidde P. van der Ploeg, Tim Takken, Marjolein Visser, Wanda Wendel-Vos, and Eco J. C. de Geus. 2018. The 2017 Dutch Physical Activity Guidelines. *International Journal of Behavioral Nutrition and Physical Activity* 15: 58. https://doi.org/10 .1186/s12966-018-0661-9.

Willis, Devon Paige, Kevin Manaugh, and Ahmed El-Geneidy. 2015. Cycling Under Influence: Summarizing the Influence of Perceptions, Attitudes, Habits, and Social Environments on Cycling for Transportation. *International Journal of Sustainable Transportation* 9(8): 565–579. https://doi.org/10.1080/15568318.2013.827285.

Winters, Meghan, Joanie Sims-Gould, Thea Franke, and Heather McKay. 2015. "I Grew Up on a Bike": Cycling and Older Adults. *Journal of Transport and Health* 2(1): 58–67. https://doi.org/10.1016/j.jth.2014.06.001.

World Health Organization. 1986. *The Ottawa Charter for Health Promotion*. Geneva: World Health Organization.

World Health Organization. 2002. *Active Ageing: A Policy Framework*. Geneva: World Health Organization.

Zander, Alexis, Erin Passmore, Chloe Mason, and Chris Rissel. 2013. Joy, Exercise, Enjoyment, Getting Out: A Qualitative Study of Older People's Experience of Cycling in Sydney, Australia. In The Effect of Active Transport, Transport Systems, and Urban Design on Population Health. Special issue, *Journal of Environmental and Public Health* 2013. https://doi.org/10.1155/2013/547453.

# 14   Social Justice and Cycling

Karel Martens, Aaron Golub, and Andrea Hamre

Since the advent of the motorcar and the enormous investments enabling its uptake and use, the transport landscape has become highly uneven. People with a car typically enjoy a high level of freedom and access to destinations, while carless households are typically relegated to a patchy public transport system and an even less complete network of sidewalks and bicycle lanes—if they are available at all. The increasing efforts to promote cycling, as described in many chapters of this book, take place in the context of this uneven landscape. Bicycle investments and policies can increase or reduce existing disparities.

The notion of social justice directs attention to these disparities, to what philosophers term the distribution of goods and bads (i.e., the benefits and burdens) over different groups of people. The notion also refers to the way people are treated and recognized by fellow citizens, companies, and public agencies. Finally, social justice is about people's ability to be involved or represented in decisions that concern them.

In this chapter, we provide examples of these different dimensions of social justice as it relates to cycling and planning for cycling. We will not provide lengthy philosophical arguments here (see, e.g., Martens et al. 2016). Rather, we discuss a range of equity issues as they relate to cycling, with the intention of making clear that bicycle planning cannot and should not be solely an engineering exercise.

In what follows, we will discuss a range of equity issues related to distribution, recognition, and representation. We will take most of our examples from the US and Dutch contexts, with which we are most familiar. They also represent two contrasting cases of low and high overall levels of bicycling and policies to support it.

## The Goods People Receive (Or Do Not)

Transport planning and bicycle planning shape the distribution of a range of goods and bads, in direct and indirect ways. In relation to cycling, it includes tangible goods, such as bicycle infrastructure, bicycle parking facilities, or bikesharing systems, but also less tangible goods, such as the bikeability of one's neighborhood. Examples of bads include exposure to air pollution while cycling, exposure to traffic danger caused by motorized traffic, or barriers or disconnections in the bicycle network.

The distribution of goods and bads is always shaped by complex processes in society. This holds true even if goods are distributed by an identifiable actor, such as a local government. Take, for instance, bicycle infrastructure or bikesharing systems. While these are provided along particular streets or at particular points in a city and are thus literally distributed over space, they are not handed over directly to individuals. Indeed, who will actually benefit from these facilities depends on complex processes, such as whether somebody has learned to cycle, whether someone owns a bicycle, or whether someone's work or school is within cycling distance. Each of these is shaped, in turn, by other processes, such as income policies, land-use regulations, the functioning of the housing market, economic processes, and service provision policies. Thus, the simple question "who receives what and why" never has a simple answer.

Precisely because of these interdependencies, much research into the social dimension of cycling often starts with the measurement of *disparities*. The notion of "disparities" refers to differences in what people receive and, in particular, to the differences between differently positioned people. Studies of disparities are typically interested in differences between population groups along dimensions of socioeconomic status, race or ethnicity, gender, age, and residential location. Such studies often highlight large differences between population groups, for instance in bicycle use, exposure to air pollution while cycling, bikeability of neighborhoods, exposure to traffic risks, or the availability of bicycle infrastructure. These studies employ the concept of disparities in a descriptive sense.

There are, however, also many studies employing the word "disparities" to implicitly or explicitly identify observed differences as "unfair," "inequitable," or "unjust." Such studies often fall short of presenting an explicit standard for assessment and fail to acknowledge that the mere observation of disparities does not, by itself, raise issues of equity or necessarily indicate injustices (Lucas et al. 2019). Whether disparities are actually unjust or unfair requires a moral, or normative, assessment of those disparities. For instance, the observation that men cycle more than women in many countries points to a recurring pattern of disparity, and this raises concerns. It only becomes a matter of injustice if it can be argued that these differences result from unwanted circumstances, such as lack of suitable infrastructure, different access to bicycles, or (fear of) harassment, rather than choice. Research into equity and cycling could clearly benefit from a willingness by researchers to explicitly engage with normative questions such as: what constitutes a fair distribution of the benefits generated by bicycling policies and investments?

While the causes for injustices are often complex, governments still have many avenues by which to address and reduce them. For instance, if excessive exposure to air pollution is considered unjust, governments can intervene in multiple ways: they can set higher restrictions on vehicle emissions, make driving less attractive, or provide bicycle infrastructure away from major road arteries. Likewise, when low-income groups have less opportunity to cycle, because of a lack of destinations in and around their neighborhoods, governments can intervene by bringing services such as schools or health clinics closer to neighborhoods, building bicycle infrastructure across major barriers to reduce cycling distances, or even subsidizing housing in dense, high-opportunity areas with a variety of destinations within cycling distance so that low-income people may also have the opportunity to enjoy the benefits of such neighborhoods.

These two examples underscore how many of the disparities and injustices related to cycling have causes well beyond the domain of transport and can often only be successfully addressed by a coherent set of mutually reinforcing policies, many of which are the responsibility of other domains of government intervention. Bicycle planning and policy is only part of the social justice equation and only one way to promote a fair transport system—in whatever way it is defined (Golub et al. 2016; Martens 2017; Stehlin 2019).

In what follows, we will discuss these complexities and explicitly address whether and how bicycle planning—in a narrow sense—can contribute to alleviating unjust disparities and promoting justice. We do not promote a particular interpretation of bicycle justice but rather base our arguments on intuitive interpretations of what bicycle justice may mean in relation to various kinds of disparities.

## Bicycle Use

As already briefly mentioned, many countries and cities with poorly developed bicycle infrastructure show substantial differences in the use of the bicycle for transport between different population groups. Differences can be observed along dimensions of gender, ethnicity, and socioeconomic position, among others.

### Disparities and Inequities in Bicycling by Gender

The share of women bicycling typically reflects the overall bicycle mode share of travel, as has been discussed in detail by Garrard (chapter 11, this volume). In countries or cities with high levels of cycling, women cycle as much as men, and often even more. In contrast, women tend to be underrepresented in car-oriented countries and cities where cycling levels are low. For instance, in the United States, women are underrepresented in bicycling across all trip purposes, including commuting (table 14.1). In contrast, in the Netherlands, women on average make slightly more bicycle trips than men do and make a larger share of their trips by bicycle (Garrard, chapter 11, this volume; see also, e.g., Martens 2013).

Gendered patterns in child care and household responsibilities (Garrard, chapter 11, this volume), as well as in employment and private car access, affect the cycling levels of women in places with both high and low levels of cycling. Prati (2018) reviewed women's cycling for transport in relation to

**Table 14.1**
Percentage of trips, commuters, and commute mode share by gender in the United States in 2017

|                              | Female | Male |
| ---------------------------- | ------ | ---- |
| All trips                    | 52     | 48   |
| *All bicycle trips*          | *29*   | *71* |
| All commute trips            | 45     | 55   |
| *All bicycle commute trips*  | *27*   | *73* |
| All regular commuters        | 46     | 54   |
| *All regular bicycle commuters* | *26* | *74* |
| Bicycle commute mode share   | 0.6    | 1.6  |

*Note*: Totals may not sum to 100% because of rounding to the nearest percent.
*Source*: *National Household Travel Survey*, US Department of Transportation (2017).

the Gender Equality Index for the 28 member states of the European Union and found evidence that women were inhibited from cycling by disparities in available time because of child care, cooking, and other housework duties. Many women in the Netherlands have lower access to a private car, but they are able to routinely make short bicycle trips to schools and shops. In contrast, these same types of household errands are often infeasible by bicycle for women in places such as the United States. In low-cycling countries, the transport and land use system therefore compounds the disparities related to gender roles, preventing more women from cycling for everyday errands.

Gendered patterns in cycling point to injustice in transport. A lack of high-quality bicycle infrastructure poses a barrier to cycling for everyone but a greater barrier for women because of their consistently greater concern about traffic risk (Garrard, chapter 11, this volume). The same pattern is observed in India (see Pucher et al., chapter 15, this volume) and Latin America (see Pardo and Rodriguez, chapter 16, this volume). These concerns over safety have a larger effect on women's accessibility because women tend to make more short trips that could be made by bicycle if conditions were safe. The stark differences between India and China illustrate this. India features little dedicated bicycle infrastructure and a large cycling gender gap, while China has more dedicated bicycle infrastructure and fewer differences in bicycling rates between men and women (see Pucher et al., chapter 15, this volume).

Recent evidence indeed suggests that more and better infrastructure could reduce gender disparities in cycling (see also Garrard, chapter 11, this volume). For example, a recent analysis of bicycling frequency in Seattle found no gender gap in the propensity to bicycle in residential locations with more than 4 mi of bicycle lane infrastructure within a 1 mi radius (Bhat, Astroza, and Hamdi 2017). In addition, the growth in women commuting to work by bike was greater between 2007 and 2011 in designated Bicycle Friendly Communities (80%) than the national average (56%), suggesting that communities investing in bicycling will attract more female riders (League of American Bicyclists and Sierra Club 2013).

### Disparities and Inequities in Bicycling by Race and Ethnicity

There are also differences between racial and ethnic groups in bicycling for transport. The issue has been studied for the United States in particular. Data from the 2017 *National Household Travel Survey* indicate that Asian

Americans and white Americans are overrepresented in making trips by bicycle, while African Americans and Hispanic or Latino Americans are underrepresented (figure 14.1).

Even in a high-cycling country such as the Netherlands, disadvantaged ethnic minorities show a slightly lower level of cycling than the national average (Martens 2013), with native Dutch making 26% of their trips by bicycle compared to 24% for Western immigrants and 19% for non-Western immigrants.

From an equity perspective, the question is whether these disparities are the result of choice or circumstance. The Dutch experience suggests multiple reasons for lower levels of cycling among ethnic minorities. First-generation immigrants often were not accustomed to cycling in their countries of origin or were not able to cycle at all. In some countries of origin, cycling is perceived as an inappropriate activity for women, as it is associated with masculinity, speed, danger, and (inappropriate) freedom of movement (Van der Kloof 2015).

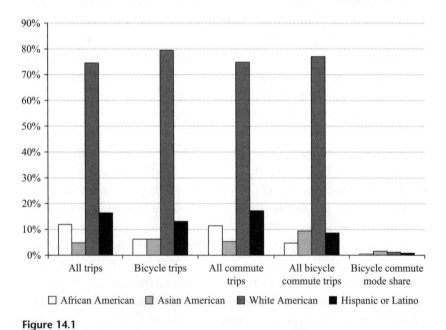

**Figure 14.1**
Share of all trips, work trips, and bicycle trips by race and ethnicity in the United States, 2017. *Note*: African American, Asian American, and white American are racial categories; Hispanic or Latino is a separate ethnic category. Totals across all four categories do not sum to 100% because of omitted racial and ethnic categories. *Source*: *National Household Travel Survey*, US Department of Transportation (2017).

The low levels of bicycle use among first-generation immigrants may subsequently affect the younger generation, as parents lack the skills to teach their children how to cycle, especially on roads with traffic. The resulting lower levels of cycling skills may subsequently interact with the quality of the bicycle infrastructure. For instance, whereas "well-trained" native Dutch cyclists may feel confident cycling on narrow bicycle lanes, immigrants may feel insecure and may avoid cycling on such routes or even avoid cycling altogether.

Socioeconomic and residential status may also interact to shape bicycle use among immigrants. For example, non-Western immigrants in the Netherlands typically live in small apartments in dense urban neighborhoods or more spacious apartments in car-oriented neighborhoods, both of which lack safe, secure, or convenient bicycle parking. This in turn could help explain the lower levels of bicycle ownership among ethnic minorities in the Netherlands. For instance, while in Amsterdam 85% of native Dutch households own one or more bicycles, the share ranges between 55% and 65% for the main ethnic minority groups (Van der Kloof, Bastiaanssen, and Martens 2014).

As we will discuss below, patterns of residential segregation among socioeconomic and racial or ethnic groups also shapes bicycle use in the United States. Cycling infrastructure investments have tended to occur outside areas with high concentrations of Americans of color, and African Americans and Hispanic or Latino Americans experience higher crash rates while bicycling than white Americans (Zimmerman et al. 2015). In a series of recent studies of bicycling among Latino immigrants, Barajas (2016, 2018, 2019) found that many face safety barriers in terms of wayfinding and access to bicycle infrastructure as well as social barriers because they lack a sense of belonging as cyclists and have low levels of cycling in their social networks.

These factors suggest a complex interplay between circumstances and choice. Irrespective of the exact reasons, the disparities may be reduced through interventions such as the provision of secure bicycle parking in lower-income neighborhoods, investments in high-quality bicycle infrastructure that is considered safe by a broad range of users, and bicycle lessons targeted at immigrant groups (Van der Kloof 2015; Lusk et al. 2017).

### Disparities and Inequities in Bicycle Use across Income Groups

Bicycling for transport also varies across income groups. In the United States, the trend across income groups is "U-shaped," where bicycling tends to be highest among those with the lowest and highest incomes (figure 14.2). The

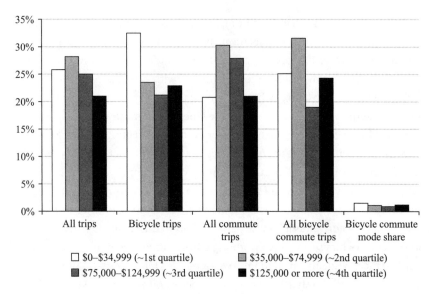

**Figure 14.2**
Share of all trips, work trips, and bicycle trips by income in the United States in 2017. *Note*: Totals may not sum to 100% because of rounding. Income quartiles were approximated by aggregating the 11 NHTS income categories into only 4. *Source*: *National Household Travel Survey*, US Department of Transportation (2017).

higher reliance on bicycling among individuals in the lowest income quartile may relate to their common experience of travel as a financial burden. Among these households, 56% report that travel poses a financial burden and 13% bicycle to save money (based on both "strongly agree" and "agree" responses), according to the 2017 *National Household Travel Survey*.

The high share of households grappling with transport expenses and a substantial share reporting that they use cycling as a coping strategy suggest that the low bicycle share itself is a more fundamental injustice. It points to the unfairness embodied in current transport landscapes, which force people into expensive driving as the only practically feasible way to travel to desired destinations. Obviously, this situation particularly burdens the lowest income groups, and investments to make bicycling safer and easier could improve transport justice by enabling more low-income households to bicycle and reduce their transport costs.

This fundamental understanding that low cycling levels are determined by a nonsupportive context (lack of cycling facilities) rather than free choice (not wanting to cycle) can be seen as a key equity argument in favor of investments in cycling. From the perspective of equity, these investments provide safer and more convenient cycling options for everyone, including those who do not currently cycle. This equity argument is relevant not only to the United States but to other countries with low levels of cycling as well.

## Bicycle Infrastructure

A renaissance of cycling is only possible through a change in the use of the right-of-way, as multiple chapters in this volume show (see Pucher and Buehler, chapter 1; Buehler and Pucher, chapter 2; Garrard et al., chapter 3; Elvik, chapter 4; Furth, chapter 5; Pardo and Rodriguez, chapter 16; Pucher, Parkin, and de Lanversin, chapter 17; Koglin, te Brömmelstroet, and van Wee, chapter 18; Geller and Marqués, chapter 19). Governments around the world have started to take up this challenge. Obviously, governments cannot provide infrastructure across their entire jurisdiction in one sweeping move, and thus disparities will be inevitable. From an equity perspective, the debate is whether investment patterns in bicycle infrastructure are biased in favor of certain groups over others during this (often lengthy) period of transition toward a complete bicycle network (Houde, Apparicio, and Séguin 2018; Lee, Sener, and Jones 2017).

The limited evidence on this topic so far shows mixed results. A study for Melbourne found that separated bicycle infrastructure is somewhat more prevalent in high-income areas, although much of this infrastructure is located in parkland and may have a predominantly recreational function. The study concluded that cycling infrastructure "is generally distributed equitably with respect to area-level socioeconomic position" (Pistoll and Goodman 2014). Meanwhile, a recent study on the expansion of the Montreal cycling network between 1991 and 2016 found that low-income individuals enjoyed relatively advantageous accessibility to cycling facilities but also noted that this advantage in access eroded over time (Houde, Apparicio, and Séguin 2018). In contrast, a study analyzing bicycle investments between 1990 and 2010 in Chicago and in Portland, Oregon, found increased cycling infrastructure investment in areas of high or increasing socioeconomic privilege (Flanagan, Lachapelle, and El-Geneidy 2016; see also Herrington and Dann 2016). Tucker and Manaugh (2018) found the same imbalance in bicycle infrastructure provision for the Brazilian cities of Rio de Janeiro and Curitiba. Zimmerman et al. (2015) found that investment patterns favoring high-income neighborhoods can also be found in infrastructure that tends to support bicycling (and walking), such as sidewalks, marked crosswalks, street lighting, and features to slow traffic.

There are several explanations for why bicycle investments may not be distributed equally across a jurisdiction or may not target population groups that can benefit most from them (e.g., carless households). First, planning for the bicycle may well follow the traditional practice of transport planning, with its strong focus on peak-hour commuter trips and prioritizing investments to reduce congestion. Since intracity congestion occurs especially along the main corridors leading into central employment districts, such a focus may result in bicycle investments on the main urban arterial roads. These are also typically high-density areas, implying both a potentially high number of cyclists and multiple destinations within cycling distance, suggesting high demand for cycling and thus a high return for bicycle investments in these areas. This practice of justifying bicycle investments based on the potential for addressing congestion is not uncommon in many countries. For instance, in the Netherlands, investments in bicycle highways can only obtain national funding if they help alleviate highway congestion (Martens et al. 2016).

Second, and related to the planning issue, governments tend to collect data for areas that are assumed to warrant interventions. Thus, data collection on the number of cyclists may be heavily biased toward city centers and other employment centers, while cyclists in other areas of the city may go unobserved. For instance, bicycle counts may be conducted only on key links heading into and out of city centers, while dedicated travel surveys may only be conducted among the employees of large companies and institutions. The result is that other types of bicycle travel, especially at the neighborhood level, are seen as less important to the transport planning process. Those sorts of travel are sometimes referred to as "shadow travel" made by "invisible cyclists" (Fuller and Beltran 2010; Zavestoski and Agyeman 2015; Barajas 2016).

Trips that are implicitly valued less include "reverse commuting" from inner-city neighborhoods out to suburban job centers, trips from one suburb to another suburb, short-distance trips within residential areas, or trips that occur at off-peak times. These trips are often made by marginalized population groups, such as caretakers, children, blue-collar workers, low-income workers, racial and ethnic minorities, the homeless, or unregistered workers, and may go unobserved (Zimmerman et al. 2015; Parker 2019). In the United States, guidance for counting bicyclists and pedestrians and generating standardized count data was included for the first time in the 2013 *Traffic Monitoring Guide* issued by the Federal Highway Administration (Alliance for Biking and Walking 2018). Efforts are under way to expand state-level estimates of bicycle and pedestrian travel (Nordback, Sellinger, and Phillips 2017) and develop a nonmotorized national traffic count archive (Nordback et al. 2015). These efforts may support increased funding and improved policy, planning, and engineering for bicycling (Nordback et al. 2015; Nordback, Sellinger, and Phillips 2017; see also Hankey et al. 2017 for research on sampling of nonmotorized travel).

Third, local governments often invest in cycling infrastructure in response to pressure from citizen and advocacy groups (see Pucher et al., chapter 20, this volume). Since stronger socioeconomic groups often are much more effective in exerting pressure on decision makers, this may also lead to bias in bicycle investments.

## Exposure to Pollution and Traffic Risk

Travel by any means opens a person to the possibility of air and noise pollution as well as the risk of suffering an injury or fatality, especially when near roadways. In recent years, increasing attention has been devoted to assessing these outcomes for bicyclists. Disparities may occur in two dimensions: between cyclists and travelers using other modes (in particular, motorists) and among different groups of cyclists.

### Disparities and Inequities between Cyclists and Motorists

As documented by Garrard et al. (chapter 3, this volume), there is evidence that bicyclists are exposed to higher doses of air pollution than motorists. Studies measuring noise pollution have also found higher exposure levels for bicyclists (Apparicio et al. 2018; Okokon et al. 2017).

As shown by Elvik (chapter 4, this volume), there is extensive evidence that the fatality and injury rates of cyclists are much higher than for motorists, especially in countries with low cycling rates. For example, in New Zealand, motorists enjoy an injury risk per million hours traveled that is 75% less than for cyclists (Chieng, Lai, and Woodward 2017). In the United

States, the bicyclist share of traffic injuries is over seven times the overall bicyclist share of all trips (Alliance for Biking and Walking 2018). Even in the relatively safe Netherlands, there are 5.5 times more traffic deaths per kilometer traveled by bicycling than by car (de Hartog et al. 2010).

These disparities in health-related impacts for air and noise pollution, as well as traffic injuries and fatalities, raise clear questions of justice. First, while cyclists generate negligible risks for other travelers, they have to contend with the elevated doses of inhaled pollutants, the noise levels, and the safety risks described here. Suárez et al. (2014) highlight the "inequity" of how "the most sustainable commuters" experience traffic emission burdens at least as large as those of car commuters. In addition, the tendency for infrastructure that facilitates safe cycling to be lacking in low-income or minority communities means that it is more difficult for these populations to take advantage of the significant health benefits offered by cycling. Lacking safe bicycle infrastructure, the net balance between exposure to pollution and risk on the one hand, and the health benefits of active travel on the other, may well be negative for these disadvantaged groups in spite of a positive net benefit on average across an entire population.

The potentially negative net health impacts for some groups raise a third kind of injustice. Because disadvantaged groups have less access to personal motorized vehicles, they may find themselves forced to bike and walk in conditions that more advantaged groups would not tolerate. This situation is not unique to the United States. For example, a recent study of air pollution and cycling in Beijing found medium- and high-income workers were more likely than low-income workers to switch to driving when pollution is high (Zhao et al. 2018).

### Disparities and Inequities among Groups of Cyclists

Disparities between groups of cyclists exist in exposure to both pollution and traffic risk, with the latter risk being particularly relevant from an equity perspective in light of the severe impacts of traffic injuries and fatalities on the victims and their families and friends. Research for the United States has repeatedly shown that disadvantaged groups are particularly affected by traffic risk. While African Americans and Hispanic or Latino Americans are underrepresented in making both work and nonwork bicycle trips, they often experience higher injury and fatality rates while biking than white Americans. For example, data from the Centers for Disease Control from

2001 indicated that African Americans experienced a fatality rate while bicycling that was 30% higher than the rate experienced by white Americans, while Hispanic or Latino Americans experienced a fatality rate while bicycling that was 23% higher than for whites (Zimmerman et al. 2015).

However, these disparities are not the same across the entire United States. State-level data for bicyclist fatalities from 2005 to 2013 indicate that Hispanic or Latino Americans are overrepresented in 18 US states but underrepresented in the remaining 32 states (Alliance for Biking and Walking 2018). Some of these differences may be explained by differences in bicycle use between population groups. However, this cannot always explain observed patterns. A study of Wisconsin found elevated traffic injury risk for African Americans and Asian Americans but not for Hispanic or Latino Americans, even after controlling for gender and travel mode (McAndrews et al. 2013). Yet, in many states, minority and disadvantaged population groups are at a higher risk. For instance, in Oregon, 56% of high-injury corridors are in areas with a larger share of people of color and low-income individuals, and both vehicle-related deaths and hospitalization rates are lower for white Americans than for other racial groups (Oregon Metro 2017).

As already discussed, these disparities in traffic injuries and fatalities may result from uneven access to high-quality bicycle infrastructure; in many places, the distribution of cycling facilities tends to be inversely related to the share of Americans of color (Zimmerman et al. 2015). This lack of access to safe bicycling conditions is reflected in surveys about riding preferences. One survey found that 26% of Americans of color would like to ride more but worry about safety, compared to 19% of white Americans (League of American Bicyclists and Sierra Club 2013), while another survey found that 48% of African Americans and 53% of Hispanic or Latino Americans would bicycle more if a physical barrier separated them from motor vehicles (compared to 44% of white Americans) (People for Bikes and Alliance for Biking and Walking 2015).

As in the United States, the evidence for Canada is mixed. A study of traffic injuries and road conditions in Montreal found "significantly more injured pedestrians, cyclists, and motor vehicle occupants at intersections in the poorest than in the richest areas" and concluded that "roadway environment can explain a substantial portion of the excess rate of road traffic injuries in the poorest urban areas" (Morency et al. 2012; see also Zimmerman et al. 2015). However, a study of Montreal's cycling network found advantageous accessibility to cycling facilities for low-income individuals (Houde, Apparicio, and Séguin 2018).

## Urban Displacement and Gentrification

While investments in safe bicycle infrastructure in low-income neighborhoods can generally be seen as an intervention that can promote equity, there is mounting evidence that such investments are connected to the displacement or gentrification of the existing working class and communities of color in central neighborhoods in US cities (Hoffman and Lugo 2014; Lubitow and Miller 2013; Stein 2011; Zavestoski and Agyeman 2014). While it is not easy to show that investments in bicycle infrastructure *cause* these changes, they are often part of the physical changes promoted and implemented as part of the neighborhood "upgrading" that accompanies shifts in residential and commercial characteristics. Studies confirm that "complete streets" investments (a broader set of design and engineering standards that often includes improvements for cycling) are accompanied by increases in real estate values, and that street improvements were an ingredient in economic development strategies (Alliance for Biking and Walking 2018; Smart Growth America 2015). Smart Growth America also noted that 8 out of 10 complete streets investment projects resulted in increased property values. Stehlin's (2015) research connected street investments in San Francisco neighborhoods to significant shifts in the racial and class compositions of those neighborhoods (see similar results for Portland in Herrington and Dann 2016). It is often the timing of investments that causes concern: neighborhoods that have been asking for safety interventions for decades suddenly receive attention only as those safety issues arise as a barrier to cycling (Lubitow and Miller 2013). Even if causation is impossible to show, and low-income groups cycle more than others according to census and other larger travel surveys, the perceived connection to urban redevelopment processes suggests that the bicycle planning process is problematic for equity. Melody Hoffman goes so far as to label bike lanes "white lanes" (Hoffman 2016) and emphasizes the role bicycle investments play in reinforcing social "markers" used to clarify who belongs in neighborhoods, who does not, and who is expected to occupy these neighborhoods in the future, whether or not those are the explicit aims of the investments.

## Social Status, Recognition, and Respect

The social significance of cycling, or lack thereof, gives rise to two interrelated issues of status and recognition: the individual rights of cyclists and the collective right to safe cycling. Bicycling is still a "second class" mode in most places in the world (see, e.g., Chieng, Lai, and Woodward 2017). For most adults in most of the developed world, bicycling is a recreational activity and not a serious solution to transport needs. As a result, one could argue that cyclists themselves are marginalized, take on a "second-class status," and are commonly excluded or discriminated against in policy, road design, traffic rules, and treatment by fellow road users (see, e.g., Prati, Puchades, and Pietrantoni 2017). Homeless bicyclists in particular are condemned for perceived disruptions to norms surrounding speed, cargo loads, and use of dedicated infrastructure (Parker 2019). An equity approach to planning should attempt to integrate cycling into the decision-making process, communication campaigns to promote safer (car) driver behavior, and transport investment packages.

While cyclists as a group may feel excluded, the societal impact of their exclusion pales in comparison to discrimination based on race, class, or gender (e.g., in areas such as housing, jobs, and education), which are therefore likely to receive higher priority in societal debates and on policy agendas. In light of the sometimes vast disparities along these dimensions of discrimination, addressing cycling safety or improving cycling infrastructure to address cyclists' second-class status while ignoring the often deep-seated discrimination along these other lines is likely to generate antagonism toward cycling advocacy (Golub 2016). From this perspective, the debate on cycling justice should be concerned not with closing the gap between cycling and car driving but with cycling's possible contribution to addressing the wider social inequities in society.

Given cycling's potential role in this sense, and given the increase in cycling's social significance over the past two decades, questions about fair access to safe cycling as a social justice issue may emerge. Bicycling has been growing in popularity and in some large cities has doubled or tripled over recent decades (see Pucher and Buehler, chapter 1, and Buehler and Pucher, chapter 2, this volume). Many places around the world, from Mexico City to South Korea, from India to parts of Africa, and in an increasing number of cities in the Global North, have greatly increased their funding for bicycling over the past two decades. If cycling becomes mainstream, safe access to it and a fair distribution of bicycle investments across neighborhoods and population groups may grow to become an object of social justice claims.

When cyclists are members of traditionally marginalized groups, there can sometimes be compounding issues of discrimination, neglect, or endangerment by drivers, and the rights to safety for these cyclists or pedestrians can be elevated as social justice issues. Several studies of driving behavior show lower rates of yielding behavior to pedestrians of color (Coughenour et al. 2017; Goddard, Kahn, and Adkins 2015). While similar studies do not exist for the attitude toward cyclists, qualitative work with cyclists of color confirms a similar fear that they will be disregarded by drivers (Lubitow 2017). Research shows that transport planners can address these biases. Traffic engineer Marisa DeMull (2018) reviews this research in her blog post, highlighting how "infrastructure drastically reduced the sample size of non-compliant behavior, demonstrating that the easiest solution is the most obvious. Provide better infrastructure for everyone, and everyone's safety increases." DeMull proposes that traffic engineers who design facilities for bicyclists and pedestrians should be more aware of the demographics near new infrastructure. They should collaborate more with planners, community members, and others who are also working in the same neighborhood. They should also work harder to leverage design decisions to reduce the tendency for bias among drivers by slowing down drivers, improving visibility, and taking other measures. Just like new policing practices that require officers to slow their decision making in the wake of a series of police killings of unarmed African American men in the United States, slowing drivers will mean they have more time to process information and make decisions based less on their gut instincts (which are shown to be biased racially) and more on the prevailing traffic norms or signage.

Meanwhile, advocates lament these "invisible cyclists" as those minority and low-income cyclists often overlooked by the planning process. They are often undercounted by bicycle cordon counts because they travel in patterns not considered important to mainstream planners, who focus on trips into and out of city centers. Minority and low-income cyclists are also undercounted in surveys of the journey to work, such as the annual *American Community Survey* (US Census Bureau 2012). They are probably undercounted in the periodic censuses and general travel surveys of other countries as well.

Public spaces and rights-of-way, including those traversed by bicycle infrastructure, are places of vulnerability to minority and low-income communities and women in most places around the world (see, e.g., Alliance for Biking and Walking 2018). Within the US context, Americans of color cope with racial profiling by law enforcement, racial harassment, and discriminatory treatment by drivers (Zimmerman et al. 2015). The continued targeting of residents of color in public spaces by the police and others is an ongoing human rights challenge, and planners should be cognizant of this reality. The question for bicycle planning then becomes, how can bicycling infrastructure help to reduce this public vulnerability?

### Participation and Recognition in Decision Making

The transport planning process is an important arena for making decisions about the future of cities and regions at a variety of jurisdictional scales. For decades in the United States, racial discrimination prevented minorities from effectively participating in the urban development process. Access to decision-making for disadvantaged groups continues to be a key goal of the transport justice movement in the United States (Golub, Marcantonio, and Sanchez 2013; Grengs 2004; Sanchez and Brenman 2007) and is also an issue in many other countries. Remaining barriers may no longer be the deliberate result of official discrimination but rather the implicit result of the closed nature of the cycling advocacy apparatus (Lugo 2015) as well as technocratic practices of (bicycle) planning. In the US context, this manifests as a predominantly white and middle-class professional class of cycling planners that creates unseen barriers to participation and influence by cyclists of color (Lugo 2015).

The situation may slowly be changing, however, with advocacy organizations becoming more aware of the need to make cycling relevant for a wider range of constituents. As described by Pucher et al. (chapter 20, this volume), the California Bicycle Coalition (CalBike), a bicycle advocacy organization, has put equity firmly on its agenda, as evidenced in its latest strategic plan, adopted in 2018. The key aim of the plan is to leverage the impact of better bicycling facilities to correct the historic transport inequities that have disproportionately impacted low-income communities and people of color for decades.

### Conclusions

Concerns over equity have long been at the bottom of the transport agenda, both in research and in practice. While the topic is receiving increasing attention in the academic community, it is still a marginal issue in most debates about transport policies and infrastructure investments. This is reflected in much of the bicycling discourse promoting cycling as a green, healthy, and urban-friendly mode of transport and often as an alternative to car use, without much attention to the equity issues at stake.

This chapter has sought to sensitize transport planners and scholars to the complex interrelationships between cycling and equity. With cycling, there are two interweaving justice issues—the marginalization of cyclists in the roadway, in transport planning, and in broader policy discourse in many countries, and marginalized peoples who may or may not be cyclists but are affected by investments in cycling. The first perspective starts from a transport mode, the second from people. Transport planners have been trained to think about modes rather than people. It is therefore no wonder that transport planners across the world have started to embrace the rallying cry to improve conditions for cyclists. Proponents of cycling often perceive the increase in cycling facilities and cycling mode share as the measure of success of their efforts to change the urban transport landscape. In doing so, they tend to ignore the broader transport equity issue at stake—providing every person with access so they can fully participate in society (Martens 2017)—and the even broader social issues in play where traditionally marginalized communities may be left out or even bear burdens from investments in cycling.

In many communities, those existing dimensions of marginality are far more important than the marginalization of cycling as a mode of transport. This is particularly evident in countries that have vast disparities in income and segregation along socioeconomic or ethnic lines, such as the United States and a number of emerging economies, but it also bubbles below the surface in countries with a still relatively strong welfare state and even in those with well-developed bicycle infrastructure. Even in the Netherlands and the Scandinavian countries, there is not enough consideration of the particular needs of minorities and disadvantaged groups. For instance, in the Netherlands, huge sums have been dedicated to the provision of free bicycle parking at train stations, disproportionately serving travelers with higher education and income levels, while little action has been taken to provide secure bicycle parking in low-income neighborhoods.

We urge advocates and practitioners to look carefully at these issues as they move their bicycle agenda and bicycle projects forward. The bicycle agenda could benefit from a broader social perspective, embracing the increasingly vocal call for an inclusive transport system across all transport modes. From that perspective, investment in cycling is only one possible means of promoting equity and not always the most effective one (compared to, for instance, improvement in pedestrian safety or public transport). At the level of bicycle projects, practitioners should consider a variety of perspectives, especially those of the affected communities, in the decision-making process so that cycling investments occur where they can really improve people's lives.

Cycling equity cannot be the sole responsibility of well-meaning planners sensitive to equity issues. Cycling provides many benefits and can be an especially attractive mode of transport for youths and low-income population groups because of its low barrier to entry. Ensuring that these groups are not ignored in planning and can enjoy the benefits of cycling, without being stigmatized or feeling threatened on the road, will require more fundamental changes—in funding mechanisms, planning approaches, policing and enforcement practices, and regulations. Transport planning should follow the lead of fields such as housing, education, and health care, which in many countries around the world are based on principles of justice, where implementation is guaranteed through minimum standards of quality and safety, guarantees of universal access, and mechanisms guaranteeing affordability for all. As in these fields, justice should become a foundational principle and hence a goal of transport planning rather than an add-on. Bicycle policy and planning should be part of that larger effort, so cycling will over time become a normal, safe, and comfortable means of transport that is available to everyone everywhere, irrespective of gender, age, ethnicity, or income, and without being seen as a symbol of either poverty or privilege.

### References

Alliance for Biking and Walking. 2018. *Bicycling and Walking in the United States: 2018 Benchmarking Report*. Washington, DC: Alliance for Biking and Walking.

Apparicio, Philippe, Jérémy Gelb, Mathieu Carrier, Marie-Ève Mathieu, and Simon Kingham. 2018. Exposure to Noise and Air Pollution by Mode of Transportation during Rush Hours in Montreal. *Journal of Transport Geography* 70: 182–192.

Barajas, Jesus. 2016. Making Invisible Riders Visible: Motivations for Bicycling and Public Transit Use among Latino Immigrants. PhD thesis, University of California, Berkeley.

Barajas, Jesus. 2018. Supplemental Infrastructure: How Community Networks and Immigrant Identity Influence Cycling. *Transportation.* https://doi.org/10.1007/s11116 -018-9955-7.

Barajas, Jesus. 2019. Perceptions, People, and Places: Influences on Cycling for Latino Immigrants and Implications for Equity. *Journal of Planning Education and Research.* https://doi.org/10.1177/0739456X19864714.

Bhat, Chandra, Sebastian Astroza, and Amin Hamdi. 2017. A Spatial Generalized Ordered-Response Model with Skew Normal Kernel Error Terms with an Application to Bicycling Frequency. *Transportation Research Part B: Methodological* 95: 126–148.

Chieng, Michael, Hakkan Lai, and Alistair Woodward. 2017. How Dangerous Is Cycling in New Zealand? *Journal of Transport and Health* 6: 23–28.

Coughenour, Courtney, Sheila Clark, Ashok Singh, Eudora Claw, James Abelar, and Joshua Huebner. 2017. Examining Racial Bias as a Potential Factor in Pedestrian Crashes. *Accident Analysis and Prevention* 98: 96–100.

de Hartog, Jeroen Johan, Hanna Boogaard, Hans Nijland, and Gerard Hoek. 2010. Do the Health Benefits of Cycling Outweigh the Risks? *Environmental Health Perspectives* 118(8): 1109–1116.

DeMull, Marisa. 2018. It's Time for Transportation Engineers to Address Racial Equity. *Alta Planning + Design* (blog), August 2. https://blog.altaplanning.com/its-time-for-transportation-engineers-to-address-racial-equity-edfd183798f6.

Flanagan, Elizabeth, Ugo Lachapelle, and Ahmed El-Geneidy. 2016. Riding Tandem: Does Cycling Infrastructure Investment Mirror Gentrification and Privilege in Portland, OR and Chicago, IL? *Research in Transportation Economics* 60: 14–24.

Fuller, Omari, and Edgar Beltran. 2010. The Invisible Cyclists of Los Angeles. *Progressive Planning*, July 14. https://www.plannersnetwork.org/2010/07/the-invisible-cyclists-of-los-angeles/.

Goddard, Tara, Kimberly Barsamian Kahn, and Arlie Adkins. 2015. Racial Bias in Driver Yielding Behavior at Crosswalks. *Transportation Research Part F: Traffic Psychology and Behaviour* 33: 1–6.

Golub, Aaron. 2016. Is the Right to Bicycle a Civil Right? Synergies and Tensions between the Transportation Justice Movement and Planning for Bicycling. In *Bicycle Justice and Urban Transformation: Biking for All?*, edited by Aaron Golub, Melody Hoffmann, Adonia Lugo, and Gerardo Sandoval, 20–31. London: Routledge.

Golub, Aaron, Melody Hoffmann, Adonia Lugo, and Gerardo Sandoval, eds. 2016. *Bicycle Justice and Urban Transformation: Biking for All?*. London: Routledge.

Golub, Aaron, Richard Marcantonio, and Thomas Sanchez. 2013. Race, Space, and Struggles for Mobility: Transportation Impacts on African Americans in Oakland and the East Bay. *Urban Geography* 34(5): 699–728.

Grengs, Joe. 2004. The Abandoned Social Goals of Public Transit in the Neoliberal City of the USA. *City* 9(1): 51–66.

Hankey, Steve, Tianjun Lu, Andrew Mondschein, and Ralph Buehler. 2017. Spatial Models of Active Travel in Small Communities: Merging the Goals of Traffic Monitoring and Direct-Demand Modeling. *Journal of Transport and Health* 7B: 149–159.

Herrington, Cameron, and Ryan Dann. 2016. Is Portland's Bicycle Success Story a Celebration of Gentrification? A Theoretical and Statistical Analysis of Bicycle Use and Demographic Change. In *Bicycle Justice and Urban Transformation: Biking for All?*, edited by Aaron Golub, Melody Hoffmann, Adonia Lugo, and Gerardo Sandoval, 32–52. London: Routledge.

Hoffman, Melody. 2016. *Bike Lanes Are White Lanes: Bicycle Advocacy and Urban Planning*. Lincoln: University of Nebraska Press.

Hoffmann, Melody, and Adonia Lugo. 2014. Who Is "World Class"? Transportation Justice and Bicycle Policy. *Urbanities* 4(1): 45–61.

Houde, Maxime, Philippe Apparicio, and Anne-Marie Séguin. 2018. A Ride for Whom: Has Cycling Network Expansion Reduced Inequities in Accessibility in Montreal, Canada? *Journal of Transport Geography* 68: 9–21.

League of American Bicyclists and Sierra Club. 2013. *The New Majority: Pedaling towards Equity*. Washington, DC: League of American Bicyclists.

Lee, Richard, Ipek Sener, and S. Nathan Jones. 2017. Understanding the Role of Equity in Active Transportation Planning in the United States. *Transport Reviews* 37(2): 211–226.

Lubitow, Amy. 2017. *Narratives of Marginalized Cyclists: Understanding Obstacles to Utilitarian Cycling among Women and Minorities in Portland, OR*. Portland, OR: Portland State University.

Lubitow, Amy, and Thaddeus Miller. 2013. Contesting Sustainability: Bikes, Race, and Politics in Portlandia. *Environmental Justice* 6(4): 121–126.

Lucas, Karen, Karel Martens, Floridea Di Ciommo, and Ariane Dupont-Kieffer. 2019. Introduction. In *Measuring Transport Equity*, edited by Karen Lucas and Karel Martens (with Floridea Di Ciommo and Ariane Dupont-Kieffer), 3–12. Amsterdam: Elsevier.

Lugo, Adonia. 2015. Unsolicited Advice for Vision Zero. *Urbanadonia.com* (blog), September 30. http://www.urbanadonia.com/2015/09/unsolicited-advice-for-vision -zero.html.

Lusk, Anne C., Albert Anastasio, Nicholas Shaffer, Juan Wu, and Yanping Li. 2017. Biking Practices and Preferences in a Lower Income, Primarily Minority Neighborhood: Learning What Residents Want. *Preventive Medicine Reports* 7: 232–238.

Martens, Karel. 2013. Role of the Bicycle in the Limitation of Transport Poverty in the Netherlands. *Transportation Research Record* 2387: 20–25.

Martens, Karel. 2017. *Transport Justice: Designing Fair Transportation Systems*. New York: Routledge.

Martens, Karel, Daniel Piatkowski, Kevin J. Krizek, and Kara Luckey. 2016. Advancing Discussions of Cycling Interventions Based on Social Justice. In *Bicycle Justice and Urban Transformation: Biking for All?*, edited by Aaron Golub, Melody Hoffmann, Adonia Lugo, and Gerardo Sandoval, 86–98. London: Routledge.

McAndrews, Carolyn, Kirsten Beyer, Clare Guse, and Peter Layde. 2013. Revisiting Exposure: Fatal and Non-fatal Traffic Injury Risk across Different Populations of Travelers in Wisconsin, 2001–2009. *Accident Analysis and Prevention* 60: 103–112.

Morency, Patrick, Lise Gauvin, Céline Plante, Michel Fournier, and Catherine Morency. 2012. Neighborhood Social Inequalities in Road Traffic Injuries: The Influence of Traffic Volume and Road Design. *American Journal of Public Health* 102(6): 1112–1119.

Nordback, Krista, Mike Sellinger, and Taylor Phillips. 2017. *Estimating Walking and Bicycling at the State Level*. Portland, OR: Portland State University.

Nordback, Krista, Kristin A. Tufte, Morgan Harvey, Nathan McNeil, Elizabeth Stolz, and Jolene Liu. 2015. Creating a National Nonmotorized Traffic Count Archive: Process and Progress. *Transportation Research Record* 2527: 90–98.

Okokon, Enembe, Tarja Yli-Tuomi, Anu Turunen, Pekka Taimisto, Arto Pennanen, Ilias Vouitsis, Zissis Samaras, Marita Voogt, Menno Keuken, and Timo Lanki. 2017. Particulates and Noise Exposure during Bicycle, Bus and Car Commuting: A Study in Three European Cities. *Environmental Research* 154: 181–189.

Oregon Metro. 2017. Regional Transportation Plan Safety Work Group Meeting Materials. Oregon Metro, October 19. https://www.oregonmetro.gov/public-projects/2018-regional-transportation-plan/safety.

Parker, Cory. 2019. Homeless Negotiations of Public Space in Two California Cities. PhD thesis, University of California, Davis.

People for Bikes and Alliance for Biking and Walking. 2015. *Building Equity—Race, Ethnicity, Class, and Protected Bike Lanes: An Idea Book for Fairer Cities*. Boulder, CO: People for Bikes.

Pistoll, Chance, and Anna Goodman. 2014. The Link between Socioeconomic Position, Access to Cycling Infrastructure and Cycling Participation Rates: An Ecological Study in Melbourne, Australia. *Journal of Transport and Health* 1(4): 251–259.

Prati, Gabriele. 2018. Gender Equality and Women's Participation in Transport Cycling. *Journal of Transport Geography* 66: 369–375.

Prati, Gabriele, Víctor Marín Puchades, and Luca Pietrantoni. 2017. Cyclists as a Minority Group? *Transportation Research Part F: Traffic Psychology and Behaviour* 47: 34–41.

Sanchez, Thomas, and Marc Brenman. 2007. *The Right to Transportation: Moving to Equity*. Chicago: Planners Press.

Smart Growth America. 2015. *Safer Streets, Stronger Economies: Complete Streets Project Outcomes from around the Country*. Washington, DC: Smart Growth America.

Stehlin, John. 2015. Cycles of Investment: Bicycle Infrastructure, Gentrification, and the Restructuring of the San Francisco Bay Area. *Environment and Planning A: Economy and Space* 47(1): 121–137.

Stehlin, John. 2019. *Cyclescapes of the Unequal City: Bicycle Infrastructure and Uneven Development*. Minneapolis: University of Minnesota Press.

Stein, Samuel. 2011. Bike Lanes and Gentrification: New York City's Shades of Green. *Progressive Planning* 188: 34–37.

Suárez, Liliana, Stephanie Mesías, Verónica Iglesias, Claudio Silva, Dante D. Cáceres, and Pablo Ruiz-Rudolph. 2014. Personal Exposure to Particulate Matter in Commuters Using Different Transport Modes (Bus, Bicycle, Car and Subway) in an Assigned Route in Downtown Santiago, Chile. *Environmental Science: Processes and Impacts* 16(6): 1309–1317.

Tucker, Bronwen, and Kevin Manaugh. 2018. Bicycle Equity in Brazil: Access to Safe Cycling Routes across Neighborhoods in Rio de Janeiro and Curitiba. *International Journal of Sustainable Transportation* 12(1): 29–38.

US Census Bureau. 2012. Census Bureau Releases Estimates of Undercount and Overcount in the 2010 Census. News Release, May 22. https://www.census.gov/newsroom/releases/archives/2010_census/cb12-95.html.

US Department of Transportation. 2017. *National Household Travel Survey*. Washington, DC: Federal Highway Administration. https://nhts.ornl.gov.

Van der Kloof, Angela. 2015. Lessons Learned through Training Immigrant Women in the Netherlands to Cycle. In *Cycling Cultures*, edited by Peter Cox, 78–105. Chester: University of Chester Press.

Van der Kloof, Angela, Jeroen Bastiaanssen, and Karel Martens. 2014. Bicycle Lessons, Activity Participation and Empowerment. *Case Studies on Transport Policy* 2(2): 89–95.

Zavestoski, Stephen, and Julian Agyeman, eds. 2014. *Incomplete Streets: Processes, Practices, and Possibilities*. New York: Routledge.

Zavestoski, Stephen, and Julian Agyeman. 2015. *Invisible Cyclist* (blog), November 3. https://invisiblecyclist.wordpress.com.

Zhao, Pengjun, Shengxiao Li, Peilin Li, Jixuan Liu, and Kefan Long. 2018. How Does Air Pollution Influence Cycling Behaviour? Evidence from Beijing. *Transportation Research Part D: Transport and Environment* 63: 826–838.

Zimmerman, Sara, Michelle Lieberman, Karen Kramer, and Bill Sadler. 2015. *At the Intersection of Active Transportation and Equity: Joining Forces to Make Communities Healthier and Fairer*. Oakland: Safe Routes to School National Partnership.

# 15 Cycling in China and India

John Pucher, Geetam Tiwari, Zhong-Ren Peng, Rong Cao, and Yuan Gao

For many decades, the bicycle has been an important means of travel in both China and India, accounting for much higher shares of travel than in almost all other countries. In the 1980s, China was known as the "bicycle kingdom" because over two-thirds of trips in urban areas were made by bicycle (Zhang, Shaheen, and Chen 2014). Since the 1990s, however, rapidly rising car ownership and vastly improved public transport have reduced cycling levels to an average of about 30% of trips in Chinese cities from 2015 to 2017. The bicycle share of trips in Indian cities averaged 18% in 2011, having fallen in recent decades as in China (Mohan 2013; Pucher et al. 2007).

Despite sharp declines since the 1980s, cycling levels in both China and India are still much higher than in most other countries for which travel survey data are available. The most important exception is the Netherlands, Europe's most bicycle-oriented country, which has a bicycle mode share of 28% (see Buehler and Pucher, chapter 2, figure 2.1, this volume). The Netherlands is followed in bicycle share by Denmark (14%) and Germany (11%). In most European Union countries, however, less than 5% of trips are made by bicycle. The United States, Canada, and Australia have even lower bicycle mode shares, at only 1%–2%.

This chapter examines cycling in China and India, two important countries that are crucial for providing a global perspective on cycling (United Nations 2018; World Bank 2018). Severe data limitations prevent time-trend analysis of some important indicators even within each country, and they make statistical comparisons between China and India difficult. Consequently, this chapter combines the limited statistical data with published research to portray key differences between the two countries in cycling trends and government policies, and how these have changed over time.

## Trends in Cycling Levels and Variation within Countries

Neither China nor India has a national travel survey that would allow consistent time-trend data on levels of cycling and the mode share of cycling. The 2011 *Census of India* is the country's first comprehensive survey of the means by which people travel to work, but as with census data for other countries, it does not include information on nonwork trips (Ministry of Home Affairs 2011). Nevertheless, the Indian census uses a consistent survey methodology that permits disaggregation along various dimensions, and thereby comparisons among different Indian cities, as well as between rural and urban areas.

The census reveals that in 2011 the bicycle was used by 19.2% of all Indian workers to reach their jobs, but with a higher cycling mode share in rural areas (21.9%) than in urban areas (17.6%). As shown in table 15.1, the bicycle mode share was roughly the same 20% of work commuters in the four city population size categories from 100,000 to 1.9 million, but dropped to 15.7% for cities with 2.0 million to 4.9 million residents and to 8.8% for cities with 5 million or more residents (Tiwari and Nishant 2018). In short, bicycle mode share was about twice as high in small and medium-size cities as in cities with 5 million or more residents. The lower bicycle mode shares in the two largest population categories probably result from the longer average trip distances generated in cities that cover a larger land area. In addition, public transport is much more available in larger cities than in smaller ones (Pendakur 2011; Pucher et al. 2007; Rathi 2017; Tiwari, Arora, and Himani 2008).

**Table 15.1**
Bicycle share of work commuters in Indian cities by population size for 2011

| City population size category | Number of cities | Bicycle share of work commuters (%) |
|---|---|---|
| 100,000 to 250,000 | 141 | 20.4 |
| 250,000 to 500,000 | 144 | 19.8 |
| 500,000 to 1 million | 105 | 19.9 |
| 1 million to 2 million | 59 | 20.6 |
| 2 million to 5 million | 34 | 15.7 |
| 5 million or more | 6 | 8.8 |

*Source*: Tiwari and Nishant (2018), based on data from the *2011 Census of India* (Ministry of Home Affairs 2011).

Figure 15.1 shows bicycle mode shares in 2011 for India's 15 largest cities—in order of population size—ranging from 2.3% in Mumbai to 30.2% in both Kanpur and Nagpur (Tiwari and Nishant 2018). The variation in cycling levels among cities with similar populations is probably caused by differences in topography, demographics, land-use patterns, roadway networks, and public transport. Even among the three largest Indian cities, there are large differences in bicycle mode share: 2.3% in Mumbai, 10.6% in Delhi, and 23.0% in Kolkata. The high bicycle mode share in Kolkata results from its many low-income residents; its mixed land use, which generates many short trips; and its roadway network, consisting mostly of closely spaced streets, with few highways and major arterial roads. The low bicycle mode share in Mumbai results from the large proportion of low-income people living in slums within walking distance of the central business district (Dharavi). For middle-income households, bicycle trips to work are not feasible because of the long trips necessitated by Mumbai's peninsular location, with most residential areas being located far away from the city core at the tip of the peninsula. Mumbai has India's most extensive public transport system (especially suburban rail), which is used by most commuters to reach their jobs. In spite of such variations among cities—and consistent with the aggregate data in table 15.1—there is a general tendency for bicycling mode share to increase as city size decreases, as seen for the five smallest cities shown in figure 15.1.

Cycling rates also vary by state, gender, and income level (Tiwari and Nishant 2018). The bicycle mode share was highest in the poorest states (over 30%) and lowest (but at least 10%) in the most affluent states, confirming the importance of cycling for travel throughout India, especially among low-income workers. There is a big difference by gender, however, with almost five times as many men as women cycling to work (22.2% vs. 4.8%). Cycling is viewed by some women as being too dangerous—susceptible to traffic injuries as well as verbal and physical sexual harassment by men (Pendakur 2011; Pucher et al. 2007; Rathi 2017).

Cycling rates also vary by trip distance. For trips up to 1 km, the bicycle share is only 12%, probably because 68% of such short trips are made by walking. The highest bicycle mode share is for trips of 2–5 km (30%), with successively lower bicycle mode shares as trip distance increases: 15% for trips of 6–10 km and 7% for trips of 11–20 km (Tiwari and Nishant 2018).

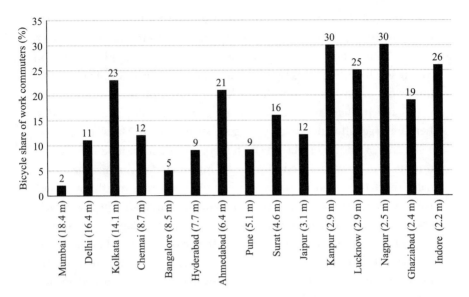

**Figure 15.1**
Bicycle share of daily work commuters in India's largest cities in 2011. *Note*: Population size (in millions) is shown in parentheses beside each city's name. *Source*: Tiwari and Nishant (2018), based on data from the *2011 Census of India* (Ministry of Home Affairs 2011).

China does not have a national travel survey of any kind, not even for the journey to work. Travel surveys conducted by large Chinese cities, roughly once per decade, are the only source of information on travel behavior. Since cities conduct their own surveys in different years, there are inconsistencies across cities in their survey timing and methodology. Thus, they should not be used for exact comparisons. Nevertheless, examined together, the surveys provide useful information on differences among cities and over time in travel behavior in large Chinese cities.

Figure 15.2 shows the bicycle mode share of trips for 10 large Chinese cities. The first bar for each city shows the bicycle mode share in the 1980s, with the survey year varying by city, as shown at the bottom of each vertical bar. The second bar shows the mode share in the 1990s, again with survey year shown at the bottom of the bar. The third bar shows the bicycle mode share for the latest year available, ranging from 2007 to 2017. Bicycle mode share fell in 9 of the 10 cities shown in figure 15.2 over the past three decades, and it fell sharply in four of those cities. For example, bicycle mode share plummeted in Beijing, China's capital and second-largest city, from 54% of all daily trips in 1984 to only 10% of all trips in 2016. Whatever the limitations of these city surveys, they are consistent with other studies reporting falling cycling rates in China in recent decades (Gao and Kenworthy 2017; Jiang et al. 2017; Pucher et al. 2007; Yang, Wang, Zhou, and Zacharias 2014; Zhang, Shaheen, and Chen 2014).

There are several reasons for the overall decline of cycling in China. Most obvious, perhaps, is the rapid growth in car ownership and use, which directly reduced cycling levels by attracting former bicyclists. Indirectly, rising motor vehicle traffic made cycling on shared roadways more dangerous, especially at intersections. In addition, the large increase in motorized travel worsened the already serious air pollution in Chinese cities to levels far exceeding standards set by the World Health Organization (Jiang et al. 2017; WHO 2018a; WHO 2018b). Forcing cyclists to breathe increasingly toxic air over the past few decades has made cycling less pleasant as well as less healthy. Finally, Chinese cities have grown rapidly both in population size and area in recent decades, generating many trips too long to make by bicycle.

To accommodate the skyrocketing demand for motor vehicle travel, many Chinese cities reallocated roadway space from bicycles to motor vehicles. As noted in the infrastructure section of this chapter, some cities reduced the supply of separate cycling facilities, even banning bicycles from key arterial roads altogether. At the same time, most large Chinese cities have vastly expanded and improved their public transport systems, especially metro and express bus systems, offering a faster yet affordable alternative to cycling. Public transport has become far more feasible than cycling for making the longer trips being generated by the geographic expansion of large cities.

It is not possible to explain all the variation among cities in figure 15.2. Different survey years and methodologies probably account for some of the variation, but some differences are the result of different city policies toward cycling infrastructure provision, car ownership, and public transport improvements (Gao and Kenworthy 2016; Pucher et al. 2007; Yang, Wang, Zhou, and Zacharias 2014; Zacharias 2002). Since 1994, for example, Shanghai has strictly limited new car registrations each year (Chen and Zhao 2013). As a result, car ownership rates in Shanghai have been much lower than in Beijing over the past two decades. In 2016, for example, Shanghai had only half as many cars per 1,000 people as Beijing (100 vs.

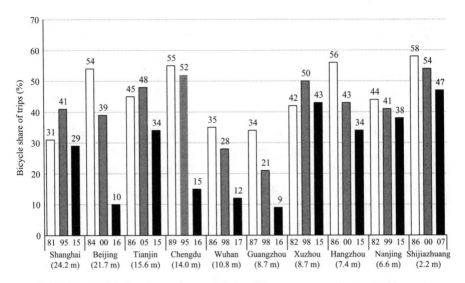

**Figure 15.2**
Bicycle share of trips in the 1980s, 1990s, and 2000s in large Chinese cities. *Note*:
Population size (in millions) is shown in parentheses below each city's name. Differences among cities in survey methodology and timing limit the comparability of
these bicycle modal share numbers. Some cities (such as Shanghai) include e-bikes in
their bicycle mode shares, while others do not. Thus, the numbers shown should be
considered approximate and not used for exact comparisons. *Source*: National Bureau
of Statistics of China (NBS China) 2019, and travel surveys conducted by each of the
cities shown.

208). In 2011—17 years later than Shanghai—Beijing finally began to limit
new car registrations (Yang, Ying, Ping, and Antung 2014). Following the
example of Shanghai, many other large Chinese cities now limit annual
car registrations, including Guangzhou, Tianjin, Hangzhou, Chengdu, Shijiazhuang, Guiyang, and Hainan.

Until recently, Beijing and many other large Chinese cities, including Shanghai, have focused mainly on massive investments in new and
expanded roadways, often eliminating or narrowing cycling infrastructure.
New expressways often divided neighborhoods, blocked cycling routes,
and caused long detours for cyclists. Roadway expansion has discouraged
cycling both directly and indirectly—by facilitating car use while making
cycling more circuitous and more dangerous (Pendakur 2011; Pucher et al.
2007; Yang, Wang, Zhou, and Zacharias 2014; Gao and Kenworthy 2017;
Zacharias 2002; Zhang, Shaheen, and Chen 2014).

Another important reason for the variation among Chinese cities, as well as the fall in bicycle mode share over time, is the extent and quality of their public transport systems. For example, the Beijing and Shanghai metro systems both started limited operations in the 1990s but evolved rapidly to become the two largest metro systems in the world by 2018. Indeed, 7 of the world's 20 largest metro systems in 2018 were in China (UITP 2018). Bus systems have also been expanded and modernized, including bus rapid transit routes with express service in many cities. In general, the best public transport is in the largest cities, where large geographic size leads to long trips ideal for public transport but too long for bicycling. Conversely, public transport is less extensive and of lower quality in medium-size Chinese cities, where trip distances are shorter and more feasible to make by bicycle. For example, almost half of daily trips (47%) are by bicycle in Shijiazhuang, the smallest of the cities included in figure 15.2.

One potential problem with the survey data in figure 15.2 is the inconsistent treatment of e-bikes in the city surveys since 2000. Some Chinese cities (such as Shanghai) include e-bikes in their reported bicycling mode shares, while others exclude them. The technology of e-bikes in China has evolved over time, starting out as pedelecs, pedal bicycles assisted by small, battery-powered motors. Currently, however, almost all e-bikes in China are not really bicycles at all but rather are battery-powered, two-wheeled motor scooters, usually without pedals. (For a detailed analysis of e-bikes, see Cherry and Fishman, chapter 9, this volume.) To a considerable extent, the decline in traditional bicycling has been offset by an increased use of e-bikes, which offer a faster but more dangerous alternative to traditional bicycles, as noted later. Travel speed has become increasingly important as Chinese cities have rapidly grown by spreading out in recent years, thus increasing trip distance. At the same time, per capita income has greatly increased, making e-bikes an affordable alternative to traditional bicycles.

In addition to overall cycling levels, there are two other important differences between cycling in China and in India: by gender and by urban versus rural areas. Cycling rates are roughly the same for women and men in China, although there is some variation by city (Cherry and Cervero 2007; Zhang et al. 2008). In contrast, men in India are almost five times as likely as women to ride a bicycle to work. Thus, there is a large gender gap in India but none in China. Differences in cycling rates between urban and rural areas in the two countries are also large. The limited evidence available

**Figure 15.3**
In spite of declines in cycling since the 1980s, most large Chinese cities have extensive cycling infrastructure, which is used both by traditional bicycles and by e-bikes, as shown here in Kunming. *Source*: Chris Cherry.

suggests that cycling rates in most rural areas of China average about 5%, roughly one-sixth the average cycling rate in Chinese cities (Wang et al. 2015; Yang, Yuan, and Feng 2014). In stark contrast, cycling rates in India are slightly higher in rural areas than in urban areas. E-bikes, motor scooters, and motorcycles are more practical for covering the long trip distances in rural areas. Although rural incomes are lower than urban incomes in both countries, they are higher in China than in India, thus enabling higher levels of motorized vehicle ownership in rural areas of China.

## Cycling Safety

Reliable comparable data on cycling safety are even more difficult to obtain for China and India than data on cycling levels. The official statistics that are available greatly understate the actual level of cyclist (and pedestrian) injuries and fatalities (Ministry of Road Transport and Highways 2011–2018; Ministry of Road Transport and Highways 2017). They are based on police reports, and many cyclist injuries are not reported to the police. For example, three independent studies showed that cyclist fatalities accounted for 6%–8% of all traffic fatalities in India in 2016, compared to only 2% reported by the Indian Ministry of Transport (Mohan, Tiwari, and Bhalla 2017). Moreover, the World Health Organization's estimate of total traffic fatalities in India in 2016 was 68% higher than the official Indian Ministry of Transport's estimate of 150,785. Thus, the actual number of cyclist fatalities is almost six times as high as the government's official estimate (17,780 vs. 3,016). There are no reliable time series data on cyclist fatalities and serious injuries in India. Even if such data were available, there are no corresponding time series data on bicycle trips and kilometers traveled. Thus, it is impossible to determine the risk of cycling fatalities and injuries relative to trends in cycling levels (exposure) over time.

Nevertheless, all existing research emphasizes the dangers of cycling in India caused by the almost complete lack of separate cycling facilities of any kind. Thus, bicyclists are forced to ride on the roads in the midst of much faster-moving motorcycles, cars, trucks, and buses (Das et al. 2012; Mohan 2013; Mohan, Tiwari, and Bhalla 2017; Pendakur 2011; Pucher et al. 2007; Rathi 2017; Singh 2005).

Cycling appears to be considerably safer in China than in India, probably because of the greater availability of separate cycling infrastructure in Chinese cities—in spite of cutbacks from about 1990 to 2010. From 2001 to 2016, bicyclist fatalities in China fell by 82%, from 16,190 to 2,946, partly because of the steep decline in cycling on traditional bicycles (Jiang et al. 2017; National Bureau of Statistics of China 2019). In sharp contrast, e-bicyclist fatalities soared. They were first recorded in official government statistics in 2004 and were divided into two categories. In 2004, there were 293 fatalities of riders of pedelecs and 589 fatalities of riders of battery-powered e-bikes (without pedals). By 2016, only e-bicyclist fatalities of the second category (essentially, electric scooters) were reported, and they had risen from 589 in 2004 to 7,201 in 2016 (Jiang et al. 2017). Including e-bicyclists with bicyclists, total fatalities fell from 16,190 in 2001 to 10,147 in 2016. That represents a 37% decline, compared to the 82% decline excluding e-bikes. Since most e-bike riders are former bicycle riders, the main modal shift has been from traditional bicycling to e-bicycling. In Shanghai, for example, the modal share of e-bicycling rose from 2% in 1995 to 22% in 2014, while the modal share of traditional bicycling fell from 39% to 7%.

The topic of e-bikes is examined in detail by Cherry and Fishman (chapter 9, this volume). It is highly debatable whether a battery-powered scooter without pedals can legitimately be called a bicycle at all, but most e-bikes in China are of this sort, and they comprise over 90% of the world's e-bikes. Because such e-bikes are allowed to use bicycle lanes and paths, they create safety hazards. They can travel much faster than conventional bicycles, thus creating large speed differentials of vehicles using the same rights-of-way. Nevertheless, even including e-cyclists, bicyclist fatalities in China have fallen considerably since 2001.

It is noteworthy that the estimated 17,780 bicyclist fatalities in India in 2016 exceeded the total of 10,147 bicyclist fatalities in China in the same year, even when e-cyclists are included in the fatality total for China. Moreover, at roughly 30%, the bicycle mode share in Chinese cities is much higher than the 18% mode share in India's cities. This suggests that—relative to mode share—bicycling fatalities in India are roughly twice as high as in China. Such comparisons of bicycle mode share and fatalities between countries are dubious, especially based on statistics of unknown reliability. Nevertheless, the measurable cycling safety gap is so large that it would be difficult not to draw the conclusion that cycling in China is indeed much safer than in India.

Clearly, the goal should be to make cycling safer in both countries. India, however, has a much greater need for improvements, indeed starting from almost zero, both in terms of cycling infrastructure and traffic safety programs focused on increasing cyclist safety. The following sections examine cycling infrastructure, programs, and policies in the two countries.

## Cycling Infrastructure

Perhaps the most important policy difference between China and India is the availability and quality of cycling infrastructure. Until recently, Indian cities had almost no separate cycling infrastructure at all, forcing cyclists to ride on overcrowded, deteriorating roads together with a wide variety of motorized traffic: cars, trucks, buses, vans, auto rickshaws, motorcycles, and motor scooters. Motorized traffic, especially large vehicles such as trucks and buses, poses the greatest traffic danger for bicyclists in India. Bicyclists must also maneuver around high volumes of nonmotorized traffic, such as pedestrians, animal-drawn carts, cycle rickshaws, and other bicyclists. The mixing of so many different modes of transport, with large variations in size and speed, causes severe problems of traffic safety, especially for bicyclists and pedestrians (Mohan 2013; Mohan, Tiwari, and Bhalla 2017; Pendakur 2011; Pucher et al. 2007; Singh 2005). Such mixing of transport modes also exacerbates the problem of roadway congestion, which would be considerable at any rate because of the high volume of traffic.

The first bicycle-friendly policies in India were proposed by the central government's National Urban Transport Policy in 2006. Both walking and cycling were officially endorsed as socially and environmentally sustainable modes of transport. Moreover, this policy document acknowledged the dangers of walking and cycling in Indian cities and proposed vast improvements in the provision of walking and cycling infrastructure separated from motor vehicle traffic (Ministry of Urban Development 2006). The central government has expressed further support for improved walking and cycling infrastructure in numerous studies, policy statements, and plans since then (Mohan, Tiwari, and Bhalla 2017; Tiwari 2014; Tiwari et al. 2015). In 2014, the Transportation Research and Injury Prevention Program (TRIPP) of the Indian Institute of Technology in Delhi proposed fully modernized, state-of-the-art design guidelines for the construction of cycling facilities (Tiwari 2014). In response, the Indian Roads Congress, which sets standards for roadway design and construction, issued new guidelines in 2015 that included suggestions for building separate cycling infrastructure when new roads are constructed, widened, or modernized. Some large Indian cities, such as Delhi, have also adopted bicycle plans that foresee construction of separate cycling facilities (Tiwari 2005).

Despite all these policy statements, plans, and guidelines, most Indian cities still have little if any separate cycling infrastructure (Mohan 2013; Mohan, Tiwari, and Bhalla 2017; Tiwari 2014; Tiwari et al. 2015). Delhi, Pune, Nanded, Gandhi Nagar, Bangalore, Chennai, Lucknow, Bhuvneshwar, Chandigarh, and a few other cities have some separate cycling infrastructure. Even in those cities, however, cycling facilities are few and far between. Many such facilities are substandard, narrow bicycle lanes that are frequently blocked by motor vehicles. No Indian city has an integrated network of connected facilities. In short, official government policies indicate awareness of the problem, and extensive plans and guidelines are now available, but so far the actual provision of cycling infrastructure in Indian cities is minimal and substandard.

The situation in China is very different but hard to pin down because it has been changing rapidly over time and varies greatly from city to city. From the 1970s to 1990s, all large and medium-size Chinese cities had extensive cycling infrastructure to handle the high volume of bicycle traffic during those decades. That generally included wide, separate bicycle lanes or paths on both sides of every major road, often with physical barriers to protect cyclists from motor vehicles, and sometimes with special bicycle traffic signals at important intersections. Even on local neighborhood streets without separate bicycle lanes, bicycling was convenient and safe because of the much lower levels of motorized traffic, particularly during the 1970s and 1980s. Cycling infrastructure was generally best in large and medium-size cities, worse in small cities, and worst or nonexistent in rural areas.

Starting around 1990, car ownership and use increased rapidly. The existing roadways were not nearly adequate for the sharply rising volume of motor vehicle traffic. Given limited space, most Chinese cities eliminated some of their extensive cycling infrastructure, either narrowing existing lanes and paths or removing them completely. In general, the highest-income cities with the most growth in motor vehicle use cut back cycling facilities the most. In addition, massive construction of new and widened roadways (including expressways and ring roads) led to the destruction of many traditional inner-city neighborhoods and disrupted the flow of local traffic, especially bicycle and pedestrian trips.

Since about 2008, a few large Chinese cities have been partially reversing their antibicycling policies implemented from about 1990 to 2010 (Pan 2011). Where cycling facilities had been removed, they are sometimes being reinstalled using more advanced design standards. So far, this encouraging new development is limited to only a few cities, but there are indications that more cities might follow their lead. Several recent studies report that the serious congestion, pollution, and safety problems caused by rapidly rising car and motorcycle use have led many city governments to slow down their accommodation of car use (Gao and Kenworthy 2017; Jiang et al. 2017; Yang and Zacharias 2016). As discussed in the next section, the recent bikesharing boom in China has increased cycling levels in some cities since about 2014 and has sparked new interest in government policies to facilitate cycling.

## Bikesharing

As noted by Cherry and Fishman (chapter 9, this volume), bikesharing has been the most prominent development in bicycling throughout the world in recent years. Because those authors examine bikesharing in great detail, this chapter focuses only on the specifics of bikesharing in Chinese and Indian cities.

Although bikesharing first evolved in Europe, it was adopted by various Chinese cities around 2008 to 2012. The first systems in China were based on docking stations (third-generation bikesharing), with approximately 1.5 million docking system bikes in 2018. Hangzhou has China's largest third-generation bikesharing system (85,000 bikes), which is also the largest in the world (see Cherry and Fishman, chapter 9, this volume). Around 2015, a new type of dockless bikesharing emerged. Now referred to as the fourth generation of bikesharing, such systems operate without docking stations and with so-called floating bikes. They are owned and operated by private firms, dominated in China by Ofo, Mobike, and Hellobike. By late 2018, only a few years after the introduction of floating bikesharing, there were an estimated 10 million to 12 million bikes in such dockless systems throughout China, in addition to about 1.5 million bikes in systems using docking stations (Li 2018; Zhou et al. 2018). As of late 2018, over 60% of the bikes in the world's bikesharing systems were in China (see Cherry and Fishman, chapter 9, this volume).

Bikesharing in China is changing rapidly, varies from city to city, and is difficult to quantify. The skyrocketing growth of floating bikesharing has led to an oversupply in many large cities, which in turn has caused serious financial problems and even bankruptcy for some operators. The oversupply of floating bikes has also led to dangerous or illegal bike parking as well as bike vandalism and abandonment. The haphazard parking of floating bikes clutters bike lanes and sidewalks (especially near metro stations), impedes the flow of cyclists as well as pedestrians, worsens safety problems, and detracts from the visual appeal of public spaces. In response, many Chinese cities have imposed limitations on the number of floating bikes allowed to operate and are increasingly requiring that bikes be parked at designated parking areas or hubs to be established and maintained by the private operators.

Whatever the drawbacks of the current oversupply of bikesharing bikes in China, the limited evidence available suggests that floating bikes in particular have been extremely popular with users, leading to increases in cycling levels in some cities between 2015 and 2018. Cherry and Fishman (chapter 9, this volume) discuss the pros and cons of floating bikesharing, but the experience in China suggests that the fourth generation of dockless bikesharing is more effective at increasing cycling levels than the third generation of docked bikes.

Bikesharing started much later in India than in China—only in the past few years—and on a much smaller scale than in China. As of late 2018, the largest dockless bikesharing system was in Pune, with 3,000 floating bikes, followed by Ahmedabad with 1,000 bikes and Kolkata, Chennai, and Bangalore, each with 500 floating bikes. The largest docking station systems were in Bhopal with 500 bikes, Mysuru with 450 bikes, and Karnal with 210 bikes. In total, there were about 7,000 bikes in India's bikesharing systems in 2018, and many of those are trials and not permanent. India has less than one-thousandth as many bikes in its bikesharing fleet as China. Thus, bikesharing has not yet had an impact on overall cycling levels in Indian cities. Given the almost complete lack of any separate cycling infrastructure in India, it seems unlikely that bikesharing will flourish as it has in China.

### Conclusions and Policy Recommendations

There are more bicycles and more cyclists in China and India than in the rest of the world combined. Thus, the future of cycling in China and India is important for the future of cycling worldwide. It is also important for global environmental issues such as climate change and the depletion of limited natural resources. Urban transport developments in China and India affect the entire world.

Most important, however, they directly affect the daily well-being of the residents of Chinese and Indian cities. Cycling policies, and urban transport policies in general, influence overall urban sustainability, including traffic safety, public health, roadway congestion, energy use, the environment, and the livability of Chinese and Indian cities. Improving the conditions for cycling benefits not only cyclists but other road users as well, indeed the entire population.

Ideally, the main policy priority for promoting safe cycling in both Chinese and Indian cities should be the provision of safe, well-designed, connected cycling facilities that form an integrated network enabling cycling throughout cities. On heavily traveled roads such as major arterials, bicycle lanes and paths should be physically protected from motor vehicle traffic—via curbs, bollards, planters, or other barriers. On less busy roads, unprotected bike lanes are the cheaper option, but, as experience shows, motorists in Chinese and Indian cities will infringe on cycling facilities unless there are physical barriers that prevent them from doing so. On-street bike lanes without protective barriers should be at least 2 m wide and, if possible, should include a diagonally striped buffer zone between bicycle and motor vehicle lanes. Because intersections are especially dangerous for cyclists, they should be reengineered to ensure safe crossings for cyclists. Such enhancements might involve advance green lights for cyclists, turn restrictions for motorists, and clear roadway markings on intersections indicating the safest routes for cyclists to take through them.

As Furth (chapter 5, this volume) discusses in detail, there are superb design guidelines for cycling facilities, based mainly on the Dutch standards (CROW 2017), which have been so successful at making cycling safer in the Netherlands than in any other country of the world (see Buehler and Pucher, chapter 2, this volume). Indian transport researchers have recently established similar design guidelines for cycling facilities specifically adapted to the situation in Indian cities (Tiwari 2014).

In addition to building a connected system of well-designed cycling facilities, Chinese and Indian cities should also consider traffic-calming their local neighborhood streets. By reducing legal speed limits to 30 km/h and enforcing them through a variety of roadway design modifications (speed humps and bumps, chicanes, diverters, raised intersections), hundreds of European cities have greatly improved pedestrian and cyclist safety, reduced traffic noise and air pollution, and increased the overall quality of life in such traffic-calmed neighborhoods (see Buehler and Pucher, chapter 2, this volume; Buehler et al. 2017).

Complementing such infrastructure measures, it is crucial to establish and strictly enforce traffic laws that give priority to pedestrians and cyclists, the most vulnerable road users. Such laws already exist in China but are rarely enforced (Gao and Kenworthy 2017; Jiang et al. 2017; Yang and Zacharias 2016; Yang, Wang, Zhou, and Zacharias 2014). Traffic laws in India do not prioritize pedestrians and cyclists despite numerous central government policy documents and plans that emphasize their need for greater protection from motor vehicles. The traffic chaos on most roads in Indian cities suggests that existing traffic laws are ignored by almost everyone anyway (Rathi 2017; Pucher et al. 2007).

The suggestions for expanding and improving cycling infrastructure mentioned earlier represent ideal solutions that might be difficult to implement for three reasons: lack of space, lack of money, and lack of public and political support. The obstacles to implementation are far greater in India than in China. Indian cities do not have China's history of extensive cycling facilities, and they have very few existing facilities to build on. In addition, most city governments are weak, ineffective, and lack funding. Roadway space in Indian cities is already overcrowded with motor vehicles. Reallocating some of that limited space from motor vehicles to bicycles and pedestrians would face public opposition from motor vehicle users and thus political opposition. Moreover, the rich and powerful in India do not ride bikes; nor does most of the middle class. They drive cars and, for obvious reasons, prefer investment in roads to investment in improved cycling and walking facilities they rarely if ever use.

Nevertheless, there is already some central government funding available in India for investment in improved cycling infrastructure. The central government could reallocate some of the massive funding currently dedicated to roadway expansion to improved cycling facilities, which would cost much less per kilometer than new roadways. As a starter, the central government could fund pilot projects in several Indian cities, focusing on improving the extent, quality, and connectivity of the facilities constructed in each city, making them models for other Indian cities to follow. As part of such pilot projects, the design of arterial roads could be revised to provide the space needed for cycling facilities. Currently, most arterials in Indian cities conform to a standard width for traffic lanes of 3.5 m, which could easily be reduced to 3.0 m without a loss of safety or traffic-carrying capacity. Combined with the narrowing of wide medians, that would free up space for separate cycling infrastructure on both sides of arterials, which is where such facilities are most needed. There is less likely to be political opposition to such pilot projects, as they would be introduced on an experimental basis and either made permanent if successful or terminated if unsuccessful.

Improving cycling infrastructure in Chinese cities would not be nearly as challenging as in India. As noted previously, Chinese cities had extensive systems of cycling infrastructure in the 1970s and 1980s, when over two-thirds of trips within cities were made by bicycle (Zhang, Shaheen, and Chen 2014). Thus, the task in Chinese cities is to restore those cycling facilities, at least in part, while updating them to meet the most advanced design guidelines. Just as in India, there is limited space on urban roads in China, which are already highly congested with motor vehicle traffic.

In some cities, there has been opposition from motorists to reallocating roadway space to cyclists, who now constitute a minority of the residents in most Chinese cities. Just as in India, the rich and powerful—who determine policy decisions at every government level in China—do not ride bicycles. Nevertheless, local government officials in large cities generally support improvements in cycling infrastructure, while officials in smaller cities prefer that roads be mainly for cars, trucks, and buses.

In addition to their past experience with cycling infrastructure, Chinese cities have five advantages over Indian cities. First, both the central and local governments in China have been investing massive amounts of funding in expanding and modernizing infrastructure, some of which is already being spent on the restoration and improvement of cycling facilities. Second, Chinese governments at every level are far more powerful, more effective, and much faster than Indian governments at getting large infrastructure projects of any kind completed. Third, the powerful central government provides explicit planning guidance that encourages and even requires local governments to build more and better infrastructure for cycling and walking. Fourth, several large Chinese cities have already limited registrations of new motor vehicles and have restricted use of those vehicles according to the last digit in license numbers (Chen and Zhao 2013; The Economist 2018; Yang, Ying, Ping, and Antung 2014). Motorists obviously oppose such restrictions, but local governments in China are more powerful than in India and not nearly as dependent on public support for implementing controversial policies. Fifth, China is seeking to be a world leader in sustainability, and reviving cycling through large investments in cycling infrastructure would enhance China's global reputation.

Finally, it is crucial to improve air quality in both Chinese and Indian cities to promote the public health of all residents but also to encourage cycling and walking, which are obviously discouraged by toxic air. In 2018, the World Health Organization (WHO 2018b) ranked 499 cities according to the severity of their pollution from ambient fine particulate matter (PM<2.5 μm). Of the world's 20 most polluted cities, 14 were in India and 1 was in China, with PM2.5 concentrations (μg per m$^3$) ranging from 173 to 105—17 to 11 times the WHO suggested maximum of 10. Of the cities ranked from twenty-first to fiftieth most polluted, 2 were in India and 21 were in China, with PM2.5 concentrations ranging from 93 to 77—nine to eight times the WHO standard. Thus, the air in Indian cities is the most toxic, but Chinese cities are not far behind. Air quality in Chinese cities has been improving in recent years, but it still has a long way to go before it meets WHO standards. It is beyond the scope of this chapter to examine strategies for reducing air pollution, but doing so is of crucial importance in both Indian and Chinese cities.

### References

Buehler, Ralph, John Pucher, Regine Gerike, and Thomas Götschi. 2017. Reducing Car Dependence in the Heart of Europe: Lessons from Germany, Austria, and Switzerland. *Transport Reviews* 37(1): 4–28.

Chen, Xiaojie, and Jinhua Zhao. 2013. Bidding to Drive: Car License Auction Policy in Shanghai and Its Public Acceptance. *Transport Policy* 27(5): 39–52.

Cherry, Chris, and Robert Cervero. 2007. Use Characteristics and Mode Choice Behavior of Electric Bike Users in China. *Transport Policy* 14(3): 247–257.

CROW. 2017. *Design Manual for Bicycle Traffic*. Ede, Netherlands: CROW.

Das, Ashis, Hallvard Gjerde, Saji S. Gopalan, and Per T. Normann. 2012. Alcohol, Drugs, and Road Traffic Crashes in India: A Systematic Review. *Traffic Injury Prevention* 13(6): 544–553.

The Economist. 2018. Why a Licence Plate Costs More Than a Car in Shanghai. *The Economist*, April 19. https://www.economist.com/china/2018/04/19/why-a-licence-plate-costs-more-than-a-car-in-shanghai.

Gao, Yuan, and Jeffrey Kenworthy. 2017. China. In *The Urban Transport Crisis in Emerging Economies*, edited by Pojani Dorina and Dominic Stead, 33–58. Cham, Switzerland: Springer.

Jiang, Baoguo, Song Liang, Zhong-Ren Peng, Haozhe Cong, Morgan Levy, Qu Cheng, Tianbing Wang, and Justin V. Remais. 2017. Transport and Public Health in China: The Road to a Healthy Future. *Lancet* 390(10104): 1781–1791.

Li, Yin. 2018. Beijing's Shared Bike Policy Analysis. *Transportation World* 30: 158–160.

Ministry of Home Affairs. 2011. *Census of India: Journey to Work by Mode of Travel*. Delhi: Ministry of Home Affairs, Office of the Registrar General and Census Commissioner.

Ministry of Road Transport and Highways. 2011–2018, annual. *Road Transport Year Book*. New Delhi: Ministry of Road Transport and Highways.

Ministry of Road Transport and Highways. 2017. *Road Accidents in India: 2016*. New Delhi: Ministry of Road Transport and Highways.

Ministry of Urban Development. 2006. *National Urban Transport Policy 2006*. New Delhi: Ministry of Urban Development.

Mohan, Dinesh. 2013. Moving around in Indian Cities. *Economic and Political Weekly: Review of Urban Affairs* 48: 40–47.

Mohan, Dinesh, Geetam Tiwari, and Kavi Bhalla. 2017. *Road Safety in India: Status Report 2016*. Delhi: Indian Institute of Technology, Transport Research and Traffic Injury Prevention Programme.

National Bureau of Statistics of China (NBS China). 2019. *China Statistical Yearbook*. Beijing: National Bureau of Statistics of China.

Pan, Haixiao. 2011. The Evolution of Bicycle Policy in Chinese Cities and Sustainability. *Urban Planning Forum* 4: 82–86.

Pendakur, V. Setty. 2011. Non-motorized Transport as Neglected Modes. In *Urban Transport in the Developing World*, edited by Harry Dimitriou and Ralph Gakenheimer, 203–231. Cheltenham, UK: Edward Elgar Publishing.

Pucher, John, Zhong-Ren Peng, Neha Mittal, Yi Zhu, and Nisha Korattyswaroopam. 2007. Urban Transport Trends and Policies in China and India: Impacts of Rapid Economic Growth. *Transport Reviews* 27(4): 379–410.

Rathi, Sujaya. 2017. India. In *The Urban Transport Crisis in Emerging Economies*, edited by Pojani Dorina and Dominic Stead, 81–106. Cham, Switzerland: Springer.

Singh, Sanjay K. 2005. Review of Urban Transportation in India. *Journal of Public Transportation* 8(1): 79–97.

Tiwari, Geetam. 2005. *Road Designs for Improving Traffic Flow: A Bicycle Master Plan for Delhi*. Delhi: Transport Research and Traffic Injury Prevention Programme, Indian Institute of Technology.

Tiwari, Geetam. 2014. *Planning and Design Guideline for Cycle Infrastructure*. Delhi: Transport Research and Traffic Injury Prevention Programme, Indian Institute of Technology.

Tiwari, Geetam, Anvita Arora, and Jain Himani. 2008. *Bicycling in Asia*. Delhi: Transport Research and Traffic Injury Prevention Programme, Indian Institute of Technology.

Tiwari, Geetam, and Nishant. 2018. *Travel to Work in India: Current Patterns and Future Concerns*. TRIPP-PR-18-01. New Delhi: Transport Research and Injury Prevention Programme, Indian Institute of Technology Delhi. http://tripp.iitd.ernet.in /assets/publication/WorkTravelReport.pdf.

Tiwari, Geetam, Alokeparna Sengupta, Dinesh Mohan, and Ruchi Varma. 2015. *Traffic and Safe Communities*. Delhi: Transport Research and Traffic Injury Prevention Programme, Indian Institute of Technology.

Union Internationale des Transports Publics (UITP). 2018. *World Metro Figures: Statistics Brief*. Brussels: Union Internationale des Transports Publics (International Association of Public Transport).

United Nations. 2018. *World Population Prospects: The 2017 Revision, Key Findings and Advance Tables*. New York: United Nations, Department of Economic and Social Affairs, Population Division.

Wang, Yu-ji, Long Cheng, Cen Feng, Xue-yu Chen, and Wei Wang. 2015. Area Difference in Travel Characteristics of Residents in Small and Medium-Sized Cities. *Traffic Planning* 3(2): 55–62 (in Chinese).

World Bank. 2018. *World Development Indicators: GDP per Capita Growth (Annual %) by Country*. Washington, DC: The World Bank.

World Health Organization (WHO). 2018a. *WHO Global Ambient Air Quality Database*. Geneva: World Health Organization.

World Health Organization (WHO). 2018b. *WHO Ambient Air Quality Guidelines*. Geneva: World Health Organization.

Yang, Jun, Liu Ying, Qin Ping, and Liu Antung. 2014. A Review of Beijing's Vehicle Registration Lottery: Short-Term Effects on Vehicle Growth and Fuel Consumption. *Energy Policy* 75(12): 157–166.

Yang, Ming, Quining Wang, Jinhua Zhou, and John Zacharias. 2014. The Rise and Decline of the Bicycle in Beijing. In *Proceedings of the 93rd Annual Meeting of the Transportation Research Board*. Washington, DC: Transportation Research Board.

Yang, Ming, and John Zacharias. 2016. Potential for Revival of the Bicycle in Beijing. *International Journal of Sustainable Transportation* 10(6): 517–527.

Yang, Qi, Hua-zhi Yuan, and Shu-min Feng. 2014. Travel Characteristics of Rural Residents under Different Economic Conditions. *Journal of Chang'an University* 1(1): 76–83 (in Chinese).

Zacharias, John. 2002. Bicycle in Shanghai: Movement Patterns, Cyclist Attitudes and the Impact of Traffic Separation. *Transport Reviews* 22(3): 309–322.

Zhang, Meng, Quan-xin Sun, Jin-chuan Chen, and Ji-fu Guo. 2008. Travel Behavior Analysis of Females in Beijing. *Journal of Transport Systems Engineering and Information Technology* 4(2): 19–26.

Zhang, Yua, Susan Shaheen, and Xingpeng Chen. 2014. Bicycle Evolution in China: From the 1900s to the Present. *International Journal of Sustainable Transportation* 8(5): 317–335.

Zhou, Zhangrong, Huixian Cai, Pin He, and Jinlong Cai. 2018. An Analysis of Shared Bicycles and Suggestions for Their Management. *Chinese Collective Economy* 33: 86–88.

# 16 Cycling in Latin America

Carlosfelipe Pardo and Daniel A. Rodriguez, with Lina Marcela Quiñones

Urban cycling was not a distinguishing characteristic of Latin American cities in the twentieth century. In contrast to many European and Asian cities, cycling has never been a dominant means of travel in Latin American cities (see in this volume Buehler and Pucher, chapter 2; Pucher et al., chapter 15; Koglin, te Brömmelstroet, and van Wee, chapter 18). Partly because of the absence of supportive policies, urban cycling was limited to a small but committed group of users, including low-income residents and members of other marginalized groups. Yet, in the past 20 years and despite large variations across cities, there has been a marked increase in cycling. In some cities, a deeply ingrained culture of exercise and recreational cycling was fertile ground for the growth of urban cycling. Cycling has recently gained importance, however, for different reasons in different cities. The factors supporting the growth of cycling include top-down political decisions, bottom-up grassroots organizing, changes in the structure of the regional economy, and the increased importance of environmental and social sustainability for policy-making.

Starting in the late 1990s, there was a significant change in the mindset of decision makers with the regional emergence of urban policies related to sustainable transport (World Bank 1996). Bogota, Colombia, was Latin America's pathbreaking leader in implementing many of these policies, creating what some have identified as a snowball effect for the rest of the region (Montero 2017). Changes occurred at various levels: funding and prioritizing policies for sustainable transport modes such as bicycles, pedestrians, and bus rapid transit over the automobile; controlling street parking and curtailing the encroachment of the automobile on spaces for bicycles and pedestrians; and dramatic expansion of separated bicycling infrastructure. These efforts to promote sustainable transport in Bogota earned the city regional and international recognition. Together, these dramatic policy advances and positive feedback provided confidence and support for local decision makers to continue prioritizing bicycle infrastructure, policies, and programming.

Although pioneering for the region, the case of Bogota is not representative of the experience of other cities in Latin America. Throughout the region, there is great variation in the growth of cycling, in the policies to encourage bicycle use, and in the social acceptability of bicycles as a means of transport. Considerable challenges to creating and sustaining a critical mass of cycling and supportive policies remain. In some cases, national and urban policies do not consistently support cycling. There is a need to strengthen institutional capacity for design, funding, implementation, and maintenance of cycling facilities. Culturally, Latin America continues to fight the stigma of the bicycle as a poor person's travel mode, even as subgroups may project an image of bicycling as avant-garde, fashionable, and healthy. Although there is increasing awareness among decision makers regarding the importance of bicycles for transport, this awareness is not shared by most residents of Latin American cities. Widespread concerns about bicycle safety continue to thwart the uptake of cycling for large segments of urban populations.

The aim of this chapter is to review the extent of urban cycling in Latin America, summarize trends, examine variations across cities and countries, and identify emerging concerns. Although data availability, reliability, and comparability are limitations, we use the best and most recent information available to review the significant gains in city cycling in Latin America, while highlighting challenges and opportunities that remain. We first discuss patterns of cycling use and demand across cities and countries. Then, we provide a summary of existing bicycling infrastructure followed by a description of policies, programs, and social supports in the region. We end the chapter with a forward-looking review of emerging opportunities. Together, we portray a heterogeneous landscape of urban cycling, with some cities and countries being exemplary leaders in the region, while others are not able or willing to prioritize cycling as a transport mode.

## Cycling Patterns among Latin American Cities and Countries

The share of all trips that are taken by bicycle varies among Latin American cities from about 1% to 5% (figure 16.1). The low bicycle mode share in

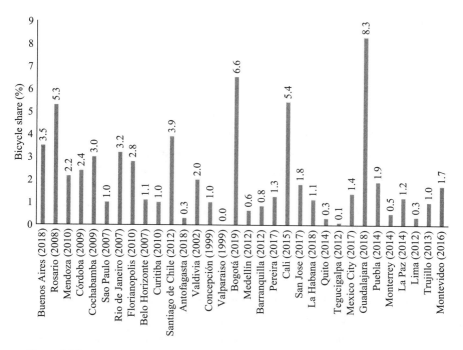

**Figure 16.1**

Recent bicycle mode share in Latin American cities. *Note*: The modal shares shown reflect travel for all trip purposes for which bicycles were the main mode of travel. Thus, trips that used a bicycle to access public transport are excluded. The modal shares are not entirely comparable because of differences in data collection methods, timing, and variable definitions. *Source*: Cities' household travel surveys.

Latin America is confirmed by a recent review of cycling statistics published by the Inter-American Development Bank (Ríos et al. 2015). Of the 20 cities included in the report, none had more than a 10% share of all trips by bicycle. More recently, the Development Bank of Latin America (CAF) surveyed a representative sample of residents in each of 11 cities of varying sizes to ask about the most frequently used travel mode, or combination of travel modes, to get from home to the main activity outside the home (work, school, or other) on a typical day (CAF 2017). The CAF data confirm the considerable variation across cities, with more than 10% of respondents in Fortaleza (Brazil) and Bogota using a bicycle for at least one leg of a trip, while in Caracas (Venezuela), Quito (Ecuador), and La Paz/El Alto (Bolivia) less than 1% use a bicycle. Although the CAF data exclude recreational trips, they are more representative of overall bicycle use, as they count bicycle use even when the bicycle is used as a feeder or a distributor for another travel mode. That also explains why the CAF data show higher cycling shares than other data for Latin America and suggests the underreported regional importance of bicycles as a first and last kilometer mode, connecting to longer-distance travel modes.

Colombia, Mexico, and Argentina are three Latin American countries that have made significant efforts to support urban cycling through major national-level initiatives. Other countries in the region have national policies that support cycling, but actual implementation has fallen short because of financial, administrative, or judicial constraints.

As shown for North America, Australia, and Britain (see in this volume Buehler and Pucher, chapter 2; Garrard, chapter 11), city cycling in Latin America also remains male dominated. Perhaps more remarkable is the variation across cities. At the high end, 40% of cycling trips in Montevideo, Uruguay, are made by women (Ríos et al. 2015). In Mexico City and the Argentinean cities of Cordoba and Rosario, women constitute more than 30% of all bicycle riders. By comparison, women account for 23% of bicycle trips in Bogota and 17% in Santiago, Chile. Medellin, Colombia, has the lowest share of bicycle trips by women (5%). Crime and traffic safety are especially important barriers to cycling for women (Ríos et al. 2015). Safety concerns also increase with age (see also Garrard et al., chapter 13, this volume), underscoring the fact that urban cycling in Latin America is largely perceived as a risky activity and thus only feasible for particular subgroups of the population, such as young men.

Climatic conditions are not a dominant barrier to city cycling in Latin America. Most cities having more than 100,000 residents are located in tropical and temperate areas where climate variation and cold weather are not obstacles to cycling. In regional surveys where obstacles to cycling are discussed, climate does not emerge as a major concern reported, although this may reflect the fact that road safety and personal security are of greater concern (Verma, López, and Pardo 2015).

By contrast, the impact of topography on cycling has not been well examined in Latin American cities. Rodriguez and Joo (2004) reported that US cyclists were about 2.5 times more sensitive to slope than pedestrians were. Thus hilly areas are likely to discourage cycling; the steeper the slope, the greater the deterrent to cycling. Larger cities in Latin America were founded in relatively flat valleys with convenient topography for laying out city streets and with good access to natural resources. Higher-income residents generally live in the more expensive neighborhoods near the centers of cities, while most lower-income residents live in peripheral, less accessible, and less expensive neighborhoods on or near hillsides.

## Leading with Bicycle Infrastructure

Many cities in Latin America have used the approach of "build it and they will come" toward bicycle infrastructure. They have focused on providing bicycle infrastructure to broaden the attractiveness and feasibility of using bicycles for daily travel (figure 16.2). Larger cities such as Bogota and the Brazilian cities of Sao Paulo and Rio de Janeiro lead in the provision of cycling infrastructure (Ríos et al. 2015). However, even the smaller Argentinean cities of Rosario and Cordoba each had about 100 km of cycling infrastructure.

Figure 16.3 shows that in just three years from 2015 to 2018 there have been important increases in bicycle infrastructure for some cities. Bogota's increase to 500 km is partly the result of having a very supportive, pro-bike mayor in office, but other cities also have increased their cycling infrastructure, including Lima (Peru), Buenos Aires (Argentina), Guadalajara (Mexico), and Medellin, among others. Exact comparisons of the extent of bicycling infrastructure are difficult because definitions of bicycling infrastructure can vary from city to city. Context also matters. A city with little or no separate cycling infrastructure but extensively traffic-calmed neighborhoods may create a more cyclist-friendly environment than a city that focuses exclusively on segregating cyclists from motorized traffic (see in this volume Furth, chapter 5; Heinen and Handy, chapter 7). Nevertheless, rapidly expanding cycling facilities indicates a particular infrastructure-focused approach in tandem with the rising prominence of bicycles for some cities in Latin America.

**Figure 16.2**
A wide variety of cycling infrastructure in (from left to right) Buenos Aires, Lima, and Quito. *Source*: Carlosfelipe Pardo.

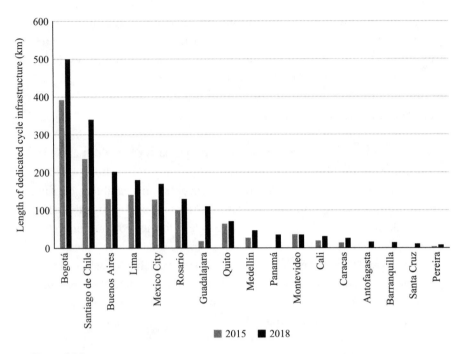

**Figure 16.3**
Extent of bicycle infrastructure in selected Latin American cities in 2015 and 2018.
*Source*: Rios et al. (2015) and data from individual cities.

Whether separated infrastructure for bicycles increases bicycle use is an empirical question (see in this volume Furth, chapter 5; Pucher, Parkin, and de Lanversin, chapter 17; Koglin, te Brömmelstroet, and van Wee, chapter 18). Figure 16.4 shows a scatterplot of bicycle share and bicycle infrastructure for a sample of 29 Latin American cities. The Pearson correlation is significant and positive ($r=0.36$; $p=0.056$). We also estimated a regression equation of bicycle infrastructure predicting bicycle mode share while adjusting for whether the city has 5 million residents or more. Every 10 km of additional separated infrastructure for bicycles was associated with a 0.14% higher bicycle mode share ($p<0.01$). This analysis is based on cross-sectional data. It would be preferable to do such an analysis with longitudinal data measuring changes over time, but such data are not currently available.

Despite the importance of bicycle infrastructure, most cities remain unwilling to consider the removal of motor vehicle lanes to provide more space for bicycles. Although this problem is not unique to Latin America, it

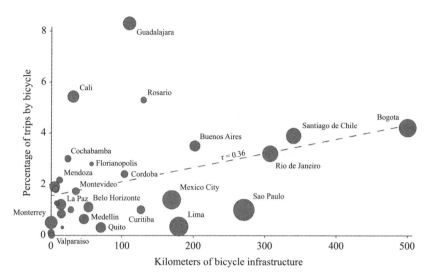

**Figure 16.4**
Statistical relationship between bicycle infrastructure and bicycle mode share for 29 Latin American cities. *Note*: The size of each city circle is proportional to the city's population. Some city names were excluded for improved legibility. As noted in the figure, there is a positive correlation ($r$=+0.36) between bicycle mode share of trips and kilometers of cycling infrastructure in the cities. *Source*: Based on data collected by the authors from government agencies, nonprofit organizations, and cycling advocacy groups in each of the cities.

severely constrains the design possibilities for bicycle infrastructure, particularly in areas where street and streetside space are limited. It also creates a vicious cycle in that limited space leads to poor cycling infrastructure, which begets few users, further diminishing support for bicycling infrastructure.

The definition of bicycle infrastructure has also expanded in the last decade to include bicycle parking and integration with other travel modes (see Buehler, Heinen, and Nakamura, chapter 6, this volume). In Maua, in southeastern Sao Paulo, a citizen-led initiative shortly after 2000 resulted in almost 2,000 bicycle parking spaces and a user cooperative that provides other services, such as bike repair and cleaning (Alcantara et al. 2009). Bogota is exemplary in its integration of cycling with public transport, with more than 4,000 bike parking spaces located directly adjacent to its bus rapid transit stations. In the past three years alone, Mexico City built 880 covered, secure bicycle parking spaces integrated with its metro system

(Gobierno de la Ciudad de México 2017). Although the level of physical integration between bicycles and other travel modes is improving, fare integration varies. For example, in 2018, Santiago's Metro Company created a parking program enabling commuters to securely park bicycles in covered areas of metro stations. However, bicycle parking fees are relatively high and are in addition to the metro fare (Metro de Santiago 2018).

In addition to bicycle parking at public transport stations, many cities have begun implementing street-level bicycle parking facilities. The inverted U bike rack is now becoming the norm. For example, Buenos Aires, Santiago, and Lima have installed them. Other cities, however, such as Bogota, have not been able to modify their street design code to accommodate the U bike rack and continue to use less convenient and secure designs. In 2017, one district in Lima (San Isidro) placed "service totems" with various bike repair tools and air pumps along the bikeway, the first time this had been implemented in the region.

Finally, bikesharing (see Fishman and Shaheen, chapter 10, this volume) has had limited implementation in Latin America. As of 2014, there were only about 12,000 bikesharing bicycles throughout Latin America (Ríos et al. 2015). In 2018, Mexico City alone had 6,800 and had an ambitious plan to add more. Despite these impressive numbers, almost all cities that have launched bikesharing systems have had substantial operating changes. Many of them have closed temporarily or permanently. For example, in the past two years, Santiago, Medellin, Rio, and Quito changed their main bikesharing operator altogether. Myriad bikesharing systems have come and gone in less than one or two years in smaller cities—including six Colombian cities that received a full subsidy for capital costs from Colombia's Ministry of Transport. Dockless bikesharing systems are becoming increasingly popular because they involve lower financial commitment and risk to local governments, although the financial sustainability of the dockless bikesharing business model remains uncertain. Mexico City and Santiago have each allowed at least one dockless operator. Somewhat surprisingly, Bogota, normally seen as the foremost example of cycling in the region, has not yet successfully implemented a citywide bikesharing system despite various failed attempts. (For a detailed account of the first pilot project in Bogota, see Pardo and Moreno Luna 2011.)

## The Emerging Importance of Cycling Policies
## and Programming in Latin America

Bicycle infrastructure design guidelines are particularly important policy attempts at ensuring that infrastructure is of consistent quality within each city, or in some cases within entire countries. They also encourage some level of uniformity in facility design expertise and management. Salient examples in the region include:

- Chile created its own cycling design guide (Ministerio de Vivienda y Urbanismo 2015), and Bolivia published a guide for cycling infrastructure design (Wiskot 2015).
- In Mexico, a local nonprofit (Instituto de Políticas para el Transporte y el Desarrollo, ITDP) published a five-volume design manual for cycling called *Ciclociudades* (ITDP México and I-CE 2011).
- Lima published the fourth version of its cycling design guide (Municipalidad Metropolitana de Lima 2017).
- An increasing number of cities have adopted bicycle plans. Medellin, Buenos Aires, Bogota, and Bucaramanga have published or updated bicycle plans and strategies for their cities or metropolitan areas (Área Metropolitana del Valle de Aburrá, Elejalde López, and Martínez Ruíz 2015; Área Metropolitana del Valle de Aburrá 2015; Bisiau 2017; Secretaría de Movilidad de Bogotá 2018).

In addition to bicycle infrastructure design, perhaps the most interesting Latin American contribution to cycling has been the growing support for and importance of cycling policies and programs (see Heinen and Handy, chapter 7, this volume). Led by public institutions at the national and city levels, and with significant support from nongovernmental organizations and informal groups, many policy and programming changes have been implemented in the last decade. Colombia's example is notable, where the following laws and regulations have been approved recently:

- In 2014, the four-year government plan allowed national-level financing for cycling improvements, established deadlines for municipalities to clarify the legal status of bicycle taxi operations, and increased support for local cycling policies.
- A national "bicycle law" was passed in 2016 but is still awaiting additional rulemaking before full implementation (Congreso de la República

de Colombia 2016). The law increases what is deemed a "safe passing distance" between motorized vehicles and bicycles. It also requires government agencies to provide one half day of paid leave for every month of riding a bicycle to work as an employee incentive.

- In 2017, the "e-bike law" clarified the specific conditions under which electric bicycles can use bicycle facilities: only those electric pedal-assist bicycles limited by design to a maximum speed of 25 km/h are allowed on bicycle facilities. All others are considered motorcycles and must operate according to motorcycle regulations (Ministerio de Transporte de Colombia 2017).

- National mandatory design guidelines published in 2016 (Ministerio de Transporte de Colombia 2016; Ministerio de Transporte de Colombia 2018b).

- A preliminary regulatory framework to allow bicycle taxi operations throughout the country (Ministerio de Transporte de Colombia 2018a).

To quantify the prevalence of policies and programs to support cycling, we conducted a survey of 30 cities. Almost 80% of cities reported that they had policies in place to support cycling (figure 16.5). An example of policy

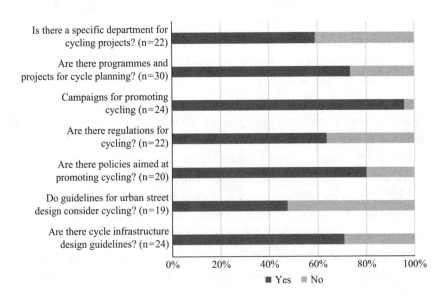

**Figure 16.5**
Policies and programs to support cycling-inclusive planning in a sample of Latin American cities. *Source*: Survey research conducted by the authors.

changes includes allowing bicycles inside public transport vehicles, such as in larger cities including Santiago, Mexico City, Medellin, and Bogota. Some cities allow folding bicycles only, while others allow full-sized bicycles, although with some restrictions to certain times of the day and days of the week. Similarly, several cities have requirements for the provision of bicycle parking for new development or redevelopment projects. In general, buildings must provide roughly one bicycle parking space for every 10 car parking spaces (Banco Interamericano de Desarrollo 2013).

Institutional staffing dedicated to cycling is another example of city-level policies supporting bicycling. More than 60% of cities have a specific department in charge of bicycling projects (figure 16.5). Since 1999, Bogota has had a professional staff focusing specifically on bicycling (Pardo 2013), and it currently has a "cycling manager" who coordinates policies and programs with other city agencies involved with cycling. Other cities, such as Lima, have created a "cycling office" (CicloLima) to advocate for and implement cycling policies. Buenos Aires has at least three dedicated staff members for implementing cycling policies. Although having a dedicated staff and some funding are positive outcomes, there is still an important need to increase the visibility and institutional presence of cycling within local governments remains. Cycling has yet to be fully integrated into urban planning, policies, and funding at the local government level. There are few cycling planners relative to other transport modes, and cycling is notoriously underfunded.

An iconic Latin American effort involving cycling programming and promotion is the car-free Sunday (commonly known as "ciclovia," "ciclorecreovia," or "open streets"). Car-free Sundays temporarily restrict the use of city streets to nonmotorized users, primarily cyclists and pedestrians. The program is often complemented by promotional campaigns and site-specific activities (in parks and plazas, for example). Car-free Sundays began in 1974 in Bogota with a citizen-led initiative. In recent years, almost every Sunday from 7 a.m. to 2 p.m., 120 km of city streets have been closed to motor vehicle traffic for the "ciclovia." More than 1 million users enjoy the benefits of the ciclovia, which has been successfully replicated in several dozen cities in the region, with the ciclovia in Guadalajara being noteworthy for its extent and success. Although their emphasis is on recreational cycling, ciclovias have raised awareness, changed attitudes, legitimized the bicycle as a way of getting around, and created incentives for further cycling. There is considerable evidence of the multiple health, personal, and community benefits of these programs (Sarmiento et al. 2017).

An important but less acknowledged recent change in Latin America is the upsurge in citizen groups and advocacy for bicycling (see Pucher et al., chapter 20, this volume). Citizen groups that promote cycling through organized rides, advocates that encourage cycling policies in city council meetings, and groups that use tactical urbanism tools to work with community members are now common in many cities. This burgeoning social support has been accompanied by more formally established groups of "bicycle networks," such as Bicirred Mexico. The World Bicycle Forum has been a trigger for the emergence of strong social advocacy. The Forum is a no-cost event created in Porto Alegre as a reaction to an infamous incident on February 25, 2011, when a car driver ran over dozens of cyclists protesting for better conditions in a Critical Mass ride. Since then, the World Bicycle Forum has been held in Porto Alegre (2012 and 2013), Curitiba (2014), Medellin (2015), Santiago (2016), Mexico City (2017), Lima (2018), and Quito (2019). The Forum has greatly strengthened and organized cycling advocacy, inspiring cyclists to advocate for improved cycling conditions and cycling-friendly policies. Despite the use of the word *World* in the title, the meeting has only been held in Latin American cities, while contenders from overseas have lost their bid in the committee selection process every year.

Finally, a rising phenomenon is the linkage of cycling to private-sector business opportunities. Company initiatives for employees to cycle to work are being supported by local nonprofits. Tourists use bikesharing systems or rented bicycles for sightseeing while enjoying the increased safety afforded by separated bicycle facilities. Even mobile applications for bicycle-based small freight deliveries have focused the private sector's attention on bicycles. Gamified applications, such as Biko and Kappo, have also found a niche for cyclists who want to boast about their riding while redeeming prizes in return and generating social interest among their networks. In some cases, car parking lot operators have increased revenues by including bike parking without sacrificing much car parking space, since a bicycle requires only a fraction of the space required by a car (Pardo, Caviedes, and Calderón Peña 2013).

## Continuing and Emerging Challenges for Cycling in Latin America

Despite the major steps taken to improve cycling conditions in Latin American cities, cycling faces many challenges. First, bicycle theft is a serious problem. Parked bicycles are not the only targets; while riding, cyclists have been assaulted by criminals attempting to steal a bicycle or even to rob the cyclist. Because records of bicycle theft are poor, reliable estimates of the number of stolen bikes are not available. The public perception, however, is that bicycle theft is a serious problem that deters cycling. Most Latin American cities have done little to prevent theft. Identifying stolen bicycles is challenging because bicycles are easy to disassemble and parts are interchangeable and easily resold. Bogota attempted to deal with the bicycle theft challenge by launching a voluntary bicycle registration system in December 2018 and targeting shops known to sell stolen bicycles and bicycle parts.

Road safety is a critical second challenge in Latin America. Cyclists continue to be among the most vulnerable road users in Latin America (see also Elvik, chapter 4, this volume). Together with pedestrians and motorcyclists, they represent the majority of traffic fatalities in Latin American cities. The police and courts often assign the blame for crashes to cyclists instead of motorists, citing the failure of cyclists to wear protective helmets, clothing, and accessories. To be politically and publicly acceptable, it is important to find safety solutions that preserve the rights of all road users while recognizing the special vulnerability of cyclists to traffic injuries. Traffic-calming opportunities in Latin American cities have emerged as an important approach to improve safety for all road users, including cyclists, while reducing noise and air pollution in residential areas. The experience of Sao Paulo is encouraging. The introduction of traffic calming reduced the number of traffic crashes, fatalities, and serious injuries (Biderman and Lima 2019). Mexico City has recently implemented similar speed-reduction measures in some of its neighborhoods.

A third challenge is the lack of reliable and comparable data. Although an increasing amount of data on cycling in Latin American cities have been collected in recent years, good data on cycling are less available than for motorized transport modes. A regional survey showed that data on cycling levels, policies, and infrastructure were only available for about one-third of cities sampled (Ríos et al. 2015). This confirms our personal experience and our research colleagues' observations that staff from municipal and national governments in Latin America sometimes claim that cycling is not considered a transport mode and therefore there is no need to include it in travel surveys. Similarly, cycling is often excluded from household travel surveys, even when trips cross the boundaries of zones used in analyses. Compounding this challenge are differences among countries and cities in their definitions of trips, how cycling trips are recorded, and the definition and measurement of cycling infrastructure. Some cities include streets shared with cars as part of their cycling network, even though they do not include any separate provisions for cycling, such as bike lanes. That overstates the total length of their cycling infrastructure (see Furth, chapter 5, this volume, for definitions and examples of different kinds of cycling infrastructure).

A fourth challenge is institutional and concerns the limited capacity that local, regional, and national governments have to support bicycling. Competition for scarce funds is high, while the cost of planning, designing, and implementing infrastructure, policies, and programs continues to increase. Because of the patchwork way in which cycling facilities have been built and the limited awareness of the importance of an integrated cycling network, cycling facilities are incomplete and disconnected, even in cities with hundreds of kilometers of facilities, such as Bogota, Santiago, and Rio de Janeiro. Prioritizing connectivity would create significant network benefits. The needed connectivity of cycling routes cannot come exclusively from separated cycling facilities. A truly comprehensive bicycling network can only be achieved through a carefully planned and integrated combination of separated and shared facilities, including traffic-calmed residential neighborhoods. Unfortunately, almost no cities in Latin America have this crucial comprehensive view of integrated cycling networks.

A fifth and most recent challenge arises from the emergence of new vehicles (such as electric scooters) that contest spaces originally designed for cycling. Similar to the Global North, private and shared electric scooters are becoming increasingly common in higher-income, central areas of Latin American cities. These scooters frequently use the limited bicycle infrastructure that exists, but the infrastructure was not designed with these new vehicles in mind. Their smaller wheels, lower ground clearance, varying speeds, and different operating characteristics set up inevitable conflicts with bicyclists. With the limited institutional capacity noted earlier, competition for currently designated bicycle infrastructure space is likely to grow (see Cherry and Fishman, chapter 9, this volume).

The final challenge has to do with cycling and equity. Cyclists continue to be stigmatized as a fringe mode undeserving of road space. The lower-class stigma of cycling arises from the view that most cyclists are low-income individuals without enough money to use public transport, cars, or motorcycles. Although there is important variation across cities, many regular bicycle users are indeed low-income residents (see figure 16.6). The general image

**Figure 16.6**
Families cycling together in Chia (left) and Bogota (right). *Source*: Carlosfelipe Pardo.

of cycling has been changing, however, with a culture of "bicycle chic" and hipster cyclists from the Global North also reaching some Latin American cities. The location of bicycle infrastructure has tended to prioritize middle- and high-income areas (for a detailed discussion of this issue, see Martens, Golub, and Hamre, chapter 14, this volume). For example, in Bogota, one study found that the distribution of the bicycle lane network differed greatly according to income levels, with lower-income areas having fewer bicycle lanes (Parra et al. 2018). Protected, physically separated cycling facilities, in particular, are rare in low-income communities. The disparities in cycling facilities are reinforced by the lack of bikesharing facilities in lower-income neighborhoods in Brazilian and Chilean cities (Duran et al. 2018; Duarte 2016). The many hills and lack of paved streets in lower-income areas also make cycling less feasible. In addition, most cycling infrastructure is concentrated in city centers, which are the overall focus of trip destinations in the city. Regardless of the specific reasons, there is unquestionably a disparity in cycling infrastructure between high- and low-income neighborhoods that deserves additional research and policy attention.

## Conclusions and Outlook

A retrospective look at city cycling in Latin America provides encouraging news. The last two decades have witnessed positive change: expanded infrastructure and innovative policies and programs have increased the legitimacy of cycling as a means of travel, reduced the stigma of cycling, and increased cycling levels. Riding a bicycle in many Latin American cities 20 years ago was a different experience than it is today. The stories of cities such as Buenos Aires, Bogota, and Mexico City and their large investments in cycling infrastructure and programs have created useful examples for many other cities in the region. Adapting, improving, and replicating these efforts has resulted in important policy transfer and a healthy competition to be the most sustainable, bicycle-friendly city. Although progress has been uneven, national and regional policies, political will, and strengthened institutional capacity suggest significant continued progress in the near future.

Despite impressive advancement, much more remains to be done. Policy continuity has been pivotal to the success of some strategies. Building on past successes will require considerable expansion in professional staff trained in cycling planning and engineering. This will provide greater continuity in cycling policies that might otherwise be rescinded by changing governments. Data collection needs to be improved in order to better monitor performance, to evaluate the effectiveness of programs and policies, and to enable comparisons across transport modes. Data are particularly important in pinpointing areas of intervention to address the needs of those who are interested in cycling but concerned about cycling safety. As emphasized by Martens, Golub, and Hamre (chapter 14, this volume), equity remains an important concern requiring further research and policy initiatives focused specifically on disadvantaged groups—especially investments in cycling infrastructure and programs.

Future prospects for cycling in Latin American cities are promising. Past and current trends suggest that bicycles will be increasingly integrated into the mobility ecosystem of Latin American cities. Bicycles can play many roles for multiple users: as a mode to overcome first- and last-kilometer barriers for public transport, as a main mode of travel, as a recreational mode, and as a complement to app-based car ride hailing. As elsewhere in the world, bicycling in Latin America continues to emerge as an effective way to mitigate climate change while providing myriad personal benefits. This is of particular relevance for those dense Latin American cities where trip distances are short enough to cover by bicycle.

Technological innovations will continue to change the landscape of bicycles, further blurring the boundaries with scooters and motorcycles, by increasing the availability (and reducing the price) of electric bicycles and the increasingly flexible possibilities of dockless bicycles and related app-mediated services. This raises the possibility of more transport modes competing for the same already-limited urban spaces. On balance, however, we are optimistic that these innovations will increase cycling in Latin American cities by enhancing the mobility ecosystem of cities, strengthening coalitions to improve the infrastructure for travel modes different from the automobile, and improving their integration with cycling options.

## Acknowledgments

Daniel A. Rodriguez's contribution was supported by the SALURBAL project. The Salud Urbana en América Latina (SALURBAL)/Urban Health in Latin America project is funded by the Wellcome Trust (grant 205177/Z/16/Z).

## References

Alcantara, Adilson, Ana Maria Destito, Jonas Hagen, and Juliana de Campos Silva. 2009. *Manual de Bicicletários: Modelo Ascobike Mauá*. Rio de Janeiro: Institute for Transportation and Development Policy.

Área Metropolitana del Valle de Aburrá. 2015. *Manual de Ciclo-infraestructura Metropolitana (Plan Maestro Metropolitano de La Bicicleta Del Valle de Aburrá)*. Medellin: Área Metropolitana del Valle de Aburrá.

Área Metropolitana del Valle de Aburrá, Hernán Dario Elejalde López, and Juan Esteban Martínez Ruíz. 2015. *Plan Maestro Metropolitano de La Bicicleta 2030*. Medellin: Área Metropolitana del Valle de Aburrá.

Banco Interamericano de Desarrollo. 2013. *Guía Práctica: Políticas de Estacionamientos y Reducción de Congestión En América Latina*, edited by Despacio and ITDP. Washington, DC: Banco Interamericano de Desarrollo.

Biderman, Ciro, and Lycia Lima. 2019. *Unraveling the Speed Limit Reduction-Traffic Crashes Relationship: Is It the Law or the Enforcement? The Case of São Paulo, Brazil*. Working Paper. São Paulo, Brazil: Fundacao Getulio Vargas. http://www.cepesp.io /publicacoes/unraveling-the-speed-limit-reduction-traffic-crashes-relationship-is-it -the-law-or-the-enforcement-the-case-of-sao-paulo-brazil/.

Bisiau, Paula. 2017. Movilidad Segura y Sostenible En Buenos Aires. Presentation at Mobilize Santiago. https://mobilizesummit.org/wp-content/uploads/sites/2/2017/07 /Bisiau_Movilidad-Segura-y-Sostenible-en-Bueno-Aires_Espanol.pdf.

CAF (Banco de Desarrollo de America Latina). 2017. Encuesta CAF 2016—Methodological Appendix. Bogota: CAF. https://www.caf.com/media/29899/informe-metodologico-ecaf -2016.pdf.

Congreso de la República de Colombia. 2016. *Ley 1811 de 2016*. *Por La Cual Se Otorgan Incentivos Para Promover El Uso de La Bicicleta En El Territorio Nacional u Se Modifica El Código Nacional Tránsito*. Bogota: Congreso de la República de Colombia.

Duarte, F. 2016. Disassembling Bike-Sharing Systems: Surveillance, Advertising, and the Social Inequalities of a Global Technological Assemblage. *Journal of Urban Technology* 23(2): 103–115. https://doi.org/10.1080/10630732.2015.1102421.

Duran, A. C., E. Anaya-Boig, J. D. Shake, L. M. T. Garcia, L. F. M. de Rezende, and T. H. de Sa. 2018. Bicycle-Sharing System Socio-spatial Inequalities in Brazil. *Journal of Transport and Health* 8: 262–270. https://doi.org/10.1016/j.jth.2017.12.011.

Gobierno de la Ciudad de México. 2017. Biciestacionamientos CDMX. Retrieved on July 4, 2018, from https://www.cdmx.gob.mx/vive-cdmx/post/biciestacionamientos -cdmx.

Instituto de Políticas para el Transporte y el Desarrollo (ITDP) México and Interface for Cycling Expertise (I-CE). 2011. *Manual Integral de Movilidad Ciclista Para Ciudades Mexicanas: Ciclociudades*. Mexico City: Instituto de Políticas para el Transporte y el Desarrollo.

Metro de Santiago. 2018. *Línea Cero*. Santiago: Metro de Santiago.

Ministerio de Transporte de Colombia. 2016. *Guía de Ciclo-infraestructura Para Ciudades Colombianas*, edited by Carlosfelipe Pardo and Alfonso Sanz. Bogota: Ministerio de Transporte de Colombia.

Ministerio de Transporte de Colombia. 2017. *Resolución 160 de 2017*. Bogota: Ministerio de Transporte de Colombia.

Ministerio de Transporte de Colombia. 2018a. *Resolución 3256 de 2018*. Bogota: Ministerio de Transporte de Colombia.

Ministerio de Transporte de Colombia. 2018b. *Resolución 3258 de 2018, "Por La Cual Se Adopta La Guía de Ciclo-infraestructura Para Ciudades Colombianas."* Bogota: Ministerio de Transporte de Colombia.

Ministerio de Vivienda y Urbanismo. 2015. *Vialidad Ciclo-inclusiva*. Santiago: Ministerio de Vivienda y Urbanismo.

Montero, Sergio. 2017. Study Tours and Inter-city Policy Learning: Mobilizing Bogotá's Transportation Policies in Guadalajara. *Environment and Planning A: Economy and Space* 49(2): 332–350. https://doi.org/10.1177/0308518X16669353.

Municipalidad Metropolitana de Lima. 2017. *Manual de Criterios de Diseño de Infraestructura Ciclo-inclusiva y Guía de Circulación Del Ciclista*, edited by Patricia Calderón, Carlosfelipe Pardo, and Juan José Arrué. Lima: Municipalidad Metropolitana de Lima.

Pardo, Carlosfelipe. 2013. Bogotá's Non-motorised Transport Policy 1998–2012: The Challenge of Being an Example. In *Aspects of Active Travel: How to Encourage People to Walk or Cycle in Urban Areas*, edited by Werner Gronau, Wolfgang Fischer, and Robert Pressl, 49–65. Mannheim: Verlag MetaGISInfosysteme.

Pardo, Carlosfelipe, Álvaro Caviedes, and Patricia Calderón Peña. 2013. *Estacionamientos Para Bicicletas: Guía de Elección, Servicio, Integración y Reducción de Emisiones*, edited by Despacio and ITDP. Bogota: Despacio and ITDP.

Pardo, Carlosfelipe, and Carlos Augusto Moreno Luna. 2011. *Recomendaciones a La Prueba Piloto Del Sistema "BiciBog."* Bogota: Despacio and ITDP.

Parra, Diana C., Luis F. Gomez, Jose D. Pinzon, Ross C. Brownson, and Christopher Millett. 2018. Equity in Cycle Lane Networks: Examination of the Distribution of the Cycle Lane Network by Socioeconomic Index in Bogotá, Colombia. *Cities and Health* 2(1): 60–68. https://doi.org/10.1080/23748834.2018.1507068.

Ríos, Ramiro Alberto, Alejandro Taddia, Carlosfelipe Pardo, and Natalia Lleras. 2015. *Ciclo-inclusión En América Latina y El Caribe: Guía Para Impulsar El Uso de La Bicicleta.* Washington, DC: Banco Interamericano de Desarrollo.

Rodríguez, Daniel A., and Joonwon Joo. 2004. The Relationship between Nonmotorized Mode Choice and the Local Physical Environment. *Transportation Research Part D: Transport and Environment* 9(2): 151–173. https://doi.org/10.1016/j.trd.2003.11.001.

Sarmiento, Olga L., Adriana Díaz del Castillo, Camilo A. Triana, María José Acevedo, Silvia A. Gonzalez, and Michael Pratt. 2017. Reclaiming the Streets for People: Insights from Ciclovías Recreativas in Latin America. *Preventive Medicine* 103: S34–S40. https://doi.org/10.1016/j.ypmed.2016.07.028.

Secretaría de Movilidad de Bogotá. 2018. *Plan Bici Bogotá.* Bogota: Secretaría de Movilidad de Bogotá.

Verma, Philip, Jose Segundo López, and Carlosfelipe Pardo. 2015. *Bogotá 2014 Bicycle Account.* Bogota: Despacio. Retrieved on April 13, 2020, from https://www.despacio.org/portfolio/bogota-bicycle-account-2014/.

Wiskot, Alexa. 2015. *Manual de Diseño de Calles Para Las Ciudades Bolivianas.* La Paz: Cooperacion Suiza en Bolivia.

World Bank. 1996. *Sustainable Transport: Priorities for Policy Reform.* Washington, DC: International Bank for Reconstruction and Development, World Bank.

# 17 Cycling in New York, London, and Paris

John Pucher, John Parkin, and Emmanuel de Lanversin

New York, London, and Paris are three of the largest and most important cities in the world, serving as financial, political, cultural, and media centers. In contrast to many cities of northern and western Europe, neither London nor Paris has a history of cycling for daily travel. Throughout most of the twentieth century, New York was typical of North American cities in relegating cycling to a marginal role. All three cities instead have relied mainly on their extensive public transport systems, combined with walking to access rail and bus stops and to travel short distances.

Over the past few decades, however, New York, London, and Paris have each implemented transport policies and infrastructure aimed at making cycling a feasible, convenient, and safe way to travel. The result has been a large increase in cycling levels and a rise in cycling's share of total trips. The promising growth of cycling in New York, London, and Paris has made headlines across the globe and inspired other cities to follow their example.

This chapter examines how the three cities have overcome the challenges facing cycling in large cities. For example, heavy traffic on noisy, congested streets makes cycling more stressful, more intimidating, and riskier than in smaller cities. Moreover, the geographic extent of large cities generates many long trips that are difficult to make by bicycle, especially for the time-sensitive commute to work. High trip volumes in narrow travel corridors create intense competition for space among cyclists, motor vehicles, and pedestrians. Space constraints make it more expensive and politically difficult to install separate cycling facilities, although they are needed in such situations.

The population density and geographic size of large cities generally favor walking for short trips, combined with public transport for longer trips. New York, London, and Paris all have extensive public transport systems, but they are often overcrowded. As public transport reaches capacity, cycling can help relieve that pressure, especially in the most crowded, central areas of public transport systems, where many trips are short enough to be taken by bicycle. Moreover, in the outer areas, cycling can provide a cheap and effective way to expand the service area of metro and suburban rail systems beyond the walking distance to stations. Public transport and cycling have the potential to work together especially well in large cities.

In this chapter, we examine recent trends in cycling levels and policies in each city and draw lessons for other large cities seeking to promote cycling.

## Demographics and Climate

In 2018, New York's metropolitan area (20.3 million residents) had a population 42% greater than that of the London metropolitan area (14.3 million) and 66% greater than that of the Paris metropolitan area (12.2 million) (see table 17.1). City populations are smaller because they are based on the boundaries of the central political jurisdiction. London (8.9 million) and New York City (8.6 million) had almost the same populations in 2018, both about four times greater than the population of Paris (2.2 million). The much smaller population of Paris is the result of its much smaller land area: only 105 km$^2$ versus 1,570 km$^2$ for London and 791 km$^2$ for New York. Indeed, Paris includes less than one-fifth of its metropolitan area's population. The smaller size of Paris distorts some of the comparisons among the cities—accounting, for example, for Paris's much higher population density. Conversely, the large area included in the central political jurisdiction of London (misleadingly designated as Greater London) accounts for its low population density, which is half of New York's and one-fourth of Paris's.

As documented by national travel surveys, cycling rates tend to be higher among students and car-free households (USDOT 2019; Buehler, Pucher, and Bauman 2020). New York (7%) and Paris (6%) both have higher percentages of university students than London (4%). Moreover, New York (54%) and Paris (55%) have a majority of households without cars, compared to 43% of London's households, which is probably related to London's lower population density. These two factors may provide a greater incentive to cycle in New York and Paris than in London.

The populations of New York and London grew considerably from 1990 to 2018: by 18% in New York and by 29% in London (vs. 4% in Paris). That

**Table 17.1**

Demographic and climatic characteristics of New York, London, and Paris in 2018

| | Population (1,000s) | | Population per km$^2$ | University students (%) | Car-free households (%) | Annual precipitation (cm) | Annual days ≤ 0° C | Annual days ≥ 30° C |
|---|---|---|---|---|---|---|---|---|
| | City | Metro | | | | | | |
| New York | 8,623 | 20,321 | 11,001 | 7 | 54 | 127 | 81 | 16 |
| London | 8,800 | 14,187 | 5,609 | 4 | 43 | 58 | 29 | 5 |
| Paris | 2,243 | 12,194 | 21,280 | 9 | 64 | 63 | 25 | 11 |

*Note:* The metro areas listed in this table refer to the metropolitan areas functionally related to the central city in terms of housing, jobs, shopping, and travel patterns.

*Sources:* Based on information collected by the authors for each city.

growth reflects an urban revival not only in their centers but also in many formerly deteriorating, low-income neighborhoods within a few kilometers of the city center. For example, there are many trendy neighborhoods in northern Brooklyn and western Queens, just across the East River from millions of jobs in Manhattan and thus within an easy bike commute. The outward-spreading gentrification of such neighborhoods has been driven by an influx of young, affluent professionals keen on an urban lifestyle. That demographic trend obviously favors growth in cycling. Indeed, cycling has become an iconic indicator of gentrification in New York neighborhoods.

Climate can influence cycling levels, with heavy precipitation as well as extreme heat or cold discouraging cycling (Meng et al. 2016; Pucher, Buehler, and Seinen 2011). Of the three cities, London has the least average annual precipitation but the most days with precipitation (109). New York has the most precipitation but the fewest days with precipitation (96). London has the fewest days above 30°C; New York has the most days with temperature extremes, averaging 81 days below freezing and 16 days 30°C or hotter (table 17.1). Topography can also influence cycling levels. With the exception of a few hilly districts, all three cities are mostly flat and thus conducive to cycling.

### Trends in Cycling

Figure 17.1 summarizes trends in bicycle mode shares derived from travel surveys in each of the three cities. The fastest growth in cycling has been in Paris, with bicycle mode share increasing by 1,150% (twelvefold) from 1991 to 2019, compared to 267% growth (more than tripling) in New York from 1990 to 2017 and 100% growth (doubling) in London from 1996 to 2017. In the most recent survey years (2017–2019), the highest bicycle share of trips was in Paris (5.0%), more than twice as high as in New York (2.2%) and London (2.4%). There are problems of comparability however, because the area of Paris is roughly equivalent to the central areas of New York and London, where bicycle mode share is also higher than in the outer areas. The most important message of figure 17.1 is that the bicycle share of trips has been rising substantially in all three cities.

Using survey data from the US Census Bureau, the New York City Department of Transportation (NYCDOT) reported an overall increase of 207% in the number of annual bicycle trips between 2000 and 2016, from 54.8 million to 167.9 million (NYCDOT 2019). Survey data for Greater London indicate

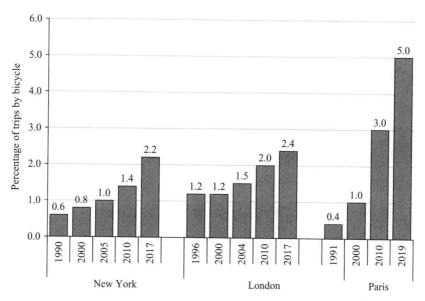

**Figure 17.1**

Trend in bicycle mode shares in New York, London, and Paris, 1990–2019. *Note*: Bicycle mode shares shown are percentages of all daily trips for all kinds of trip purposes, as derived from travel surveys for each city. *Sources*: NYCDOT (2019); Transport for London (2000–2019b); Transport for London (2000–2019d); City of Paris (2000–2019b).

a 154% increase in the number of annual bicycle trips in London from 2000 to 2016 (Transport for London 2019b; Transport for London 2019d). The number of annual bicycle trips in Paris increased by 483% from 30 million to 175 million from 2000 to 2018 (City of Paris 2019b). Cordon counts of cycling volumes on major roads or crossing points in all three cities reveal that cycling has at least doubled and in some cases increased more than tenfold, thus confirming the strong growth in cycling shown in figure 17.1.

As mentioned, cycling levels are not uniform throughout the metropolitan areas. In New York City, the bicycle mode share of daily work commuters ranges from about 5% in Lower Manhattan and northwestern Brooklyn to 0.2% or less in eastern Brooklyn and Queens, the Bronx, and Staten Island, just about the same low rate as in the suburbs (USDOC 2019). This spatial pattern holds in almost all North American cities, with bicycle mode shares falling sharply from the center to the periphery of urban areas (Pucher, Buehler, and Seinen 2011). Similarly, the bicycle mode share in

Paris is five times as high as in the larger Île de France region (5.0% versus 1.0%) (City of Paris 2019b; STIF 2015; IFM 2019). Inner London has a 3.6% bicycle mode share compared with 1.7% in outer London (Transport for London 2018b). As this spatial variation in bicycle mode share suggests, Paris is more comparable with the inner areas of New York and London than with those two cities as a whole.

The purposes of cycling trips vary among the three cities. Sports, recreation, and exercise account for over half of bicycle trips in New York (NYC-DOT 2019). By comparison, about two-thirds of bicycle trips in London and Paris are for utilitarian purposes such as commuting, shopping, and accessing services (City of Paris 2019b; Transport for London 2018a).

The demographics of cyclists also vary among the three cities. Women account for only 29% of bicycle trips in New York and 30% in London, compared to 40% in Paris (City of Paris 2019b; NYCDOT 2019; Transport for London 2018b). In London, cycling accounts for 3.8% of trips made by men, almost three times the 1.3% of trips made by women. In all three cities, the highest rates of cycling are for ages 25–44. Cyclists aged 65 and older account for 9% of bicycle trips in Paris, compared to 5% in London and only 3% in New York (City of Paris 2019b; NYCDOT 2019; Transport for London 2018b).

## Cycling Safety

As shown in figure 17.2, rates for cycling fatalities and serious injuries have fallen in all three cities. Ideally, fatality and injury rates would be calculated relative to kilometers cycled per year, but data on distance cycled are not available. Moreover, the official annual trip statistics are not fully comparable among cities, thus making per trip rates misleading. As the best feasible alternative, the fatality rates in figure 17.2 were calculated by dividing annual average fatalities (over a five-year period bracketing the year of the rate) by the population of the city in each year, as adjusted for cycling exposure, measured as the proportion of daily trips by bicycle.

The same method was used to calculate rates of serious cyclist injuries. Because there are at least 10 times as many serious injuries as fatalities, fatality rates are expressed per 100,000 adjusted population units, while serious injury rates are expressed per 10,000 adjusted population units.

The cyclist fatality rate fell by 57% in New York (from 28 in 2000 to 12 in 2017), by 63% in London (from 19 in 2000 to 7 in 2016), and by 84%

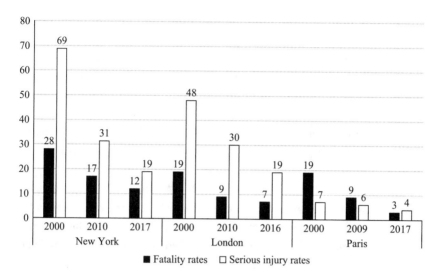

**Figure 17.2**

Decline in cyclist fatality and serious injury rates, 2000–2017. *Note*: The fatality rates shown are calculated as average annual fatalities over a five-year period per 100,000 people weighted by bicycle mode share (percentage of all daily trips). The serious injury rates are calculated as serious cyclist injuries per 10,000 people weighted by bicycle mode share. Thus, they are per capita rates but with population weighted by bicycle mode share of trips to control for exposure. The fatality rates are much more comparable among cities than the serious injury rates. Each city has somewhat different definitions of serious injuries, while all three cities have the same definition for fatalities. *Sources*: NYCDOT (2019); City of Paris (2000–2019a); City of Paris (2000–2019b); Transport for London (2000–2019a); Transport for London (2000–2019b); Transport for London (2000–2019d).

in Paris (from 19 in 2000 to 3 in 2017). The serious injury rate in New York fell by 72% (from 69 in 2000 to 19 in 2017), by 60% in London (from 48 in 2000 to 19 in 2016), and by 43% in Paris (from 7 in 2000 to 4 in 2017). Although these rates are approximations, they suggest significant reductions in death and serious injury rates since 2000 in all three cities.

There are several possible reasons for the observed reduction in fatality and injury rates. The theory of "safety in numbers" (see Elvik, chapter 4, this volume) suggests that as cycling levels rise, cycling becomes safer because more cyclists are more likely to be expected and seen by motorists, who consequently adjust their driving behavior to avoid collisions with cyclists. Moreover, as a higher percentage of the population cycles, more motorists are likely to be cyclists themselves and thus sensitive to the safety concerns of cyclists, and to drive in a way that avoids endangering cyclists on the roadway. "Safety in numbers" might help explain the reduction in cycling fatality and serious injury rates in the three cities. In New York, a tripling in annual bicycle trips (+207%) from 2000 to 2016 was accompanied by a 74% decline in cyclist fatalities and serious injuries (combined) per 10 million bicycle trips (NYCDOT 2019). In Paris, the sixfold increase in annual bicycle trips (+483%) from 2000 to 2017 was accompanied by an 84% reduction in cyclist fatalities per 10 million bicycle trips (City of Paris 2000–2019a; City of Paris 2000–2019b). Although the improvement in cycling safety in all three cities is consistent with the theory of "safety in numbers," the most important explanation for declining cyclist fatality and injury rates is the vastly expanded and improved cycling infrastructure over the same period, as described in the next section.

## Cycling Infrastructure

The expansion and improvement of cycling infrastructure have been central to the cycling promotion strategies in New York, London, and Paris. Table 17.2 summarizes the supply of cycling infrastructure in the three cities, both in total and on a per capita basis. (For a detailed analysis of different kinds of cycling infrastructure, see Furth, chapter 5, this volume.)

The network of on-street bike lanes and off-street bike paths in New York expanded sevenfold between 2000 and 2019, increasing from 284 km to 2,103 km (NYCDOT 2019). Most of the new cycling facilities feature improved design. Since 2006, New York has built 244 km of cycle tracks, which are protected on-street bike lanes (illustrated in figure 17.3), separated from motor vehicle traffic by physical barriers such as bollards, curbs, and raised concrete barriers. In some cases, cycle tracks in New York are completely separate bikeways immediately adjacent to the road. Most cycle tracks are further protected at intersections by improved offset crossing designs and by traffic signals that prohibit motor vehicle turns during a special crossing phase for cyclists. In late 2019, NYCDOT adopted an accelerated construction plan to build an additional 30 mi (45 km) of protected cycle tracks every year.

Many regular bike lanes in New York have been converted to buffered bike lanes (see Furth, chapter 5, this volume), which are separated from traffic lanes with diagonally striped buffer zones, offering more protection

**Table 17.2**
Supply of bike lanes and paths and bike parking, 2018–2019

| | Bike lane and path network | | | | | Bike parking | |
|---|---|---|---|---|---|---|---|
| | On-street lanes (km) | Protected on-street lanes (km) | Off-street paths (km) | On-street lanes and off-street paths (km) | Bike paths and lanes (km) per 100,000 people | Bike parking spaces | Bike parking spaces per 100,000 people |
| New York | 1,528 | 244 | 582 | 2,103 | 24 | 59,872 | 695 |
| London | 1,021 | 89 | 1,158 | 2,179 | 25 | 145,993 | 1,658 |
| Paris | 860 | 115 | 81 | 941 | 59 | 68,500 | 3,055 |

*Note*: The kilometers of protected on-street lanes are included in the total kilometers of on-street lanes. The extent of on-street bike lanes is counted on the basis of directional lane kilometers. Thus, if there is a bike lane on both sides of a two-way street, it counts as twice the kilometers of that street. In Paris, 148 km of on-street bike lanes are shared, extrawide bus and bike lanes with prominent roadway markings and vertical signs indicating their use by cyclists as well as buses.
*Source*: Information collected by the authors directly from each of the three cities.

from motor vehicles than usual bike lanes. Even ordinary, unprotected bike lanes have generally been widened, and many have been painted bright green to enhance visibility. The city government has committed to building 75 mi (121 km) of new bicycle infrastructure of various designs in 10 Bicycle Priority Districts by 2022 (NYCDOT 2019).

Improving intersection design is crucial for reducing bike conflicts with motor vehicles (see in this volume Elvik, chapter 4; Furth, chapter 5). Most cyclist fatalities and serious injuries occur at or near intersections. New York has implemented a multifaceted approach to making intersection crossings safer for cyclists: advance stop lines (3–5 m ahead of the stop line for cars); separate turning lanes for cyclists; mixing zones; offset crossings; painted intersection islands; special bike crossing markings within the intersection; and special bike traffic signals that give cyclists an advance green light, a special signal phase where motorists are prohibited from making turns crossing the bike lane (NYCDOT 2019).

The network of separate cycling facilities in London (including on-street lanes and off-street paths) increased fivefold from 2001 to 2017, from 435 km to 2,179 km. That expansion was accompanied by a dramatic change in the conceptual approach to the provision of cycle infrastructure in London, as reflected in changes to the London Cycling Design Standards—first published in 2005 and updated in 2016 (Transport for London 2016)—and the much greater commitment by the city government to promote cycling. Mayor Ken Livingstone's London Cycling Action Plan (Transport for London 2004) identified 900 km of a London Cycle Network Plus, which was designed to provide greater connectivity within the cycle network. Livingstone's successor, Mayor Boris Johnson, developed a number of cycle superhighways, but these were no more than blue bike lanes on main roads. Those blue lanes offered only a minimal enhancement in level of service to cyclists, and their shortcomings were soon recognized (see Furth, chapter 5, this volume).

To remedy those shortcomings, Transport for London developed a comprehensive vision for cycling (Greater London Authority 2013). The vision called for bicycling networks that connect key origins and destinations. Four programs of work were promoted: segregated cycle superhighways, Quietways, Livable Neighborhoods, and the Central London Grid. These distinctions were initially used in branding different types of cycling facilities. In practice, however, they are sometimes overlapping, and have since been integrated with each other into a single unified brand for the London-wide cycle network.

The 11 segregated cycle superhighways are routes planned to connect public transport stations, town centers, and key destinations. Two of the most significant are the East-West (#3) and the North-South (#6) cycle superhighways. They comprise either a two-way cycle track on one side of a road or a one-way cycle track on each side of a road. Enhancements at signal-controlled junctions include a range of methodologies for dealing with the left-turn conflict (left-hand rule of the road), such as separately staged left turns, and the right-turn conflict, such as two-stage right turns.

There are 12 Quietways constructed as routes along networks of less busy roads in London. They use various kinds of infrastructure: bike lanes; light segregation (bike lanes with some form of physical protection, such as intermittent raised ribs, planters, or flexible bollards); stepped cycle tracks (cycle tracks at an intermediate level between the carriageway and the footway, separated from each by a level difference created by a curb); and one-way or two-way cycle tracks.

There are three Livable Neighborhoods, which used to be called "mini-Hollands." They are comprehensive area-wide traffic management schemes to control motor traffic volume and speed in a neighborhood to create routes that are suitable for cycling. Some of the measures used for such

**Figure 17.3**
Protected bikeways in (from left to right) Paris, London, and New York. *Sources:* Christophe Belin (City of Paris), Tom Bogdanowicz (London Cycling Campaign), and New York City Department of Transportation.

traffic calming include road closures to motor traffic, banned turns, and streetscape enhancement.

The Central London Grid is a transport plan that mixes different types of physical infrastructure to create comprehensive routes for bicycle traffic. These include bike lanes, cycle tracks, and intersection enhancements.

In December 2018, London's mayor Sadiq Khan announced the city's new Cycling Action Plan, which foresees a tripling in the bikeway network by 2024, an expansion of 450 km, which would put 28% of London residents within 400 m of a separate cycling facility (Transport for London 2018a). It will be by far the city's largest investment in cycling infrastructure and is certain to encourage future growth in cycling.

Paris expanded its network of cycling facilities sixfold between 1999 and 2019, from 151 km to 941 km (City of Paris 2000–2019a). First established in 2001, extrawide, prominently marked, shared bike and bus lanes accounted for 147 km of the on-street cycling routes in Paris in 2019. Other facilities include bike-only lanes on the street and bike lanes on sidewalks. In 2010, the city began installing contraflow bike lanes on a few streets. Paris did not begin to build protected bikeways (as shown in figure 17.3) until 2017, but by the end of 2019, only three years later, Paris had built 115 km of protected bikeways, and it plans to expand them in the coming years. Thus, in all three cities, there is a strong trend toward physically separated, protected bike lanes and paths.

**Bike Parking and Integration with Public Transport**

Despite quadrupling the number of bike parking spaces between 2008 and 2019, New York still lags behind London and Paris. As shown in table 17.2,

New York has less than half the supply of public bike parking spaces per 100,000 residents as London and less than a fourth as many as Paris. New York provides insufficient, insecure, and unsheltered bike parking at the city's most important suburban and long-distance train and bus terminals. In 2018, however, NYCDOT began pilot testing small-scale secure, enclosed bike parking at three major public transport hubs. The New York City subway system itself provides no bike parking at any of its 467 subway stations, but NYCDOT provides some bike racks on nearby sidewalks (NYCDOT 2019). London and Paris have done far more than New York.

Since 2000, for example, London has added about 5,000 bike parking spaces at its underground and suburban rail stations (Transport for London 2018a). In 2019, Paris had 68,500 bike parking spaces, including 11,886 mixed parking places shared with motorbikes. Most of these bike parking spaces were created by removing car parking places. Paris is steadily expanding and improving bike parking, and by 2025 there will be about 75,000 spaces (City of Paris 2019b).

In all three cities, integration of cycling with public transport is mainly through the provision of bike parking at stations and not by facilitating the carriage of bikes aboard vehicles. None of the cities provide bike racks on their buses. London and Paris prohibit bikes on their metro and suburban trains at most times. New York's official policy is to allow bikes on the subway, even during rush hours. As of 2018, however, less than one-fourth of its 472 subway stations had elevators, and, on average, each elevator was out of service 53 times a year (Barron 2018). The many steps from street level to the platform make it difficult to get bikes on board, thus sharply limiting cyclists' actual usage of the seemingly generous policy.

As noted in the next section, many of the bikesharing docking stations in New York, London, and Paris are located at or near public transport stations. Many public transport riders rely on the cheap and convenient availability of these public access bikes. In that respect, bikesharing systems are also a way to integrate cycling with public transport.

## Public Bikesharing Systems

Public bikesharing systems have been rapidly proliferating in cities around the world (see Fishman and Shaheen, chapter 10, this volume). Paris's Vélib' was the world's first large bikesharing system and remains one of the most famous. It began in 2007 and was generally considered a huge success. As of 2016, Vélib' offered 20,600 bikes and 1,595 docking stations, with 300,485 annual subscribers and 38 million annual trips (City of Paris 2016c). Vélib' was used mainly for short trips. A 2009 survey found that 79% of Vélib' trips were for accessing metro and suburban rail stations, confirming the important role of bikesharing as a complement to public transport (TNS Sofres 2009).

Although Paris had undertaken other measures to increase cycling since the mid-1990s, Vélib' provided an extraordinary stimulus to cycling, generating widespread interest in cycling and attracting users who had rarely cycled before. Between 2006 (before Vélib') and 2008 (one year after its inception), the city's cordon counts of cycling volumes on major streets registered a 56% increase (City of Paris 2008b). Electronic usage records indicate that in 2016 Vélib' generated 78,000–120,000 bike trips per day, depending on the season of the year and day of the week, with 34% of annual subscribers using it to commute to work.

Great disruption was caused in 2017–2018 during the transition from JC Decaux to Smovengo as operators of Vélib' (City of Paris, 2010–2019c; City of Paris, 2010–2019d). It took over a year to implement the new contract while the previous one was discontinued. As of November 2018, only 988 of 1,595 Vélib' stations were open for use. Meanwhile, several private companies began operating free-floating, dockless bikesharing programs, with a total of 8,250 bikes as of November 2018. Vélib' has since been renamed Vélib' Metropole, still operated by Smovengo, and has expanded from Paris to include 64 surrounding towns. As of December 2019, however, Paris's bikesharing system still had not recovered from the 2017–2018 disruption, with fewer bikes (14,500 vs. 20,600) and fewer docking stations (1,100 vs. 1,595) than in 2016 (Meddin 2020). It remains to be seen how successful the regionalization of Paris's bikesharing system will eventually be and to what extent the disaster of 2017–2018 can be overcome.

Since opening in July 2010 in the central area of London, bikesharing has expanded to the east and southwest. In fiscal year 2016–2017, there were 10.5 million trips made with 11,500 bikes at 750 docking stations (Transport for London 2018b). There were 305,000 members at the end of 2017, with almost half of bikesharing trips made by nonmembers in the summer months (Transport for London 2018b). At the most popular docking stations, most bikes are used for work commuting journeys to and from key rail stations.

In 2013, New York introduced Citi Bike, North America's largest bikesharing system, starting with 6,000 bikes and 330 docking stations in Manhattan below 59th Street (Midtown and Downtown) and northern Brooklyn. Citi Bike has been funded by private capital, sponsorship agreements, and revenue from users, avoiding the need for direct public subsidy of the program (NYCDOT 2019). By December 2019, the system had expanded to 12,000 bikes and 750 docking stations, covering most of Manhattan and expanded portions of Brooklyn and Queens. Citi Bike trips more than doubled from 8.1 million in 2014 (the first full year of operation) to 17.6 million in 2018. In 2018, Citi Bike's operator, Motivate, was bought by Lyft, which entered into an agreement with the city to double Citi Bike's service area and triple the number of bikes to almost 40,000 by 2023 (NYCDOT 2019). The city also started conducting demonstration projects to evaluate the feasibility of dockless bikesharing technology in outer areas of the city not yet served by Citi Bike (NYCDOT 2019). By August 2018, NYCDOT had permitted the trial operation of 1,200 dockless bikes (in total) by four private bikesharing companies. Some of the dockless bikes include electric pedal assist to test which e-bike version would be appropriate to permit for operation on city streets.

## Cycling as an Alternative to Public Transport

As noted previously, cycling and public transport can be complementary or compete with each other. The surge in cycling in New York in recent years has been partly the result of the severe deterioration in subway (metro) and bus services. Public transport in New York has been plagued by overcrowding, delayed or canceled trains, buses slowed by traffic congestion, and outdated vehicles, stations, and rights-of-way. Although there have not been any scientific studies to measure the role of bad and worsening public transport services on cycling levels, there has probably been some shift of trips from public transport to cycling, especially among younger riders, for trips short enough to make by bike. This "push" effect from bad public transport has combined with the "pull" effect of better cycling facilities to encourage more cycling in New York.

As in New York, public transport in London is overcrowded during peak hours in central sections, where a shift to cycling could relieve some of the excess demand. Between 2000 and 2016, the number of public transport passenger trips in Greater London increased by 85% for suburban rail, by 49% for underground (metro) and light rail, and by 70% for buses. Although public transport services in London are being expanded, they still struggle to satisfy rising passenger levels. Expanding infrastructure for cycling is less expensive and quicker than expanding public transport services, which require enormous capital and operating subsidies.

## Restrictions and Charges on Car Use

Reducing the speed and volume of motor vehicle traffic on residential streets can greatly increase the safety, comfort, and convenience of cycling (see in this volume Buehler and Pucher, chapter 2; Furth, chapter 5; Heinen and Handy, chapter 7). Of the three cities examined in this chapter, Paris has done the most to restrict car traffic in residential neighborhoods. Starting in 2002, Paris established 38 "quartiers verts" ("green districts"), extensive traffic-calmed neighborhoods with speed limits of 30 km/h or less, car-free zones, narrowed roadways, and widened sidewalks. Roughly one-third of Paris streets now incorporate traffic-calming measures. There is also a special program for weekends and holidays called "Paris respire" ("Paris breathes"), when districts close their streets to cars and reserve them for nonmotorists. Paris's car-restrictive policies have been reinforced by a new policy to further reduce car traffic along the Seine River and to implement a citywide traffic-reduction area, called the Low Emission Zone, involving a variety of additional, coordinated measures to regulate car use.

London has also been introducing traffic calming in many residential neighborhoods through 20 mph (32 km/h) speed limits and roadway design modifications that include dead ends and detours (diverters) for motor vehicles but passthroughs (shortcuts) for cyclists and pedestrians to provide more convenient, faster connections. The more comprehensive Livable Neighborhoods approach now being implemented in London combines people-friendly street design with overall improvements in urban design in street corridors.

Over the past 10 years, New York has been implementing neighborhood slow zones in all five boroughs, with 26 by 2019. They have 20 mph (32km/h) speed limits that are sometimes enforced by speed cameras as well as traffic-calming road infrastructure modifications, which not only reduce motor vehicle speed but also limit traffic volumes by routing through traffic around residential neighborhoods. Traffic-calming measures have been implemented in many commercial areas of the city as well. Moreover, in 2013, New York reduced its general speed limit from 30 mph (48 km/h) to 25 mph (40 km/h) on almost all roads except highways. In order to enforce that lower limit, the city has been deploying speed cameras throughout the city, and it currently has the largest system of speed-ticketing cameras of any American city.

The most extreme restriction on car use is to ban it completely. Like hundreds of other European cities, both London and Paris have many car-free zones, squares, and portions of streets, and they sometimes connect to each other to form a network ideal for walking and cycling. By comparison, New York has only a few isolated car-free stretches of streets, such as Broadway from Herald Square to Times Square, which also includes a separate cycling facility. Throughout New York, however, key intersections have been improved by reallocating street space from cars to pedestrians and cyclists, thus enhancing the safety, convenience, and pleasantness of crossing intersections. In addition, the city has been steadily increasing the number, size, and quality of car-free plazas and squares (68 as of 2019) by building or widening walkway and cycling facilities, planting more trees and shrubs, and installing better seating, fountains, and artwork. Recently, the city has been experimenting with other kinds of car restrictions as well. In autumn 2019, for example, 14th Street (a major east-west crosstown road in Manhattan) was converted into a bidirectional busway. It prohibits private cars but permits local truck deliveries and emergency vehicles. In advance, NYCDOT installed protected bike lanes on 13th and 15th Streets (parallel to 14th Street) to give cyclists convenient and safe alternatives to 14th Street, where bicycles are not allowed.

Taxes and fees on car ownership and use in the United Kingdom and France are much higher than in the United States (see Buehler and Pucher, chapter 2, this volume). Although there are variations by type of vehicle, sales taxes on new car purchases average 20% throughout France and the United Kingdom, more than three times the 6% average in the United States and twice the 9% tax in New York City, including state and local taxes.

Congestion charging was introduced in London in 2003 in a 21 km$^2$ central area, including the financial district (City of London) and the main commercial and entertainment district (West End) (Transport for London 2003–2019c). As of 2019, the daily charge was £11.50 between 7 a.m. and 6 p.m. Monday to Friday. As the term suggests, the main purpose of congestion charging is to reduce traffic congestion, especially in central city areas. By making car users pay more to drive—while also reallocating roadway space from cars to cyclists and buses—congestion charging has also stimulated more cycling and public transport use. Cycling volumes into and out of the congestion charging zone increased by 42% from 2002 (before) to 2005 (after). Since 2005, cycling volumes have continued to increase every year, and by 2019 they had risen to almost four times as high as in 2012 (283% increase) (Transport for London 2003–2019c). Massive improvements in cycling infrastructure in and leading to London's city center have been partly responsible for that continued strong growth.

After decades of alternative proposals, consideration, approval, rejection, and reconsideration, congestion pricing was finally approved in April 2019 by the legislature and governor of New York State (Regional Plan Association 2019). The legislation requires that the congestion tolls generate at least a billion dollars a year in net revenue and entrusts a Traffic Mobility Review Board controlled by the governor with specifying the details of their implementation. As of May 2020, congestion pricing was scheduled to go into effect in late 2021—the first time in any North American city. The congestion pricing zone would have an area of 22 km$^2$ (roughly the same as London's) and include all of Lower and Midtown Manhattan (south of 61st Street) except for two peripheral highways. It would increase inbound tolls already in place on four bridge and tunnel facilities and impose inbound tolls for the first time on four currently free bridge crossings over the East River (from Brooklyn and Queens). The congestion charge would also apply to southbound crossings between Uptown and Midtown Manhattan at the northern boundary of the congestion zone at 60th Street.

Congestion pricing is needed to reduce the serious traffic congestion in Manhattan. However, the main political motivation for congestion pricing in New York was to generate revenues to finance desperately needed improvements to the city's old and deteriorated subway (metro) and suburban rail systems. One important side effect will be to stimulate more cycling as a free alternative to driving to Manhattan, especially from Brooklyn, Queens, and northern Manhattan (Uptown). The large increase in cycling volumes in London's congestion charging zone suggest large increases in cycling in New York as well.

## Enforcing Traffic Regulations That Protect Cyclists

It is crucial that the police enforce traffic regulations requiring motorists to avoid endangering cyclists (as well as pedestrians). London's police have made a long-term effort to ensure that both motorists and cyclists adhere to traffic regulations affecting cycling safety. The police have encouraged safer driving as well as safer cycling by strict ticketing of motorists and cyclists who violate traffic laws. In addition, the Metropolitan Police Service established a Cycle Task Force of 40 police officers patrolling on bicycles to reduce bike theft and enforce good road user behavior. The police work closely with cycling education programs in London schools and facilitate group bike rides (Transport for London 2018a).

As in London, the Paris police have contributed to cycling safety by strictly ticketing motorists who speed or encroach on bike lanes. From about 2000 until 2007, motorcyclists rode illegally in bike lanes, causing many crashes resulting in serious cyclist injuries. After the police began strictly ticketing motorcyclists in 2008, motorcyclists' use of bike lanes fell from 25% to 3%. In 2018, the city implemented a citywide video traffic-control system that enhances the enforcement of regulations (City of Paris 2000–2019a; City of Paris 2000–2019b).

In sharp contrast to London and Paris, relations between police and cyclists have been highly confrontational in New York, with many cyclists accusing the police of harassment, mistreatment, and ignoring their needs (NYCDCP 2005). The situation has continued in the years since then, with NYC police evidently sharing the view of many motorists that cyclists do not have the right to use the road (Komanoff and Gordon 2017). New York's police have seriously undermined cyclist safety by not strictly enforcing bike lanes, which are often blocked by cars (including police cars) and delivery vehicles. That forces cyclists to veer out of the bike lanes into busy streets to navigate around the blockages, which is very dangerous. The frequent blockage of regular on-street bike lanes has been an important motivation for NYCDOT to build a rapidly growing network of physically separated, protected bike lanes (cycle tracks), as noted earlier in this chapter.

## Cycling Advocacy

London, New York, and Paris have strong cycling advocacy organizations that support laws and funding to improve cycling conditions, offer bike training, and coordinate many of the group rides and special events noted previously. The London Cycling Campaign (LCC) had 11,000 members in 2018 (LCC 2019). The LCC has played a role in almost every recent cycling development in London. It campaigned successfully for the creation of a dedicated cycling unit at Transport for London and lobbied for increased funding for cycling, which rose from only £5 million a year in 2002 to more than £100 million in 2018. Every year, LCC organizes the dozens of guided rides to the RideLondon FreeCycle event and about 500 local rides and events, as well as promoting cycle training, running repair workshops, and providing input to hundreds of city traffic schemes and bike route projects. In 2012, LCC ran a successful campaign called Love London, Go Dutch to bring Dutch-style high-quality protected cycle lanes to the capital. With buy-in from London's mayor at the time, Boris Johnson, the outcome, in 2016, was the installation of wide cycle tracks through the center of the capital, past Big Ben and Buckingham Palace, funding for influential "mini-Holland" cycle schemes in several town centers, and new infrastructure design guidance. LCC's campaign for trucks without "blind spots" has led to a new "direct vision" standard that will mandate safer vehicles as of 2020. LCC offers cycling information through a popular website, electronic newsletter, magazine, and web-based networks such as Facebook and Twitter (LCC 2019).

Bicycling advocacy in New York City is led by Transport Alternatives, whose membership tripled from 2,600 in 2000 to 7,900 in 2018 (Transport Alternatives 2019). It works closely with local government officials, the media, and other groups advocating for the environment, sustainable transport, and social justice. It has vigorously supported the expansion of New York's bikeway network and organizes large group rides such as the Century Ride, the Tour de Brooklyn, and the Tour de Bronx (Transport Alternatives 2019). Since 2005, Bike New York has offered a range of cycling training courses for different ages and skill levels (Bike New York 2019). Since 2017, it has formally partnered with the New York City Department of Education to offer cycling training in the city's schools. It also organizes the city's largest group bike ride, the annual Five Borough Bike Ride, on the first Sunday of May. The ride attracts over 30,000 cyclists, who ride a 40 mi (64 km) route on roads that are reserved for cyclists during the event.

The two main cycling advocacy organizations in Paris are Mieux se déplacer à Bicyclette (MDB), with 1,000 members, and Paris en Selle, with 800 members. They engage in activities similar to those of the London Cycling Campaign and Transportation Alternatives but are much smaller and less influential (MDB 2019; Paris en Selle 2019). The Fédération Fran-çaise des Usagers de la Bicyclette (FUB), Club des Villes et Territoires Cycla-bles (CVTC), and Animation, Insertion et Culture Velo (AICV) are three other cycling organizations that have also supported pro-bike policies in Paris (AICV 2019; CVTC 2019; FUB 2019).

## Conclusions

Over the past three decades, New York, London, and Paris have greatly increased cycling levels, both in the annual number of bicycle trips and in the bicycle mode share of daily trips. Over the same period, cycling has become safer, with fatality and serious injury rates falling. What makes these cities' accomplishments especially impressive is that none of them had a history of cycling as a major travel mode. They started from a point of almost no cycling and no experience with cycling infrastructure, programs, and policies. New York, London, and Paris show that, even in large cities with no tradition of cycling, it is possible to establish cycling as a safe and feasible way to travel.

The three cities also demonstrate the importance of demographics. As in many cities throughout the world, there has been a trend toward the revival and gentrification of many inner-city neighborhoods, with an influx of young, affluent professionals keen on an urban lifestyle (Goodwin and van Dender 2013). Changing demography has provided a large pool of potential cyclists to use the vastly expanded cycling facilities that have been built.

As in most Dutch, Danish, and German cities, the key to success has been the implementation of a coordinated package of mutually reinforcing measures such as those summarized in table 17.3. Each city has had a some-what different mix of policies, suitably adapted to their different political, demographic, and transport environments.

New York has focused on massive expansion and improvement of sepa-rate cycling facilities, especially protected on-road cycling facilities that are physically separated from motor vehicle traffic by curbs, planters, or other raised barriers. Providing more and better cycling infrastructure has also been part of the policy packages in London and Paris but has not been as overwhelmingly dominant as in New York. London has been pathbreaking in the development of cycle superhighways (express bikeways) for long bike commutes, a type of cycling facility that has rapidly spread throughout Europe. Paris initially relied primarily on a large network of shared bus and

**Table 17.3**

Policy highlights in New York City, London, and Paris

| New York | • From 2006 to 2019, built 244 km of physically separated, protected bike lanes and 560 km of regular bike lanes |
|---|---|
| | • In 2013, launched Citi Bike bikesharing, which in 2017 had 12,000 bikes, 750 docking stations, 17.6 million trips per year, with planned expansion to 40,000 bikes by 2023 |
| | • In 2018, started pilot testing dockless (floating) bikesharing in outer parts of city not yet served by Citi Bike |
| | • In 2008, introduced Summer Streets, an open streets event three weekends every August, with over 350,000 participants in 2018 |
| | • Limited public bike parking, no secure bike parking at all, and no bike lockers or bike stations |
| | • In 2009, introduced requirements for provision of bike parking in commercial and residential buildings |
| | • Worst bike-transit integration of any city, with woefully inadequate bike parking at subway stations and major transit terminals |
| | • Police failure to enforce bike lanes leads to frequent blockage of lanes by motor vehicles |
| | • In 2013, general speed limit reduced from 30 mph (48 km/h) to 25 mph (40 km/h) on all roads except highways. Slow Zone Neighborhoods have 20 mph (32 km/h) speed limit. Speed limits enforced by most extensive speed camera surveillance program in the country |
| | • Congestion charging scheduled to be implemented in early 2021 for entry into the central business district, which is all of Manhattan south of 61st Street |
| London | • Congestion charging in Central London, imposing £11.50 per day fee for private cars workdays from 7 a.m. to 6 p.m. |
| | • London Cycle Network Plus aimed to provide 900 km of routes by 2009–2010 |
| | • The London Greenways provide walking and cycling routes to and through green space |
| | • By 2018, 12 of 15 planned Quietways were completed along back streets, through parks, and along waterways or tree-lined streets |
| | • By 2018, 8 of 11 planned superhighways were completed, with curb-separated cycle tracks along major cycle routes |
| | • On-street bike parking continues to be expanded, and a number of large-scale parking hubs have been constructed |
| | • Santander Cycle Hire bikesharing provides 11,000 bicycles at 760 docking stations |

(*continued*)

**Table 17.3 (continued)**
Policy highlights in New York City, London, and Paris

| Paris | • Sixfold increase in cycling network from 151 km in 1996 to 941 km in 2019, including 115 km of physically separated, traffic-protected on-street bike lanes (cycle tracks) and 148 km of extra-wide, shared bus and bike lanes |
| --- | --- |
| | • New bike plan aims to double number of kilometers of bike lanes, build an express bikeway, and add 10,000 more bike parking spaces by 2025 |
| | • From 2008 to 2019, 45,000 additional bike parking spaces were installed |
| | • World's first large bikesharing system, Velib', opened in 2007, stimulating Paris's bike boom. Velib' has been emulated worldwide in hundreds of cities. In 2016, Velib' had 20,600 bikes, serving 38 million annual bike trips. Disruption in 2017 caused by switch in operators |
| | • Start of free-floating bikesharing in 2018, operated by Ofo and Mobike |
| | • Planned expansion of 30 km/h traffic-calmed zones to most residential neighborhoods |
| | • Cyclists allowed to treat red traffic light as yield (give way) sign and to continue through intersection if clear |

*Source*: Information collected by the authors directly from the case study cities.

bike lanes but has increasingly complemented those with bike-only lanes, and from 2017 to 2019, it built 115 km of traffic-protected cycle tracks. All three cities are planning vast expansions of their cycling facilities in the coming years, with a special emphasis on traffic-protected bike lanes and paths along roads.

Both Paris and London have facilitated safe cycling on many local neighborhood streets by reducing speed limits (30 km/h in Paris; 32 km/h in London). Those speed limits are enforced through a range of traffic-calming measures, such as road narrowing, speed humps and bumps, raised crosswalks and intersections, and other road infrastructure modifications that reduce motor vehicle speeds while diverting through traffic from local streets to arterial roads. On a more limited scale, New York had 26 Slow-Zone Neighborhoods in 2019, with reduced speed limits of 20 mph (32 km/h), sometimes reinforced with the kinds of traffic-calming roadway modifications used in Paris and London. Far more important—because citywide—New York reduced the general speed limit to 25 mph in 2013 on all city streets except highways.

With its Vélib' bikesharing system launched in 2007, Paris set the example for the rest of the world to follow, including London and New York. Bikesharing has further stimulated cycling in all three cities, which are now also experimenting with the fourth generation of bikesharing, without docking stations and instead with floating bikes that can be picked up and dropped off at many more locations. The increasing use of e-bikes, whether as part of bikesharing or personally owned, is also likely to stimulate more growth in cycling.

The prospect is for continued growth in cycling in the coming years, but only if the city governments of New York, London, and Paris continue to implement the policies crucial for cycling to thrive, most importantly an integrated network of safe and convenient cycling facilities. As examined in this chapter, complementary policies and programs are also necessary, but the recent cycling booms in these three cities would not have been possible without the vast expansion and improvement of cycling facilities to ride on.

## References

Animation, Insertion et Culture Velo (AICV). 2019. Animation, Insertion et Culture Velo. https://www.aicv.net/.

Barron, James. 2018. For Disabled Subway Riders, the Biggest Challenge Can Be Getting to the Train. *New York Times*, July 27, A21.

Bike New York. 2019. *Bike Education*. New York: Bike New York.

Buehler, Ralph, John Pucher, and Adrian Bauman. 2020. Physical Activity from Walking and Cycling for Daily Travel in the United States, 2001–2017: Demographic, Socioeconomic, and Geographic Variation. *Journal of Transport and Health* 13. Forthcoming. https://doi.org/10.1016/j.jth.2019.100811.

City of Paris. 2000–2019a, annual. *Bike Accidents in Ile de France*. Paris: City of Paris, Department of Transport.

City of Paris. 2000–2019b, annual. *Travel in Paris*. Paris: City of Paris, Mobility Observatory.

City of Paris. 2010–2019c, annual. *Vélib' Statistics*. Paris: City of Paris, Department of Transport.

City of Paris. 2010–2019d, annual. *Bike Sharing in Paris*. Paris: City of Paris, Department of Transport.

Club des Villes et Territoires Cyclables (CTVC). 2019. Club des Villes et Territoires Cyclables. http://www.villes-cyclables.org/.

Fédération Française des Usagers de la Bicyclette (FUB). 2019. Fédération Française des Usagers de la Bicyclette. https://www.fub.fr/.

Goodwin, Phil, and Kurt van Dender. 2013. Peak Car—Themes and Issues. *Transport Reviews* 33(3): 243–254.

Greater London Authority. 2013. *The Mayor's Vision for Cycling in London: An Olympic Legacy for all Londoners.* London: Greater London Authority. https://www.london .gov.uk/sites/default/files/cycling_vision_gla_template_final.pdf.

Île-de-France Mobilités (IFM). 2017–2020, annual. *Mobility and Transport in Greater Paris.* Paris: Île-de-France Mobilités.

Komanoff, Charles, and Doug Gordon. 2017. NYPD Stonewalling Won't Protect Cyclists. *City Limits*, August 1.

London Cycling Campaign (LCC). 2019. *The Voice of Cyclists in London, Working to Create a Better City.* London: London Cycling Campaign. https://lcc.org.uk/pages /about-us.

Meddin, R. 2020. *World Bike Sharing Map.* Philadelphia, PA: Bikesharingmap.com. https://bikesharingmap.com.

Meng, M., J. Zhang, Y. D. Wong, and P. H. Au. 2016. Effect of Weather Conditions and Weather Forecast on Cycling Travel Behavior in Singapore. *International Journal of Sustainable Transport* 10(9): 773–780.

Mieux se déplacer à Bicyclette (MDB). 2019. *Baromètre des Villes Cyclables.* Paris: Mieux se déplacer à Bicyclette. https://mdb-idf.org/.

New York City Department of City Planning (NYCDCP). 2005. *The State of Cycling in New York City.* New York: New York City Department of City Planning.

New York City Department of Transportation (NYCDOT). 2019. *Bicycling in New York City.* New York: New York City Department of Transportation. http://www.nyc .gov/html/dot/html/bicyclists/bicyclists.shtml.

Paris en Selle. 2019. *Rapport d'Activité.* Paris: Paris en Selle. https://parisenselle.fr/.

Pucher, John, Ralph Buehler, and Mark Seinen. 2011. Bicycling Renaissance in North America? An Update and Re-assessment of Cycling Trends and Policies. *Transport Research Part A: Policy and Practice* 45(6): 451–475.

Regional Plan Association. 2019. *Congestion Pricing: Getting It Right.* New York: Regional Plan Association. http://library.rpa.org/pdf/RPA-CongestionPricingNYC_GettingItRight .pdf.

Syndicat des Transports d'Île-de-France (STIF). 2010–2016, annual. *Mobility and Transport in Greater Paris.* Paris: Syndicat des Transports d'Ile de France.

TNS Sofres. 2009. *Vélib Satisfaction Survey*. Paris: TNS Sofres.

Transport Alternatives. 2019. *Walking and Cycling in New York*. New York: Transport Alternatives. https://www.transalt.org/.

Transport for London. 1989–2019a, annual. *Pedal Cyclist Casualties, Killed and Seriously Injured*. London: Transport for London.

Transport for London. 2000–2019b, biannual. *London Travel Demand Survey*. London: Transport for London.

Transport for London. 2003–2019c, annual. *Central London Congestion Charging: Impacts Monitoring, Annual Report*. London: Transport for London.

Transport for London. 2004. *Creating a Chain Reaction: The London Cycling Action Plan*. London: Transport for London.

Transport for London. 2009–2019d, annual. *Travel in London*. London: Transport for London.

Transport for London. 2016. *London Cycling Design Standards*. London: Transport for London.

Transport for London. 2018a. *Cycling Action Plan: Making London the World's Best Big City for Cycling*. London: Transport for London.

Transport for London. 2018b. *Santander Cycles Transparency*. London: Transport for London.

US Department of Commerce (USDOC). 2019. *American Community Survey, Journey to Work*. Washington, DC: US Department of Commerce, US Census Bureau.

US Department of Transportation (USDOT). 2019. *National Household Travel Surveys 2001, 2009, and 2017*. Washington, DC: US Department of Transportation.

# 18 Cycling in Copenhagen and Amsterdam

Till Koglin, Marco te Brömmelstroet, and Bert van Wee

Amsterdam and Copenhagen are often cited as models of successful cycling cities. As shown in figure 18.1, Amsterdam and Copenhagen have by far the highest cycling shares of daily trips of any of the large European cities shown. Moreover, both cities raised their cycling mode shares by about one-third over the past three decades despite increases in per capita incomes and car ownership. In short, Amsterdam and Copenhagen have been the world's two most successful large cycling cities. This chapter examines the reasons underlying their success and, in particular, the policies that have enabled cycling to flourish, providing lessons for other cities seeking to increase cycling.

Both cities had high levels of cycling before World War II, but they were not exceptions, as many European cities had high cycling levels. After World War II, both Amsterdam and Copenhagen witnessed a decline in cycling levels as a result of increasing motorization and car use, but to a lesser extent than in most other cities in Europe. One reason for this smaller decline in cycling levels in Amsterdam and Copenhagen is that, unlike in many countries, where cycling is primarily among low-income groups, cycling in the Netherlands and Denmark has always included all income levels, all ages, and women as much as men. This may reflect the high quality of cycling conditions as well as the long egalitarian traditions in the Dutch and Danish societies. Unlike in many other cities, where cycling has only recently been getting more popular, cycling in Copenhagen and Amsterdam has been a mainstream part of their transport systems for a much longer time, indeed for many decades.

As shown later in this chapter, there are large variations in cycling levels among neighborhoods in each of the two cities. There are also differences between the two cities. For example, Amsterdam is *not* among the most cycling-friendly cities in the Netherlands, whereas Copenhagen is by far the

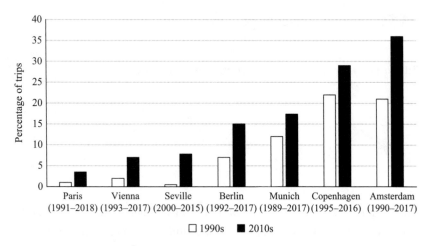

**Figure 18.1**
Increasing cycling mode share of daily trips in seven large European cities over three decades. *Note*: Years of data vary for each city as shown in parentheses under the name of each city. *Source*: Based on Pucher and Buehler, chapter 1, this volume, figure 1.1.

best cycling city in Denmark. This is because Copenhagen has a very long tradition of planning for cycling and because cycling overall does not have the same priority in Denmark as in the Netherlands (Koglin 2013). Furthermore, looking at the national cycling statistics of Denmark, one can see the differences between the nation and Copenhagen. Cycling is only 14% of the mode share on the national level, whereas it is 19% in Copenhagen (DTU 2019; City of Copenhagen 2016). Several factors help explain why Amsterdam is not among the best Dutch cycling cities. First, it has an extensive, high-quality, and inexpensive public transport system. Second, its large size (compared to most other Dutch cities) results in much longer travel distances than in smaller Dutch cities. Third, there are personal safety problems caused by the isolated location of some cycle paths (e.g., in parks), thus increasing vulnerability to assault and theft. Finally, the car is clearly the dominant mode of travel outside the city center, especially in the suburbs.

Because of these differences, Amsterdam and Copenhagen offer useful lessons for future cycling policies. We portray trends in cycling and related policies for both cities, first for Copenhagen and then for Amsterdam. For each city, we examine the relationships between cycling trends, government policies, and social and spatial developments. Based on the case study analysis for both cities, we draw lessons for other cities that want to implement cycling-related strategies. Moreover, we propose recommendations for future cycling policies for Amsterdam and Copenhagen.

## The Case of Copenhagen

Copenhagen is the capital of Denmark. Population density is high in Copenhagen; it has 602,504 inhabitants in an area of 89.78 km², resulting in a density of 6,711 inhabitants per km² (City of Copenhagen 2018; Koglin 2015). Copenhagen is also a flat city although windy. Moreover, the city has a good public transport system. However, population density and flatness alone cannot explain why cycling is so high in Copenhagen. Moreover, there are factors endangering cycling in Copenhagen, such as rising car ownership and use. This section discusses trends over time in some factors favorable to cycling and others challenging the role of cycling in Copenhagen. The current and future challenges to cycling in Copenhagen are largely unknown outside Denmark and need to be recognized.

### The History of Cycling in Copenhagen

Copenhagen has a long tradition of planning for cycling. The first traffic-protected cycle track was built in the early twentieth century because cycling on cobblestones was seen as uncomfortable; moreover, there were increasing conflicts between horses, horse-drawn carriages, and cyclists. That was the starting point for a focus on cycling infrastructure in the city of Copenhagen, which expanded from the 1900s until the 1960s. The expansion of cycling infrastructure was interrupted in the 1960s and 1970s, when motorized traffic entered Danish cities on a large scale. Copenhagen's focus in transport planning shifted increasingly toward motorized traffic, and cycling was not prioritized as before. Indeed, during the 1970s, some cycling infrastructure was demolished to create more space for motorized traffic. That led to massive public protests in Copenhagen, which were organized by the Danish Cyclist Federation and supported by many thousands of Copenhageners. The protests against reallocating roadway space from cyclists to cars were the political foundation for a return to the former policies of expanding and improving cycling infrastructure in the following years and decades (Koglin 2015). The overall history of cycling infrastructure in Copenhagen extends over a century. The network of traffic-protected cycle tracks has grown from about 25 km in the early 1900s, to about 200 km in the 1960s, and further to about 380 km in 2018 (Jensen 2013; City of Copenhagen 2018a).

## Cycling Infrastructure

The backbone of the cycling infrastructure in Copenhagen consists of cycle tracks that are separated from motorized traffic throughout the city. The cycle tracks are usually located beside the street and are separated from motorized traffic and from pedestrians by curbs. This design of cycling infrastructure increases the safety for cyclists, as noted elsewhere in this volume (Buehler and Pucher, chapter 2; Elvik, chapter 4; Furth, chapter 5). Cyclists are prioritized at intersections in Copenhagen. Traffic signals turn green for cyclists before signals for motorized traffic turn green, giving cyclists a head start through the intersection, raising their visibility to motorists, and increasing cycling safety. Copenhagen also has "green waves" of traffic signals over the length of roads, timed to be most convenient for cyclists, with traffic signals at successive intersections synchronized to turn green for cyclists riding at the average cycling speed. The green wave is set at higher speeds on longer-distance, commuter bike routes than for shorter-distance, local routes. The city has also built several new bridges for cyclists to enable more convenient, fast connections.

At some intersections, cycle tracks are converted to bicycle lanes, marked in bright colors on the street. This improves cyclist safety because cyclists ride on the street in the intersection, not widely separated from motorized traffic. That enables car drivers to clearly see and avoid endangering cyclists when traffic signals turn green for cars. Another feature of planning for cycling in Copenhagen is that bicycling against the flow of motor vehicle traffic on one-way streets is allowed in the inner city of Copenhagen. Making one-way streets bidirectional for cyclists increases the convenience and directness for cyclists while reducing trip times. As in Dutch, German, and other Danish cities, such bidirectional cycling on one-way streets has been shown to be very safe if motor vehicle speeds are limited to 30 km/h, as is usual in most traffic-calmed residential neighborhoods and city centers in Europe.

As the bicycle network has been extended, improved, and better integrated, the number of killed and injured cyclists has decreased despite increasing cycling levels (see figure 18.2). Thus, the cyclist fatality and

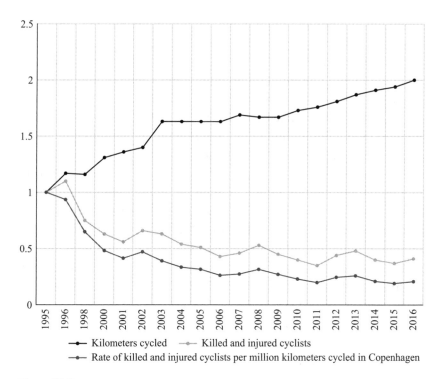

—•— Kilometers cycled    —•— Killed and injured cyclists
—•— Rate of killed and injured cyclists per million kilometers cycled in Copenhagen

**Figure 18.2**
Cyclists killed or seriously injured per million kilometers cycled in Copenhagen relative to the base year 1995 (=1). *Source*: Copenhagen Police (2018).

injury rates per trip and per kilometer cycled have fallen sharply over time. For example, the total number of cyclist crashes has been decreasing since 2004, while kilometers cycled have increased. This dramatic improvement in cycling safety is most obviously the result of the continued expansion and improvement in cycling infrastructure. It might also be related to the phenomenon of "safety in numbers," which stipulates that cycling safety improves as cycling levels rise. As emphasized by Elvik (chapter 4, this volume), however, most scientific evidence shows that improvements in cycling conditions appear to be the main explanation for increased safety as well as increased cycling levels. "Safety in numbers" does not improve cycling safety unless accompanied by improvements in cycling infrastructure and programs and by restrictions on car use such as reduced car speeds, traffic calming, and car-free zones.

## Policies in Copenhagen

Copenhagen has implemented a wide range of transport, development, and land-use policies that have promoted cycling. Although the city's first official comprehensive bicycle policy plan was not adopted and published until 2002 (Eltis 2018), Copenhagen has a long tradition of planning for cycling, with several goal-related policies and plans, such as the plans for projected expansions to the cycle track network (see Koglin 2013).

Since 1995, Copenhagen has implemented a biannual Bicycle Account, which monitors levels of cycling over time and attitudes of cyclists about cycling conditions and needed improvements. This benchmarking report is updated through a survey every two years and contains data on how Copenhageners perceive the existing cycling facilities and their maintenance, Copenhagen as a cycling city, bicycle parking, and other aspects of cycling conditions. Table 18.1 shows the results of the Bicycle Account for the 20-year period from 1996 to 2016. The accounts not only provide interesting information about cycling levels and attitudes but more importantly alert city transport planners, and bicycling planners in particular, to those aspects of cycling conditions that need to be improved as indicated by low satisfaction ratings assigned by cyclists who responded to the survey. As shown in table 18.1, cycling conditions in Copenhagen improved from 1996 to 2016, with cyclist ratings of every aspect of cycling rising continuously over the two decades of the Bicycle Account.

Copenhagen was a pioneer in establishing this regular benchmarking report for cycling, and many cities, states, and countries around the world have used Copenhagen's Bicycle Account as a model for their own cycling benchmarking efforts. For example, the League of American Bicyclists publishes a comprehensive biannual benchmarking report with information on cycling (and walking) trends, safety, infrastructure, policies, and programs for all 50 US states and the 50 largest US cities (League of American Bicyclists 2018).

## Current Trends for Bicycling and Planning in Copenhagen

There are three key trends influencing the future of cycling in Copenhagen. Perhaps most important is rising car ownership and use, with a rising car share of total trips. Moreover, as the public transport system in Copenhagen has been steadily expanded and improved, more and more Copenhageners are using public transport instead of cycling, especially as the city has spread out and trip distances have increased. With the shift to cars and

**Table 18.1**

Percentages of Copenhagen cyclists satisfied with various aspects of cycling conditions biannually from 1996 to 2016

| Indicator | 1996 | 1998 | 2000 | 2002 | 2004 | 2006 | 2008 | 2010 | 2012 | 2014 | 2016 |
|---|---|---|---|---|---|---|---|---|---|---|---|
| Copenhagen as a cycling city | 71 | 79 | 79 | 77 | 83 | 83 | 85 | 93 | 95 | 94 | 97 |
| Amount of cycle tracks | 56 | 55 | 66 | 58 | 64 | 65 | 65 | 68 | 76 | 80 | 87 |
| Cycle track maintenance | 48 | 51 | 40 | 45 | 50 | 48 | 54 | 50 | 61 | 63 | 71 |
| Cycle track width | 67 | 68 | 62 | 45 | 50 | 48 | 43 | 47 | 50 | 53 | 62 |
| Road maintenance | 24 | 27 | 23 | 28 | 27 | 28 | 26 | 31 | 32 | 36 | 44 |
| Combination of bicycle and public transport | 54 | 44 | 49 | 51 | 54 | 58 | 49 | 55 | 60 | 60 | 53 |
| Bicycle parking generally | — | — | — | — | 30 | 26 | 26 | 27 | 29 | 33 | 37 |

Source: City of Copenhagen (2016).

public transport, cycling levels have stopped their long trend of rising and indeed fell slightly between 2014 and 2016.

## The Role of the Car

Automobility is increasingly affecting Denmark and its culture. The concept of automobility is often referred to as a social concept of the automobile society and includes much more than just car traffic, also including socio-economic factors, culture, politics, and safety (Beckmann 2001; Beckmann 2004; Sheller and Urry 2000; Urry 2004). Car ownership in Copenhagen increased between 1999 and 2008, from 155 to 186 per 1,000 inhabitants, and has fluctuated since then. In 2016, it was 186 per 1,000 inhabitants (City of Copenhagen 2018b). Moreover, the percentage of trips made by car rose from 28% to 34% from 2006 to 2016 (see table 18.2). The increase in car ownership and use in Copenhagen may seem low compared to other European cities but it was problematic for Copenhagen, which for decades was used to a very low level of car ownership and use. The recent increase in motorization has probably contributed to the slight decline in bicycling mode share of trips from 31% in 2006 to 29% in 2016 (see table 18.2).

There are various possible explanations for recent increases in car owner-ship and use. Between 2000 and 2016, the real, inflation-adjusted incomes of Copenhagen residents increased by 26%. Moreover, in 2015, the Danish government lowered taxes on new car registrations from 180% on the pre-tax purchase price to 100%. The dramatic reduction in the tax rate on cars will strongly encourage rising car ownership and use in the coming years.

**Table 18.2**
Trends in modal split in Copenhagen

|                  | 2006 | 2012 | 2016 |
|------------------|------|------|------|
| Car              | 28.0 | 33.3 | 34.0 |
| Public transport | 13.9 | 18.3 | 18.0 |
| Cycling          | 31.0 | 26.9 | 29.0 |
| Walking          | 26.8 | 21.0 | 19.0 |
| Other            | 0.4  | 0.5  | 0.0  |
|                  | 100  | 100  | 100  |

*Note*: The trips included in this table refer to all trips starting in and/or ending in Copenhagen.
*Source*: City of Copenhagen (2016).

Moreover, Danish per capita incomes are forecast to rise even further, thus making car ownership and use more affordable.

Increasing car ownership levels will almost certainly affect cycling levels—as they have elsewhere in the world—because car ownership encourages car use. Copenhagen cannot influence some of the trends encouraging rising car ownership, such as increasing income or national taxes. Nevertheless, the city can implement local policies discouraging car use, such as more restrictive and expensive parking policies, lower speed limits, more circuitous car routes, traffic calming of residential neighborhoods, car-free zones, and congestion charges for the city center (as in Stockholm, Oslo, and London). In 2011, an experiment with new parking spaces for both cyclists and motorized traffic was introduced, using on-street car parking after 5 p.m. for bicycle parking. Planners have also reduced the number of car parking spaces at small-scale building projects, mainly in the inner-city parts of Copenhagen, in order to make car use less attractive.

The combination of higher incomes and lower car taxes makes it more affordable to own a car in Denmark overall, and thus in Copenhagen as well. This especially encourages more recreational car use for leisure trips outside Copenhagen, where recreational destinations are often difficult to reach by public transport and cycling. Moreover, cycling to school is also declining, as Danish parents have become increasingly concerned about perceived personal and traffic safety problems and thus drive their children to school to ensure their safety (see McDonald, Kontou, and Handy, chapter 12, this volume; Freudendal-Pedersen 2015). By comparison, car use seems least likely to increase for the commute to work in the city itself, where trips are much shorter and cycling is a convenient, inexpensive, and quick way to reach the workplace. On balance, it seems certain that overall cycling levels will fall in the coming years, while car use increases. Increased car use will endanger and discourage cycling by making it more stressful and dangerous to share roads with cars, especially at intersections (Haustein et al. 2020).

### The Role of Public Transport

As shown in table 18.2, the share of trips by public transport increased from 14% in 2006 to 18% in 2016. Rising public transport use in Copenhagen is mainly the result of the opening of the city's first metro line, which opened in October 2002. Since then, the metro network has been extended several times, with new lines and stations. The metro system tripled in size from 14 km (with 15 stations) in 2002 to 43 km (with 48 stations) in 2019. The construction and continuous expansion of the metro has not led to a decrease in the share of motorized traffic, which rose from 28% to 34% of trips. Instead it led to a sharp decrease in the share of trips by walking (from 27% to 19%) and a slight decrease in the share of trips by cycling (from 31% to 29%). Thus, the planned expansion of the metro is likely to further reduce walking and cycling mode shares while having little impact on car use. One possible strategy to increase cycling as the metro system expands would be to improve the integration of cycling with public transport. That would include much better bike parking facilities at all metro stations (sheltered, secure, and ample parking), bikesharing focused on metro stations, and inexpensive bike rentals at metro stations combined with discounted monthly public transport tickets (such as OV-Fiets in the Netherlands).

### How to Reverse the Trend in Falling Cycling Levels

The city of Copenhagen has established plans to reverse the recent trend of falling cycling levels. Indeed, the city has set the goal of increasing the share of work and education trips made by bike from 41% in 2018 to 50% by 2025 (City of Copenhagen 2016). This goal is supported by a specific plan and increased funding to expand and improve cycling infrastructure.

To maintain its position as one of the world's leading cycling cities, Copenhagen will have to make massive investments in policies, programs, and infrastructure to improve the competitiveness of cycling, especially relative to the car. As noted in two recent publications, the key problem in Copenhagen is the conflict between cars and bicycles. Handerson and Gulsrut (2019) describe political fights over limited street space in Copenhagen because of rising car ownership and use. Freudendal-Pedersen (2020) argues that restricting car use in Copenhagen is problematic because of rising public and political support for accommodating the car, as an increasing number of Copenhagen residents have cars.

Planners, cycling advocates, and cyclists themselves, however, are committed to making cycling an increasingly convenient, safe, and quick way to get around Copenhagen for daily travel. As of late 2019, cycling was still a popular means of daily travel in Copenhagen. Increasing the mode share of bicycling, however, will require significant restrictions on motorized traffic. Perhaps most important, street space needs to be reallocated from motor vehicles to cyclists to enable the needed expansion of cycling infrastructure, which would help alleviate crowding of bike lanes and paths during peak hours. Fewer car parking spaces, limited parking times, and higher parking costs must be imposed on motorists. The general speed limit for cars should be reduced, and all residential neighborhoods should be traffic calmed at 30 km/h or less. Car-free zones and home zones would also discourage car use and encourage cycling. With congestion charging already in effect in nearby Stockholm and Gothenburg, Copenhagen has excellent Scandinavian examples to follow for implementing congestion charging in Copenhagen as well. Although such measures will evoke fierce opposition from many motorists, and thus from some politicians as well, they would surely be the most effective way to discourage car use and encourage cycling as well as walking and public transport use, all three of which form an essential package of integrated sustainable transport modes. The big question is whether such measures would be blocked by motorists and politicians supporting the accommodation of more car ownership and use. If such car-restrictive measures cannot be implemented, it seems certain that cycling in Copenhagen will have passed its peak and that the decline in cycling observed in recent years will continue into the future.

### The Case of Amsterdam: An Unknowingly Skilled Cycling City

The historic, concentric structure of Amsterdam's center and the close-knit urban fabric of canals and narrow streets have played an important role in the success of cycling in Amsterdam. That, however, does not fully explain how cycling has remained so dominant in recent decades despite profound changes in the historic urban fabric of the city, newly developed city districts, dramatically increased car ownership, and other social, economic, and technological trends of the past five or six decades.

Amsterdam is a city of over 800,000 inhabitants in a metropolitan area with over two million people. It consists of a patchwork of different "urban fabrics," defined as being "the pattern form created by tying buildings, streets and open spaces to one another" (Sennett 2018, 38). Each of the identifiable "fabrics" in Amsterdam is a result of the interplay between urban development and mobility dynamics. In this sense, Amsterdam is a patchwork of typical walking fabrics, public transport fabrics, and car fabrics. This results in radically different modal splits for different neighborhoods (see table 18.3). Nevertheless, the quality of the cycling network is at a similarly high level throughout the city and its suburbs (figure 18.3).

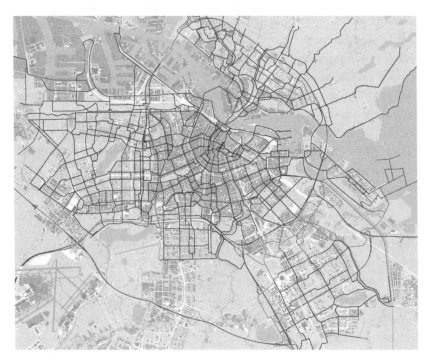

**Figure 18.3**
The main cycling infrastructure network (Hoofdnet Fiets) in the city of Amsterdam.
*Source*: City of Amsterdam.

In addition, most local streets in the central district have low speed
limits for cars and allow cyclists to share the road. That results in a fine-
grained, diversified, multilevel integrated network of bikeways and bike-
friendly neighborhoods.

As in Copenhagen, a crucial component that adds to the understanding of how and why cycling remained a mainstream transport mode in
this diverse patchwork has been the bottom-up social pressure at different
points in time and in different guises, as we will discuss. This made cycling
in Amsterdam resistant to most large-scale developments that marginalized cycling in many other cities around the world. To examine how this
worked, we look at three pivotal moments in the city's recent history that
divide its history into three distinct periods. Because these three different
phases in trends over more than a hundred years play an important role,
we cover the history of Amsterdam more elaborately than for Copenhagen.

## The History of Cycling in Amsterdam: A Head Start and a Rebound

Until the 1920s, cycling played a prominent role in many European cities, including Amsterdam. After 1920, public transport became a formidable competitor to the bike in many cities. In Amsterdam, traveling by tram was expensive, and the supply of cheap bikes from Germany enabled cycling to remain the best and cheapest means of travel (Jordan 2013). In addition, the city's many canals formed barriers that hindered the construction of public transport and car networks with convenient and direct connections. The historic structure remained largely intact during World War II. This starting point offered Amsterdam the ideal conditions for cycling to thrive until the late 1950s: a compact and mixed urban fabric with short average distances between functions.

That changed with the implementation of the General Expansion Plan that was developed by Cornelis Van Eesteren, head of the City Development Department of Amsterdam from 1929 to 1959. The plan proposed revising the land-use structure in Amsterdam so that it would be transformed into a "functional city" made up of single-function residential and employment districts separated from each other. Fast mobility, mainly individual auto-mobility, offering quick access between those districts was essential to this new approach to transport and land use. Van Eesteren, an influential architect and urbanist, was a strong proponent for strict functional separation within cities. Large new housing districts were developed in the South East and New West (garden cities). Spacious development ("light, air, space") was seen as the opposite of the densely populated city center that was facing decline as well as severe hygienic and livability challenges. As part of Van Eesteren's approach, the city center would become mainly a business and services district with few residences.

As a result, the average distances between the different urban functions increased. Increasing trip distances had to be served by large investments in extensive roadway and public transport networks. Although the neighborhoods were supposed to be self-sufficient and planned as "health machines" with lots of light, air, and space, the long-term result was large numbers of physically inactive motorists and tram passengers. In a second phase, this dynamic was even more severe, when new urban developments in so-called bundled deconcentration resulted in large new towns farther away from the central city of Amsterdam. The new suburban extensions offered a *clean slate*. Wide streets and an abundance of green space offered enough space to develop high-quality separated cycling infrastructure along the new roadways and through the parklike landscapes. Arguably, one can find the world's highest standard cycling infrastructure in these newer urban districts (table 18.3).

However, since these new suburban neighborhoods also catered to high-quality car and public transport mobility, cycling levels have been much lower than in the more densely populated city center. Their lesser accessibility to amenities requires more use of cars and public transport and reduced cycling conditions (table 18.3). It would be an oversimplification, however, to explain spatial variation in cycling levels on the basis of population density and infrastructure alone (Nello-Deakin and Harms 2019). Complex socioeconomic developments, for instance related to ethnicity and income levels, also play a role. Non-Western immigrants, for instance, have had much lower cycling rates than longtime Dutch residents. Unlike most other countries, moreover, cycling in Amsterdam (and other Dutch cities) is positively correlated with income (Harms et al. 2016). Another factor that discourages cycling in suburban neighborhoods is the isolation of cycle paths located in lush, green areas that are rather secluded and out of public sight. That creates personal safety problems by subjecting cyclists to possible physical attack and sexual assault or harassment, an especially important problem for women (see Garrard, chapter 11, this volume).

**Table 18.3**
Cycling share and accessibility of amenities across Amsterdam neighborhoods

| Neighborhood | Mode share (% of trips by residents on workdays) | | | Average number of amenities (within 3 km of home) | | | |
|---|---|---|---|---|---|---|---|
| | Cycling | Public transport | Car | Doctors | Child care | Restaurants | Supermarkets |
| Centrum | **68** | 19 | 14 | 61 | 107 | 1,067 | 61 |
| West | **61** | 21 | 16 | 53 | 103 | 711 | 50 |
| Oost | **53** | 27 | 20 | 37 | 64 | 290 | 33 |
| Zuid | **45** | 14 | **41** | 52 | 108 | 630 | 48 |
| Nieuw-West | 28 | 31 | **41** | 23 | 37 | 85 | 21 |
| Noord | 27 | 22 | **50** | 10 | 14 | 26 | 8 |
| Zuidoost | 26 | **46** | 28 | 11 | 27 | 25 | 10 |

*Note*: Bold numbers indicate the dominant transport mode in each neighborhood, defined as the means of travel accounting for the highest percentage of trips.
*Source*: Based on data from the City of Amsterdam (2016).

In these new single-function suburban areas, the car became, and to this day still is, a mainstream means of travel. Access from the motorways to both the expansion districts and the growth centers around Amsterdam was optimal. Even within the new suburban neighborhoods, cars were accorded plenty of space, with generous access routes and ample parking spaces nearby, often free. Remarkably, this strong position of the car was strengthened further by the prevailing philosophy of completely separating car and cycling networks. That provided a smooth flow of car traffic throughout the urban environment.

The increasing dominance of cars in traffic space across the city coincided with other related trends. First, the newly developed Bijlmer housing district in the South-East needed to be connected to the jobs in the central city by the first metro line of an envisioned well-developed metro network. Construction of the new metro, opened in 1977, was intended to connect Zuidoost with the city center. But the construction of a 3.5 km metro tunnel meant that a large part of the dilapidated Nieuwmarkt district had to be demolished. Once the metro was finished, this dilapidated district would rise again as a large-scale city district with urban highways, offices, and shops.

Second, the increase in car use led to a dramatic decline in cycling levels throughout the city year after year, just as in other cities in the Netherlands and across Europe. Large-scale building demolition was a common occurrence at the time, but not because of a shortage of plans to modernize the city. So, what stopped it?

After years of car-oriented policies, modernization visions for the historic city, and expansion of the public transport network, the plans for the demolition and rebuilding of Nieuwmarkt unlocked a broad countermovement. This redevelopment plan was already being implemented, with buildings demolished and the metro tunnel close to completion. With the impending results increasingly visible, the project sparked massive protests from all strata of society. Protesters included squatters fighting for affordable housing, housewives calling for more attention to the basic livability of their neighborhoods, parents who protested the alarmingly high number of child traffic fatalities ("Stop de Kindermoord," meaning "stop the child murder"), and building conservationists. The protests culminated in violent riots that required the intervention of armed forces in 1975 (Feddes, de Lange, and te Brömmelstroet 2019). The widespread and united opposition of the diverse protesters to abandon the modernistic ideals for a futuristic city and instead focus on the needs of the existing inhabitants can arguably be seen as the world's first "integrated land use and mobility" riots. The broad-based, strong public resistance ultimately succeeded by changing the planning question to: how can the vitality of cities with a small-scale, fine-grained structure be sustained without destroying their essential character? Cycling was not a central theme in this movement, but it became a powerful symbol, as expressed by the slogan attributed to Robert Jasper Grootveld: "A bike is something, and almost nothing" (de Wildt 2015).

To respond to this bottom-up resistance against the functional city concept, there were key changes in urban and mobility planning strategies in the late 1970s. Key was the rebuilding of small-scale urban fabric at the places where it had been demolished as well as preserving the remaining mixed-use districts everywhere else. At the same time, traffic calming (maximum speed 30 km/h) and the diversion of through traffic around residential neighborhoods and the city center were introduced on a massive scale throughout the Netherlands. The result was that car drivers became "guests" on most of the urban streets throughout the city, while pedestrians and cyclists were prioritized as the main intended users. Some researchers find that this dramatic increase in the priority of walking and cycling over driving was the main factor in the remarkable improvements in road safety across the Netherlands that started in 1975 (Schepers et al. 2017). Major streets such as arterials were usually redesigned to offer cyclists protected cycling infrastructure, separated from moving motor vehicle traffic by a buffer of parked cars. While never viewed as the main goal, cycling obviously benefited directly from these new planning directions.

### Recent Trends—the Hardly Visible, Slow Hand

The road to recovery was not introduced by a strong political vision or document. On the contrary, cycling got its own vision, mission documents, and long-term plans quite late, in the mid-1990s, similar to Copenhagen. Most of these cycling planning documents and policy statements were a response to growing levels of cycling instead of a specific, previously developed policy goal of advancing cycling.

In 1992, Amsterdam held a referendum to decide on the further reduction of car traffic in the city center. The people of Amsterdam voted by a small majority in favor of a policy aimed at a low-traffic city center. City center residents, in particular, came out in favor of the low-traffic plan, since they experienced the most negative effects. Measures included slowly reducing the number of on-street parking places, constructing parking garages around the center, sharply increasing parking fees and one-way traffic on main roads, mixing traffic instead of having full separation where possible, and improving public transport. None of these measures were directly and fully implemented, but 25 years later the incremental steps that were taken in this direction are clearly visible. In 1992, car traffic dominated trips within Amsterdam. By 2016, cycling and walking had become the dominant modes—especially in central neighborhoods (table 18.3).

Another, almost invisible policy also took a long time to manifest itself. On the national level, a law was passed that prohibited shopping centers from being developed outside the existing built-up areas of Dutch cities (the Weidewinkel Wet, or Big Box Retail law). This was important for preserving the limited open space around cities but also for revitalizing inner-city shopping districts. Small supermarkets are now dispersed across the city, providing a fertile ground for cycling trips, especially in trip chaining between work and home. This spatial distribution co-evolved with increases in cycling rates. As a result, everybody has at least one supermarket only a short distance from home. The same applies to other amenities, such as day care, primary schools, and cash machines.

In 2002, another important element was added to the mix. Roland Haverman, a small entrepreneur, launched an improved version of rental bicycles that were offered at Dutch railway stations. His OV-Fiets has been growing exponentially ever since; in 2018, around 20,500 bicycles served over 4.2 million rides a year. OV-Fiets played a pivotal role in unlocking the potential of integrating bicycles and trains from a user perspective. Since 2002, both the number of train trips and the percentage of people who arrive at the station by bicycle have been steadily growing. Of course, this trend cannot be attributed solely to the introduction of OV-Fiets. However, this integration of cycling with public transport offers important synergetic effects that further promote the success of cycling in Amsterdam.

With each of its 10 railway stations having a large, overlapping catchment area, cyclists can personalize their trip chaining, thus offering a very flexible and fast mobility option. As a result, the city center of Amsterdam offers a cyclist over 200 train connections within a 20-minute bicycle ride. An individual traveler can combine high-speed public transport for the longest part of this personalized chain with the highly flexible routing of cycling, which can reach virtually every location in the city. The overall speed of such a combination of public transport with cycling rivals that of free-flowing car traffic in most urban settings (Kager, Bertolini, and te Brömmelstroet 2016). Over half the 1.2 million train trips each year in the Netherlands start with a bicycle ride, and approximately one-third of all cycling trips are directly linked to a public transport trip (Jonkeren et al. 2018).

The combined use of bicycle and train has been growing rapidly since 2002, with an estimated 5% annual growth rate. Motor vehicle traffic has been slowed down and mixed with cyclists where possible. Fine-grained, highly textured urban development has been favored over large projects. In addition, the supply of car parking has been reduced and its price raised via taxes. A final factor that has supported cycling is the very slow implementation of Amsterdam's long-term metro plan. Thus, public transport is a less attractive alternative than cycling for most trips within the city. The new north-south metro line, linking the northern part of Amsterdam with the city center, did not open until mid-2018, seven years later than planned.

## A Future Cycling City? New Challenges

Amsterdam might have reached another turning point, heralding the fourth phase of its development as a cycling city. We have seen that current favorable conditions were not always planned. A complex mix of various factors contributed to the current state of cycling in Amsterdam. The most important factors include an urban structure that encourages cycling, poor alternatives to cycling, grassroots public pressure supporting cycling, and complementary policy choices, especially restrictions on car use (see also Harms, Bertolini, and te Brömmelstroet 2016).

Those favorable conditions will facilitate further growth in cycling in Amsterdam. The urban structure, enhanced by future densification and mixing of functions, is the ideal land-use pattern for cycling. Cycling has a very positive image among most of its citizens. Moreover, cycling also is a key focus of city policies and plans for an *Autoluwe Stad* ("car-light city") and the future redevelopment of large brownfields. Could this be a new "perfect storm" to usher in a new phase of cycling in Amsterdam?

A new generation of city dwellers and entrepreneurs is locating in Amsterdam, which is also attracting a growing number of tourists and expats. Planners, policy makers, and urbanists are increasingly viewing cycling as an indicator for social, healthy, and economically successful cities. To ensure future success, however, some challenges have to be overcome, as discussed in the next section.

## Lessons Learned from Both Cities

First, we draw some lessons related to cycling trends. How were Amsterdam and Copenhagen able to reverse the downward trend in the mode share of cycling? The widespread protests and related advocacy during the 1970s were an important factor. In response, both cities slowly implemented a wide variety of cycling-related policies, with cycling infrastructure playing a dominant role. Cycling-related policies were often based on strategies for safer and more livable neighborhoods. The role of the car was reduced, and land-use strategies favored urban centers over big box retail on the urban fringe. Until the 1990s, these policies were not primarily aimed at increasing cycling, but they created important necessary conditions for the wide-scale recovery of cycling levels.

This raises the question: which policies support cycling most? The stories of cycling Amsterdam and Copenhagen show that a wide range of measures is necessary. Adequate cycling infrastructure is a prerequisite but not the only relevant factor. The two cities' success in keeping cycling mainstream is the product of an intricate web of interrelated factors that span social, spatial, and technical domains. Nevertheless, an important lesson from both cities is that existing land use and land-use planning for the future supported cycling and vice versa. High to moderate population densities and mixed-use developments both favor cycling. Conversely, high cycling levels support dense and mixed-use urban developments. Thus, just as car use and suburban sprawl co-evolve, cycling and dense, mixed-use urban structures co-evolved.

Recent changes in policy-making and shifting cultural attitudes toward cycling indicate that Copenhagen and Amsterdam might be headed in different directions. In recent plans for Amsterdam, the intrinsic logic of cycling and walking is increasingly being taken as a starting point for the redesign of public spaces and roadways to deal with future traffic growth. The exception is the additional metro line, opened in 2018, connecting the north of the city to the center. Copenhagen, on the other hand, has been greatly expanding a new metro system that has contributed to the recent stagnation of cycling. The policy focus is still on cycling, but public transport is getting increasing emphasis. The shift of policies and funding in favor of public transport in Copenhagen has made public transport a much stronger competitor to cycling than in Amsterdam, where cycling remains faster, cheaper, and more convenient for most trips. While Amsterdam has clearly prioritized cycling and walking over the car and even public transport, Copenhagen has tried to accommodate all travel modes. As it turns out, however, the car and public transport have now become formidable competitors to cycling. As shown in table 18.2, the modal share of the car in Copenhagen is larger than the share of the bike and public transport. Amsterdam, on the other hand, is prioritizing pedestrians and cyclists over cars for ever-larger portions of the inner city.

So what about the future prospects for cycling in Copenhagen and Amsterdam? First, it is important to realize that increasing cycling levels is not an aim in itself. Rather, cycling is a way to provide affordable, convenient, and safe transport for everyone, to ensure high levels of accessibility, to offer an attractive alternative to car use, to make cities more livable and attractive, to improve environmental quality, and to enhance public health. Based on personal experience, it is our impression that cycling policies are easier to implement if they are not promoted solely to reduce car use but rather because of many broader benefits for society in general (i.e., accessibility, equity, health, and environment). Emphasizing that everyone benefits provides a broader base of public and political support.

What is the best strategy for promoting future growth of cycling in Copenhagen and Amsterdam? We see several challenges. The first cluster of challenges is related to infrastructure. We find that cycling infrastructure is of paramount importance, a prerequisite for the success of cycling. Both the Copenhagen and Amsterdam models for providing cycling infrastructure work well to foster cycling within a car-dominated landscape. Unfortunately, most of the physical space that has been devoted to such high-quality cycling facilities was taken from sidewalks. Parts of both cities have reached tipping points where cycling and walking dominate and where the separated cycling and walking facilities are no longer sufficient to accommodate the increasing volumes of walk and bike trips. The need for more space for cyclists and pedestrians requires radical revisions in how roadways and public spaces are designed. Cycling and walking, not motorized traffic, should dominate the design. The most effective way forward is to radically shift space away from both parked and driving motor vehicles.

Another step would be to scale up cycling success from the cities to a more regional scale. Strategies for this could include a network of regional express bikeways providing comfortable and safe infrastructure for longer trips. Such a network would facilitate cycling not only on conventional bicycles but also on faster e-bikes, which account for a rapidly increasing share of bicycle sales. Such a regional network would link cycling with high-capacity interurban public transport via high-quality transfer points with ample, sheltered, and secure bike parking.

The regional scale of cycling infrastructure will become increasingly important because e-bikes make cycling attractive even over longer distances (see Cherry and Fishman, chapter 9, this volume). Cycling infrastructure should be able to accommodate a further increase in the use of e-bikes, especially in larger metropolitan regions such as Copenhagen and Amsterdam. The most important challenge is to design bike networks that can merge both types of bikes. Next to having more space for cyclists in general, the inclusiveness of cycling should be safeguarded. Separated bicycle paths should continue to provide a safe and relaxed place to cycle for all ages and abilities. Motorized, potentially high-speed e-bikes, scooters, and other mobility innovations should not be allowed on separated cycle paths if the resulting speed differences cause serious safety problems.

Infrastructure matters, but so do complementary programs. An important challenge is to design policies for the increasing role of several forms of bikesharing in addition to, or as an alternative to, bike ownership. This applies to all cities with bikesharing systems but especially densely built, compact cities such as Copenhagen and Amsterdam, which continuously struggle to accommodate conflicting spatial claims. Dockless bikesharing systems raise concerns, particularly in the most densely populated parts of the city, which already experience high rates of parked bicycles.

As with any complex innovation in the public-private domain, it is difficult to come up with one ideal package of policies to implement. The same solution will not be appropriate for all cities. Cities worldwide should be inspired by best practices of the most successful cities, but they need to adapt what other cities have done to their own specific situation. Thus, other cities should experiment within their own specific context and learn their own lessons to determine the policies that are best for their own situations.

Both the Danish and Dutch cultures are egalitarian relative to most other countries, but not all groups cycle. Therefore, an important challenge is to get those demographic groups that currently do not cycle to adopt cycling, both for its health benefits and for its social benefits (see in this volume Garrard et al., chapter 3; Martens, Golub, and Hamre, chapter 14).

A final challenge relates to understanding possible saturation levels in cycling. Where are cycling levels stabilizing or declining in Copenhagen and Amsterdam? Is there something like peak cycling? Are there cycling levels at which cycling starts to have more adverse effects on society than positive ones (e.g., too much bike parking)? If so, what does this mean for cycling policies? It should be emphasized, however, that Copenhagen and Amsterdam have exceptionally high levels of cycling. The theoretical, potential problem of too much cycling only exists for a few of the world's most bicycling-oriented cities.

## References

Beckmann, Jörg. 2001. Automobility—a Social Problem and Theoretical Concept. *Environment and Planning D: Society and Space* 19: 593–607.

Beckmann, Jörg. 2004. Mobility and Safety. *Theory, Culture and Society* 21 (4–5): 81–100.

City of Amsterdam. 2016. *Amsterdamse Thermometer Bereikbaarheid*. Amsterdam: Gemeente Amsterdam.

City of Copenhagen. 2016. *Copenhagen Bicycle Account 2016*. Copenhagen: City of Copenhagen.

City of Copenhagen. 2018a. *Cykelregnskab 2018*. Copenhagen: City of Copenhagen.

City of Copenhagen. 2018b. *Statistikbanken*. City of Copenhagen. http://sgv2.kk .dk:9704/analytics/saw.dll?PortalPages.

Copenhagen Police. 2018. *Politiets registreringer*. Copenhagen: Copenhagen Police.

Danish Technical University (DTU). 2019. *Danish National Travel Survey*. Copenhagen: Ministry of Transport. https://www.cta.man.dtu.dk/english/national-travel-survey /hovedresultater.

de Wildt, Annemarie. 2015. *Provo's Fietsenplan*. Amsterdam: Provo. https://hart.amster dam/nl/page/49069/witte-fietsenplan.

Eltis. 2018. Cycle Policy—for the First Time in Copenhagen (Denmark). Eltis. http:// www.eltis.org/discover/case-studies/cycle-policy-first-time-copenhagen-denmark.

Feddes, Fred, Marjolein de Lange, and Marco te Brömmelstroet. 2019. Hard Work in Paradise: The Contested Making of Amsterdam as a Cycling City. In *The Politics of Cycling Infrastructure: Spaces and (In)equality*, edited by Peter Cox and Till Koglin, 133–156. Bristol: Policy Press.

Freudendal-Pedersen, Malene. 2015. Cyclists as Part of the City's Organism: Structural Stories on Cycling in Copenhagen. *City and Society* 27(1): 30–50.

Freudendal-Pedersen, Malene. 2020. Velomobility in Copenhagen—a Perfect World? In *The Politics of Cycling Infrastructure: Spaces and (In)equality*, edited by Peter Cox and Till Koglin, 179–194. Bristol: Policy Press.

Handerson, Jason, and Natalie Marie Gulsrut. 2019. *Street Fights in Copenhagen—Bicycle and Car Politics in a Green Mobility City*. Oxon: Routledge.

Harms, Lucas, Luca Bertolini, and Marco te Brömmelstroet. 2016. Performance of Municipal Cycling Policies in Medium-Sized Cities in the Netherlands since 2000. *Transport Reviews* 36(1): 134–162.

Haustein, Sonja, Till Koglin, Thomas Alexander Sick Nielsen, and Åse Svensson. 2020. A Comparison of Cycling Cultures in Stockholm and Copenhagen. *International Journal of Sustainable Transportation* 14(4): 280–293.

Jensen, Niels. 2013. Planning a Cycling Infrastructure—Copenhagen—City of Cyclists. In *Cyclists & Cycling around the World: Creating Liveable & Bikeable Cities*, edited by Lotte Bech, Juan Carlos Dextre, and Mike Hughes, 1–12. Lima: Fondo Editorial, Pontificia Universidad Católica del Perú.

Jonkeren, Olaf, Roland Kager, Lucas Harms, and Marco te Brömmelstroet. 2019. The Bicycle-Train Travellers in the Netherlands: Personal Profiles and Travel Choices. *Transportation*. Published online first, October 26. https://link.springer.com/article/10.1007/s11116-019-10061-3.

Jordan, Peter. 2013. *In the City of Bikes*. London: Harper Perennial.

Kager, Roland, Luca Bertolini, and Marco te Brömmelstroet. 2016. Characterisation of and Reflections on the Synergy of Bicycles and Public Transport. *Transportation Research Part A: Policy and Practice* 85: 208–219.

Koglin, Till. 2013. Vélomobility—a Critical Analysis of Planning and Space. Doctoral diss., Lund University, Department of Technology and Society, Transport and Roads, Bulletin 284.

Koglin, Till. 2015. Organisation Does Matter—Planning for Cycling in Stockholm and Copenhagen. *Transport Policy* 39(1): 55–62.

League of American Bicyclists (LAB). 2018. *2018 Benchmarking Report on Bicycling and Walking in the United States*. Washington, DC: League of American Bicyclists.

Nello-Deakin, Samuel, and Lucas Harms. 2019. Assessing the Relationship between Neighbourhood Characteristics and Cycling: Findings from Amsterdam. *Transportation Research Procedia* 41: 17–36.

Schepers, Paul, Divera Twisk, Elliot Fishman, Aslak Fyhri, and Anne Mette Dahl Jensen. 2017. The Dutch Road to a High Level of Cycling Safety. *Safety Science* 92: 264–273.

Sennett, Richard. 2018. *Building and Dwelling: Ethics for the City*. London: Farrar, Straus and Giroux.

Sheller, Mimi, and John Urry. 2000. The City and the Car. *International Journal of Urban and Regional Research* 24(4): 737–757.

Urry, John. 2004. The "System" of Automobility. *Theory, Culture and Space* 21: 25–39.

# 19   Implementation of Pro-bike Policies in Portland and Seville

Roger Geller and Ricardo Marqués

Portland, Oregon, in the United States and Seville, Andalusia, in Spain are cities of similar population size that have increased cycling levels roughly sixfold. While the two cities and their transport policies differ in many important aspects, the cornerstone of bicycle promotion in both cities has been the development of a high-quality, connected network of bikeways with the goal of increasing bicycling among average citizens. By 2018, both cities had created cohesive bicycling networks that served multiple major destinations and that connected to residential and mixed-use areas. Both cities were able to expand and improve their networks because of successful political leaders and strong local cycling movements that prioritized bicycling and recognized the fundamental importance of developing a network of connected bikeways to increase bicycle use both overall and for a wide spectrum of their residents.

While there are similarities between the cities, there are also differences. Seville is a medieval city with three times Portland's population density. Though both cities began with a bicycle mode share of approximately 1%, Portland achieved its highest bicycle use, 7%, in 2014, after taking more than 20 years to develop a cohesive bikeway network. Seville rapidly built a cohesive network in four years and increased its bicycle mode share of trips from 1% to 6% in that short time.

Portland's bicycle network comprises a variety of bikeway designs, which reflect more than two decades of shifting guidance and standards. At least half the city's network falls short of current infrastructure design standards for safety and comfort. By comparison, Seville built its bicycle network rapidly and in an era with widespread expertise in modern bikeway design. Most of its network of cycling facilities are similarly designed and provide physical

**Table 19.1**
Demographic and climatic characteristics of Portland and Seville

| | Population (in 1,000s) | | Population density (per km²) | | Per capita gross domestic product (US$) | Percentage of car-free households | Percentage of university students (city) | Annual climate normals | | |
| | City | Metro region | City | Metro region | | | | Precipitation (cm) | Days below 0°C | Days above 32°C |
|---|---|---|---|---|---|---|---|---|---|---|
| Portland | 605 | 2453 | 1748 | 142 | 62,606 | 14.30 | — | 91 | 0 | 0 |
| Seville | 689 | 1536 | 4887 | 313 | 22,800 | 19.76 | 14 | 53 | 0 | 80 |

*Sources:* Data obtained by the authors from each of the two cities.

separation between cyclists and motor vehicles, thus adhering to current best practice in cycling infrastructure (see Furth, chapter 5, this volume). A significant difference between the cities suggests the importance of not only creating safe and comfortable conditions for bicycling but also actively discouraging driving. Sevillians' rapid adoption of bicycling, and their achievement of higher cycling rates than in Portland, is correlated with restricting car use in its historic city center, which had been a major attractor for car trips.

This chapter demonstrates the crucial need for political support to implement bicycle-friendly policies and the results that can be achieved by the modest investments in building a high-quality and cohesive bicycle network.

## How Portland, Oregon, Became a Bicycle City

Until the early 1990s, the bicycling situation in Portland, Oregon, was similar to that in most other North American cities. There was little separate bicycle infrastructure, and levels of cycling were low. There was insufficient political support for improving cycling conditions despite small pockets of local advocacy. In sharp contrast, by 2018, Portland boasted a comprehensive, integrated bikeway network extending more than 630 km. The mode share of regular bicycle commuters to work was around 7% citywide and as high as 20% or more in neighborhoods close to the city center. Much of Portland's transport capital program is devoted to bicycling and pedestrian infrastructure projects.

This case study tells the story of Portland's transformation—and continuing evolution—based on four elements: (1) preexisting conditions conducive to bicycling that aided Portland's early and continuing efforts; (2) political support magnified by Portland's unique form of city government; (3) a dedicated and long-term cadre of experienced and committed professional staff and advocates who identified the most effective policies to promote cycling and helped implement them; and (4) research that supported Portland's introduction of European-inspired bikeway designs in a US city.

### Preexisting Conditions Conducive to Bicycling and Good Planning
The development of Portland's bikeway network was facilitated by its tight street grid, its low-speed downtown, a mixed-use planning and development pattern, and state-level "complete streets" legislation. These factors

enhanced—and continue to support—Portland's initial modest improvements to its bikeway network.

Portland's historic street grid is based on 20 blocks per mile. Where most complete, this grid creates a network of closely spaced, lengthy parallel roadways. By classification policy and design, collector streets (small arterial streets) are spaced approximately every quarter mile to half mile, leaving low-volume local streets between them. Portland used those local streets to create its grid of bicycle boulevards, which are low-volume, low-speed roadways with enhancements (traffic calming, crossing treatments, traffic diverters) that prioritize their use for people bicycling. When they meet established guidance for automobile car volumes and speeds, these low-stress, functional bikeways are attractive to riders of all ages and abilities.

Portland developed its first small subnetwork of bicycle boulevards in the late 1980s to early 1990s as part of a neighborhood traffic-calming program. This subnetwork was implemented on roadways that led directly to Portland's commercial core. For decades, the neighborhoods served by these bikeways have been among the leading neighborhoods in bicycle-to-work mode splits (figure 19.1).

Relative to other North American cities, Portland's bikeway network has relied heavily on its bicycle boulevards. Few large North American cities have the type of street grid that can be developed to create such a network. Portland's bicycle boulevards account for 25% of its existing bikeway network and half its "all ages and abilities" network. It is a combination of Portland's street spacing, especially close to the commercial center, and an early focus on traffic calming that enabled Portland's development of bicycle boulevards.

Portland's tight street grid in the inner city also resulted in its arterial streets being low speed and narrower than in most cities. Many Portland collector streets have either a two-lane cross section or now, following decades of safety improvements, a three-lane cross section (with the middle lane serving as a turning lane for both directions) on those collectors that were previously four lanes. They also operate at low speeds, typically 25–30 mph. An important benefit of slow motor vehicle traffic is that standard bicycle lanes, which have been the city's dominant bikeway type, are comfortable for many people despite not being physically separated and protected from motor vehicles.

Portland has also benefited from state-level complete streets legislation dating back to 1971. This legislation, titled "Use of Highway Funds for

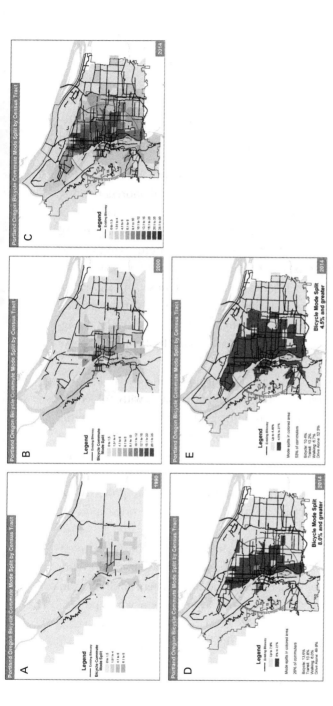

**Figure 19.1**

Spatial variation among Portland neighborhoods in bicycle mode share of daily work commuters in 1990, 2000, and 2014. *Note:* Maps A, B, and C show bicycle commuting by census tract in 1990, 2000, and 2014 (five-year average of 2012–2016). The lines show the bicycle network as it existed in each year. Maps D and E highlight those census tracts with bicycle commute rates of 8.0% and higher (map D) and 4.5% and higher (map E). Portland uses these and similar maps to make the case that "if you build it they (people bicycling) will come." On each map, the circle indicates a 4 mi radius centered on Portland's Burnside Bridge, which is roughly the center of the central city. As shown in these maps, the rate of bicycle commuting is highest in neighborhoods close to Portland's most dense employment districts in the central city. *Source:* All bicycle commuting data are from the US Census Bureau. Data from 1990 and 2000 are from those years' decennial census. Data for 2014 are from the five-year *American Community Survey* average for the period 2012–2016, which the Census Bureau reports because it is much more reliable than the often-fluctuating ACS annual estimates.

Footpaths and Bicycle Trails" (ORS 366.514, 1971), was part of a wave of environmental legislation passed in Oregon in the late 1960s and early 1970s. Colloquially known as "the bike bill," this law requires all Oregon's jurisdictions to provide appropriate bicycle and pedestrian facilities whenever a roadway is newly built or reconstructed, and that jurisdictions spend a minimum of 1% of their transport dollars for bicycling and walking.

All these factors contributed to a foundation on which future efforts built. The bike bill is especially noteworthy, as it helped address two key issues: dedicated funding and commitment to including bicycle facilities in large capital projects to improve roadways and public transport. Portland staff not only had a source of funding but also faced a requirement to spend it. Though the 1% funding provided by the legislation was low, so were the costs of improvements. When large capital projects arose, the legislation required that appropriate bicycle facilities be provided. Over the past 20 years, the Portland region has built five light rail lines that included significant construction in the public right-of-way—much of which triggered mandatory bicycle improvements.

In recent years, Portland's Bureau of Transportation (PBOT) has produced excellent bicycle infrastructure through partnerships with TriMet, the regional public transport authority, as both agencies have worked together on light rail construction projects throughout the city. The bike bill was also effectively used by the Bicycle Transportation Alliance (a statewide bicycle advocacy organization, now known as The Street Trust) to successfully sue the city for noncompliance on a capital roadway project. That action ensured that future capital investments by PBOT provided appropriate bicycle facilities.

Because of these supportive factors, even in lean times, when little investment was being made in bicycling, there remained a base level of cycling activity because of a rudimentary network and underlying conditions conducive to cycling. That foundation allowed rapid growth when political support focused on advancing bicycling. Portland's advances in bicycling have benefited from two factors. First are the strategic programmatic, operational, and capital advances made through the political leadership of three Portland elected officials. Second are the steady advances made through Portland's bureaucracy. Political leadership and effective implementation were themselves aided by effective advocacy and good data and research.

## Political Leadership

Political support for cycling is crucial for funding and implementing good bicycle infrastructure. Motor vehicle travel lanes and/or on-street car parking are sometimes replaced with bicycle lanes. Car movements at intersections are often restricted by bikeway features. On-street car parking is replaced with on-street bicycle parking in the form of bicycle corrals. Given limited street space, there must be public and political support to resist opposition to repurposing space from motor vehicles to bikes.

All these visible restrictions on motor vehicle travel and parking have resulted in pushback both from the public directly affected and from the media. Newspaper reporters, editorial boards, and television reporting have all portrayed the visual presence of bicycle lanes as a "war on cars."

Portland's commission form of government is unique in the United States. Five citywide elected officials form the government, consisting of four commissioners and one mayor. Their powers are coequal with two exceptions: the mayor assigns bureau responsibilities to the other elected officials, and the mayor proposes an annual city budget for the city council to consider. In the absence of a city manager, who would be beholden to the consensus direction of the elected body, elected officials are directly responsible for running their assigned government agencies. Each agency has a professional director in charge of running day-to-day operations in coordination with the responsible elected official.

This government structure creates a bureaucracy that is responsive to the policy directions desired by the elected official in charge. Bureau managers recognize the policy direction communicated by the political leadership and work to ensure that agency actions reflect those directions. Thus, PBOT's actions are at times highly politicized and responsive to public feedback.

Since 1991, Portland's bicycling efforts took significant advances under three elected officials: Earl Blumenauer (city commissioner, 1987–1996), Sam Adams (city commissioner and mayor, 2005–2012), and Steve Novick (city commissioner, 2013–2016). These leaders all shared two attributes. First, they recognized the benefit of bicycling to address core civic issues. Second, they facilitated structural changes in PBOT that allowed significant advances in bicycling. That, in turn, energized and empowered the professional staff to take the steps needed to build bicycle infrastructure.

**Earl Blumenauer (City Commissioner, 1987–1996)**   Earl Blumenauer had a strong personal and professional interest in bicycling. As an Oregon state representative (1972–1979), he enjoyed riding his bicycle to work at the legislature. Later, as a Multnomah County commissioner, Blumenauer set up a bicycle pool to complement the county's motor pool. He understood the benefits of bicycling and worked to promote bicycling and bicycle facilities.

Blumenauer served on Portland's city council from 1987 to 1996. He took initiative in two distinct areas that led to dramatic early growth in the development of Portland's bikeway network. His formation in 1993 of a bicycle program created a structure within PBOT that fostered the development of expertise in bikeway design and public process around bicycle projects. The bicycle program's staff became a centralized source of information and proficiency within the agency that was easy for other employees to access. It also became a source of data that allowed the city to better describe to the public and other decision makers the effects of building bicycle infrastructure. Blumenauer focused available funding to undertake these tasks. He also marshaled the agency to take advantage of the 1991 Intermodal Surface Transportation Efficiency Act, which for the first time made available significant federal funding for active transport.

Blumenauer focused on organizing the city's efforts to create a cohesive and clearly understood programmatic approach to bicycling. Among the bicycle program's more prominent and lasting accomplishments was the development of the city's Bicycle Master Plan, which was adopted by the city council in 1996. By creating a classification system for 1,080 km of city bikeways and off-street pathways, that plan provided an implementation blueprint for developing the city's bikeway network. With the plan's adoption came official city policies recognizing that Portland was to "make bicycle transport a part of daily life."

Blumenauer thought that involving citizens, elected officials, and public administrators in developing a bicycling plan would foster widespread understanding and support for the city's approach. He wanted the Bicycle Master Plan to be a blueprint around which people could rally. Under his leadership, agency staff members understood that building a bicycle network was a priority. He noted that people within PBOT were "fantastic" and found ways to identify funding and projects that began to implement in earnest Portland's bicycle network.

Portland is a city governed by policies. Portland's transport staff members honor the city's written policies and regularly refer to them when faced with difficult trade-offs. That is especially true when making operational and design decisions in the often narrow widths of Portland's public rights-of-way. Portland's 1996 bicycle policies were groundbreaking for their time in the United States. Though they lacked the clarity of the city's future bicycling policies, they nonetheless signaled to the professional staff that bicycling was important and had to be addressed in a manner that made it "a part of daily life" for Portland residents.

The tangible results of Blumenauer's efforts are shown in figure 19.2, which shows an expansion of bikeways that continued well past his tenure with the city.

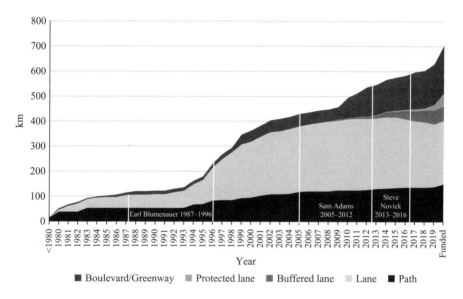

**Figure 19.2**

Portland's cumulative bikeway length (km) by facility type. *Note*: "Path" represents a shared off-street path. "Boulevard/Greenway" indicates bicycle boulevards (which Portland now calls "neighborhood greenways"). "Lane" is a conventional bicycle lane (generally 5–6 ft wide). "Buffered Lane" indicates those lanes with a buffer stripe. "Protected Lane" indicates protected bicycle lanes (also called "physically separated" or "traffic-protected" bicycle lanes or "cycle tracks"). This graph shows the significant expansion resulting from Blumenauer's foundational efforts, which focused on development of bicycle lanes. The continued growth under Adams focused on bicycle boulevards but also included the city's first efforts to install buffered bike lanes and protected bike lanes. Under Novick, the focus was on improving the quality of cycling infrastructure, including significant expansion of buffered and protected lanes, which are now considered the safest, lowest-stress, and most popular cycling facility. *Source*: Portland Bureau of Transportation.

**Sam Adams (City Commissioner, 2005–2008; Mayor, 2009–2012)**   Adams ran on a platform of making Portland the country's best city for bicycling, and indeed, it was the first large city to be awarded a platinum-level bicycle-friendly community rating from the League of American Bicyclists, the country's main national bicycle advocacy organization. As with Blumenauer, Adams's interest in bicycling was personal and political. He grew up riding a bicycle, taking long family trips across states and up and down Oregon's coast. Politically, he appreciated that bicycling was good for the environ-ment, saying it was "great for the economy" by providing a green dividend and that it would promote the kind of community and multiuse activation of the street that "Portland is rightfully obsessed about" (Cortright 2007).

Adams's efforts led to four structural changes within the city. First, he communicated to PBOT leadership his strong interest in advancing bicy-cling. His expressed interest immediately energized the staff to action. Second, he advanced planning for bicycling by updating the city's 1996 Bicycle Master Plan, developing the current *Portland Bicycle Plan for 2030*. Third, Adams provided funding to begin immediate strategic implementa-tion of bikeways as called for in the 2030 plan. Fourth, he stimulated a civic conversation about the role of bicycling in Portland.

These structural changes resulted in advances on the ground. During Adams's tenure, PBOT introduced features to its bikeway network that were considered innovative and experimental. These included the use of pedestrian-cyclist hybrid beacons, scramble signals, bicycle signal indications, and pro-tected bicycle signal phases to facilitate bicycle crossings at intersections. Portland also introduced modern intersection bike boxes (a colored, exclusive advanced waiting area for bicyclists in front of cars) and similar intersection treatments. He further energized city staff and expanded professional knowl-edge by having staff members participate in fact-finding tours in the world's best cycling cities in the Netherlands and Denmark. He also engaged Portland as a founding city member of the National Association of City Transporta-tion Officials' (NACTO) Cities for Cycling program. NACTO provides national guidance on bikeway design through regularly updated design manuals with the latest state-of-the-art standards (NACTO 2014).

The updated bicycle plan strengthened the city's policies for bicycling by introducing a "green transport hierarchy" policy, elevating mode split targets, deemphasizing automotive transport relative to bicycling, and creat-ing a more muscular functional classification for bicycling on more streets.

The civic conversation that began under Adams culminated in a highly regarded report by Portland's City Club in 2013. The report, titled "No Turning Back: A City Club Report on Bicycle Transportation in Portland," strongly supported bicycling and signaled a shift in what had been an often acrimonious conversation about the future of bicycling in Portland.

Adams led a 75 mi expansion of Portland's bikeways, launched the city's annual Sunday Parkways program, and dedicated $20 million of city funds for the expansion of bicycle boulevards. This controversial move jump-started new development of this type of facility (figure 19.1).

**Steve Novick (City Commissioner, 2013–2016)**  As with Blumenauer and Adams, Steve Novick supported bicycling because of its civic benefits. Novick saw bicycling as a key strategy for addressing climate change, reducing health care costs, and lessening pressure on family budgets. He recognized that securing funding was his most important task, as he knew that "great infrastructure" was needed to increase bicycle use. He pursued this, in part, through successfully championing a City of Portland 10 cent per gallon gasoline tax to raise $64 million between 2017 and 2020. The public education efforts by Novick and PBOT were largely responsible for this ballot initiative, which was approved by Portland voters in mid-2015. Approximately half the total is for bicycling and walking safety projects, with the rest for roadway maintenance.

Novick also gave more independence to PBOT. Energized by a new director, Leah Treat, whom he and Mayor Charlie Hales together hired, with a strong focus on safety and active transport projects, PBOT began to lay out an aggressive course to take advantage of the stronger policies and increased funding. Novick trusted PBOT's professional staff and was willing to give them much discretion. Under his leadership, and that of Treat, the agency adopted new guidance for physically separated, traffic-protected bicycle lanes, increasingly focused transport system development changes on bicycle projects, and advanced efforts to increase the price of on-street car parking, while reducing its supply.

## Professional Staff with Experience and Commitment

Portland has long been recognized as a leading bicycling city in North America. That recognition is based on Portland's far-reaching and cohesive bikeway network, adoption of European bicycle infrastructure, achievement of high cycling levels, and ability to cogently report and share its results and efforts with other cities. Although political leaders foster advances by making good decisions about policy direction and funding, they often take such actions in response to initiatives introduced by agency staff. For all the good decisions made by Portland's elected leaders, it was PBOT staff members that elevated the issues and opportunities and that formed the programs and operating strategies to realize the direction provided by elected leaders.

The city's bicycle program of the early 1990s allowed the development of expertise in infrastructure design and its rapid implementation. In 2000 PBOT restructured its bicycle staff to promote bicycling. The bicycle program was disbanded in favor of a bicycle coordinator and a group of project managers who were assigned to the full range of bureau projects. While the bicycle coordinator became the go-to subject matter expert, the knowledge that had developed within the bicycle program was disseminated throughout the bureau. This reorganization created an enriched PBOT in which exposure to bicycle issues is widespread and where expertise in bicycling is well developed throughout the department.

Portland's bicycling efforts benefited from four features associated with the professional staff:

1. *Personal experience bicycling in the city.* Almost all the key staff members in the city's bicycle program (including project managers, program managers, the bicycle coordinator, traffic engineers, civil designers, and the city traffic engineer) are themselves bicycle riders.

2. *Longevity and resultant institutional knowledge.* Many of the planners, engineers, and senior bureau managers have been with PBOT since the early days of the bicycle program and have continually learned and improved their skills as standards and guidelines changed.

3. *A user-oriented approach.* Portland staff members were early adopters of the PBOT-developed "Four Types of Transportation Cyclists" typology and thus began designing to a "design user" (explored further later) standard rather than to accepted national standards.

4. *A mission- and policy-driven approach.* Portland planners, managers, and engineers are well versed in city and regional land and transport planning policy. Portland's bicycling policies are among the strongest in North America, and the agency staff's response to these policies is reflected in a capital program that is top-heavy with bicycle projects, including many that repurpose car-carrying capacity and on-street car parking in favor of bicycle infrastructure.

These elements, combined with knowledge about international best practices in bicycle design and frustration with national guidance for the design of bicycle facilities, contributed to a spirit of doing not what is allowed but what is the right thing to do when designing bicycle facilities.

## Design Approach

Portland's approach to bicycle facility design changed around 2005 with acceptance of the PBOT-developed typology about bicycle riders. This typology has influenced bicycle planning and design in cities around the world and posits that most of any city's population is interested in bicycling but are not willing to ride because of concerns about their safety in standard bicycle facilities (Geller 2009). Early on, city staff recognized that achieving Portland's advanced goals for bicycling would require addressing the needs of this "interested but concerned" group by designing bikeways and networks with them as the "design user." This recognition is reflected in the city's *Portland Bicycle Plan for 2030*, which states that "a major premise of this plan is that the residents who are described as 'interested but concerned' will not be attracted to bicycle for transport by the provision of more bicycle lanes, but may become tempted by a dense network of low-stress bikeways" (PBOT 2010).

This understanding, together with the goal of achieving a 25% bicycle commute mode split, dramatically strengthened bicycling policies. These policies, first recommended in the *Portland Bicycle Plan for 2030*, were later adopted as city policy in 2018 updates of the city's Comprehensive Plan and Transportation System Plan.

Portland planners, engineers, policy staff members, and senior managers are highly responsive to current city policies that direct Portland to "create conditions that make bicycling more attractive than driving" to "create a bicycle network that is safe, comfortable, and accessible to people of all ages and abilities," and to prioritize "transport system decisions" in this order: walking; bicycling; transit; fleets of electric, fully automated, multiple-passenger vehicles; other shared vehicles; and finally "low or no occupancy vehicles, fossil-fueled non-transit vehicles" (City of Portland 2020). These policies are discussed internally in planning and design meetings, as well as externally as part of public engagement.

Supporting these policies are subsequent PBOT directives and a programmatic focus on roadway safety. Leah Treat, PBOT's director from 2013 to 2018, did three things that helped translate broad policies into concrete action. First was the agency's focus on safety as a Vision Zero city. Second was a directive she issued, in a September 15, 2015, email, to PBOT staff members to "make protected bicycle lanes the preferred design on roadways where separation is called for." Third was her directing Portland's city engineer to make 10 ft travel lanes the preferred width under almost all conditions. The intent of narrowing the car lanes was to secure width needed for protected bicycle lanes. Figure 19.3 displays the facilities that contribute to Portland's bicycle transport network.

The sum total of these policies and directives has been a dramatic shift in PBOT's focus and designs that strongly support the movement of bicycles rather than cars and the development of the highest-quality bikeways.

## The Role of Advocacy

Several advocacy efforts in Portland contributed to its advances in bicycling. In 1993, the Bicycle Transportation Alliance (BTA, now The Street Trust), a professional bicycle advocacy organization in Portland, successfully sued the city based on state law related to the provision of bicycle facilities as part of new roadway construction (*Bicycle Transportation Alliance v. City of Portland*). BTA then leveraged the resulting respect now offered it by PBOT to get prominent seats at the table for every significant planning, policy, and implementation effort related to bicycling subsequently undertaken by the city. The BTA participated effectively on many stakeholder advisory committees, stakeholder working groups, project advisory committees, the bureau budget advisory committee, and citizen steering committees. Through this multifaceted participation, BTA successfully influenced the outcome of many efforts undertaken by PBOT.

BikePortland.org, a local bicycling blog, had a multiplier effect on bicycle advocacy by providing widespread coverage to every substantive issue related to bicycle funding, planning, policy, design, and culture in Portland. This successful blog often broke bicycle stories that were then reported by mainstream Portland news outlets. Media coverage then pressured PBOT and elected officials to solve the issues raised. BikePortland.org provided arguments, information, and illustrations that could be used by every advocate to promote cycling.

Better Block PDX is the Portland chapter of a national movement of advocates working to improve public spaces for people walking, bicycling, and using transit. Better Block PDX worked with city transport officials and

**Figure 19.3**

Variety in cycling facility types in Portland's bicycle network. *Note*: The main types of cycling facilities are conventional bicycle lanes (A), off-street pathways (C), and bicycle boulevards (D). Increasingly, however, Portland is updating its bicycle lanes to wider, buffered lanes (B and E) and protected lanes (F and G). New protected lanes are often retrofits with temporary devices (F) and only occasionally are constructed with high-quality finishes (G). *Sources*: B and G: Jonathan Maus/BikePortland; F: Shane Valle, Portland Bureau of Transportation; all others by Roger Geller.

through the city's permit system to initiate tests of world-class bikeway designs. Their efforts initiated public discussion and design efforts for what are becoming some of Portland's signature bikeway facilities and facility types.

Bike Loud PDX is a more recent, volunteer, grassroots advocacy group. It focuses on improving cycling conditions on some of Portland's bicycle boulevards. Using direct political action, public demonstrations, and grassroots political organizing and campaigns, Bike Loud PDX successfully lobbied for planning efforts and subsequent facility improvements by PBOT. Its ongoing involvement at public meetings, council hearings, and in social and other media in support of better bicycling conditions continues to influence city efforts.

## The Role of Research

Portland's bicycling efforts have benefited from a close association with researchers at Portland State University (PSU). Researchers working in the PSU Transportation Research and Education Consortium (TREC), the National Institute for Transportation and Communities (NITC), and the Initiative for Pedestrian and Bicycle Innovation (IBPI) have contributed to Portland's practitioners' understanding of effective facility design as well as to their confidence when introducing new and nonstandard designs. The data and analysis of Portland's designs coming from these programs contributed to the determination of successive city traffic engineers to proceed with and expand the use of such "innovations," including bicycle signals, intersection bicycle boxes, buffered bicycle lanes, protected bicycle lanes, and other bicycle intersection treatments. This collaboration between practitioners and researchers has been a critical link in allowing Portland to advance its designs beyond those officially approved in federal guidance to designs considered best practice internationally.

## The Future of Bicycling in Portland

As of 2019, Portland was in a better position to implement world-class bicycling infrastructure than at any time in the past. PBOT now has a stronger focus, stronger policy support, and better guidance to develop low-stress bikeways. Funding is now available to implement networks of protected lanes in Portland's central city as well as in Portland's outer east post–World War II suburban-style neighborhoods. Over the next five years, these projects promise to transform the city. Planning efforts are under way to tackle the hilly suburban neighborhoods of southwest Portland as well as the densely populated inner neighborhood of northwest Portland. PBOT leadership and staff have developed a level of expertise in bicycle design that rivals or exceeds that of any other North American city. That knowledge is being effectively transferred to a new generation of PBOT traffic engineers, civil designers, planners, and project managers.

PBOT is poised to create the conditions to make bicycling more attractive than driving by building a bicycle system that appeals to people of all ages and abilities, in accordance with its policies. Whether that potential will be fully realized, and the pace at which investments may be made, depends largely on the actions of local leaders who recognize the benefits of increased bicycle use and who will work to further bicycling in Portland.

## Promoting Bicycling in Seville, Spain

Seville is a medium-size southern Spanish city with flat topography, located in the valley near the mouth of the Guadalquivir River. Its main demographic and climatic characteristics are shown in table 19.1. Table 19.2 shows key aspects of cycling trends and policies in Seville over time.

The evolution of cycling in Seville can be roughly divided into four main periods. Before 2006, cycling accounted for less than 1% of trips in Seville. From 2007 to 2011, cycling boomed, reaching a peak in 2011, with around 72,000 bicycle trips per day, representing an estimated 5.6% of total trips and 9.0% of vehicular trips (i.e., excluding walking trips). Starting in 2011, there was a slow decline or stagnation in cycling because of the new city government's lack of support for cycling. With a more pro-cycling city government since 2016, cycling is again on the rise.

Urban mobility in the period before 2006 was oriented toward the car and suburban sprawl, fueled by strong real-estate speculation in the metropolitan ring surrounding the central area of the conurbation. The population of the municipality of Seville (the area shown in figure 19.4) has remained roughly the same since 1990, at around 700,000 inhabitants. In sharp contrast, however, the population in the metropolitan ring rose from 200,000 to 800,000 inhabitants between 1990 and 2006. Mobility surveys showed that more than 170,000 people living in the metropolitan ring commuted daily to the municipality of Seville, most of them by car

**Table 19.2**

Cycling infrastructure, bikesharing, cycling levels, and cyclist safety in Seville, 1990–2017

| | Protected cycle paths (km) | Bikesharing rentals (millions) | Bikesharing (% of bicycle trips) | Bicycle trips (millions) | Women's share (%) | Cyclists killed or severely injured | Cyclists killed or severely injured per million trips |
|---|---|---|---|---|---|---|---|
| 1990 | 0 | | | 3.0 | 13* | n.a. | 0.98 |
| 2006 | 12 | | | 3.1 | 25 | 3 | 1.71 |
| 2000–2006 (average) | | | | 3.1 | | 5.3 | 0.46 |
| 2007 | 77 | 0.1 | | 6.5 | | 3 | 1.08 |
| 2008 | 92 | 3.9 | | 9.3 | | 10 | 0.79 |
| 2009 | 105 | 6.1 | | 14.0 | | 11 | 0.19 |
| 2010 | 120 | 4.7 | | 16.2 | | 3 | 0.06 |
| 2011 | 133 | 4.9 | 29 | 17.0 | 32 | 1 | 0.23 |
| 2012 | 144 | 5.0 | | 17.7 | | 4 | 0.39 |
| 2013 | 152 | 4.2 | 27 | 15.6 | | 6 | 0.22 |
| 2014 | 158 | 3.8 | | 13.9 | | 3 | 0.37 |
| 2015 | 162 | 3.5 | 26 | 13.4 | | 5 | 0.37 |
| 2007–2015 (average) | | | | 13.7 | | 5.1 | |
| 2017 | 175 | 3.4 | 24 | 14.4 | 36 | n.a. | |

*Note:* Data from 1987.

*Sources:* Marqués et al. (2015a), Marqués and Hernández-Herrador (2017), Ayuntamiento de Sevilla (2017), and additional information provided by the staff of the municipality of Seville.

(Marqués et al. 2015b). This caused problems of traffic congestion and air pollution in the city, which were the background for the pro-bike policies that gave rise to the strong urban cycling boom after 2006. It is notable, however, that other Spanish cities facing similar problems did not develop pro-bike policies. There were several additional reasons specific to Seville for the dramatic shift to pro-bike policies.

The period between 1990 and 2006 was characterized by the onset of a strong cyclists movement grouped around the Association of Urban Cyclists (A Contramano), founded in 1987 (Huerta and Hernández-Ramírez 2015). This group, now a member of the European Cyclists' Federation, organized cycling demonstrations, which peaked in 1993 with more than 10,000 participants, and made several proposals for a city network of bicycle paths inspired by Dutch models. In 2003, a coalition between the Partido Socialista Obrero Español (PSOE), the biggest Spanish Social Democratic party, and a left-wing federation including Communists and Greens, named Izquierda Unida, won local government elections. Izquierda Unida had strong connections with many local environmental groups and A Contramano, so some members of these groups became members of the new city government. As part of their negotiation for joining the city government, Izquierda Unida insisted on the construction of a complete network of bike paths as a key goal of the new coalition government. Building the bikeway network received additional support when Seville citizens voted for making it a budget priority for the city within the framework of the participatory budgeting processes developed in the city in 2004 and 2005 (Huerta and Hernández-Ramírez 2015).

Between 2004 and 2011, the city government, led by the Social Democratic mayor Alfredo Sánchez-Monteseirin and the vice-mayors of Izquierda Unida, Paula Garvín (2003–2007) and Antonio Rodrigo-Torrijos (2007–2011), promoted a series of pro-bike policies. Their most salient feature was the planning and implementation of a complete network of separated, traffic-protected bicycle paths (cycle tracks), with a total budget of €32 million (Marqués et al. 2015a). In just four years, the city went from having 12 km of unconnected bikeways to having a network of 120 km of protected bike paths and lanes, primarily for utilitarian travel within the city. Most locations in the city were to be within about 500 m of such a bike path network.

The evolution of the bikeway network is shown in figure 19.4. This network was complemented by the implementation of a bikesharing system in

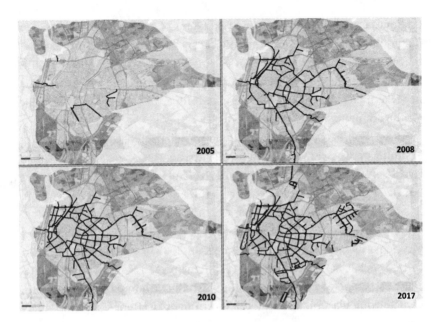

**Figure 19.4**
Evolution of the network of protected bicycle paths in Seville between 2003 and 2017. *Source*: Ayuntamiento de Sevilla (2017).

July 2007. The public bikesharing system had 2,500 bicycles at 250 docking stations by the end of 2008 and expanded slowly until it had 2,600 bicycles in 260 stations in 2019, with an average distance of about 300 m between stations (Marqués et al. 2015a). In addition, the city installed 5,700 free bicycle parking spaces along streets (Marqués et al. 2015a).

To implement such policies, a new Bicycle Office was created in Seville's Department of Urbanism. It was directed by Pepa García-Jaen, an officer of the Area of Urbanism, under the supervision of Jose A. García-Cebrián from Izquierda Unida. In addition, severe car traffic restrictions were imposed on nonresidents in the historical center (the inner area without bicycle paths in figure 19.4), whose very narrow streets prevent the implementation of any specific bicycle infrastructure. Simultaneously, important areas of the city center were pedestrianized, and some important new public transport infrastructure was built, such as the first metro line of the city and a tram through the main avenue of the historical center. Those improvements in public transport helped alleviate traffic congestion in the city center (Castillo-Manzano, López-Valpuesta, and Marchena-Gómez 2015).

The network of bicycle paths was designed according to the following key principles (Marqués et al. 2015a), which were supported by the local bicycle advocacy association A Contramano and strongly inspired by the Dutch design principles (CROW 2007), as described in detail in Furth (chapter 5, this volume):

- *Continuity and connectivity.* The network was designed with the aim of connecting, through the direct linkage of bicycle paths, the main trip destinations and residential areas of the city.
- *Uniformity and visibility.* The design of the bicycle paths was consistent so that cyclists could easily recognize and follow it. A uniform green pavement was chosen in order to enhance the visibility of the bicycle paths.
- *Directness and relevance.* The network followed the main streets of the city so that most trip destinations were along it.
- *Separation and comfort.* The entire network was separated from motorized traffic and mainly built on sidewalks that were expanded for this purpose, with the aim of providing an enhanced sensation of safety to the cyclists, while still providing ample space for walking.

Figure 19.5 shows a typical protected bike path in Seville along the Calle Parlamento de Andalucía near the Arco de la Macarena, a big avenue near the city center. On such an avenue, car parking lanes along the sidewalks were replaced by bicycle paths on the widened sidewalks, separated from the pedestrian areas by flowerbeds and a row of trees. Presently, most bicycle paths in Seville (83% of the total length of the network) share this basic design. They are on widened sidewalks but with different pavements for pedestrians and cyclists, with cyclists and pedestrians often physically separated by trees or shrubs. The remainder of the city's bikeway network consists of on-street bike lanes (10% of the length of the bicycle network) and shared-use paths (8%) (Ayuntamiento de Sevilla n.d.). The installation of bike paths on sidewalks did not cause much loss of pedestrian space, because of the widening of sidewalks to accommodate the bicycle paths. Indeed, the net loss of pedestrian space resulting from the implementation of bicycle paths was only 9% of the total surface dedicated to such bicycle facilities (Hernández-Herrador and Marqués 2017). At intersections, protected bike lanes were usually designed to be parallel to pedestrian crossings but with separate crossing paths and different bike signage for cyclists and pedestrians.

**Figure 19.5**
A typical protected bike path in Seville. *Source*: Ricardo Marqués.

The impacts of the new bicycle infrastructure on cycling levels, public health, and the economy were dramatic. As can be seen in table 19.2, the volume of bicycle traffic grew sixfold between 2006 and 2011. Moreover, the purpose of cycling trips shifted from mainly recreational to utilitarian. In fact, on-street surveys of cyclists in 2011 showed that most bicycle trips during a typical weekday are to work or school (Marqués et al. 2015a). Moreover, cyclists included all age groups, but with younger people predominating.

Table 19.2 shows a dramatic increase in the percentage of bike trips by women, almost tripling, from only 13% in 1990, to 25% in 2006, and to 36% in 2016. That increased rate of women cycling in Seville is consistent with the findings of many studies reporting that women greatly prefer separation from motor vehicle traffic through protected cycling infrastructure, which significantly raises the share of bike trips made by women (see Garrard, chapter 11, this volume). Moreover, the experience of Seville also confirms the strong relationship between rising overall cycling levels (as a share of all trips) and the increasing share of bike trips made by women. As

Garrard (chapter 11, this volume) notes, women have been identified as an "indicator species." Where a high percentage of bike trips are by women—such as in the Netherlands, Denmark, and Germany—overall cycling levels and mode share are also much higher than in countries where cycling is male dominated, such as the United States, Canada, and Australia (see in this volume Buehler and Pucher, chapter 2; Garrard, chapter 11; Koglin, te Brömmelstroet, and van Wee, chapter 18).

An ethnographic analysis of cycling in Seville showed that, as cycling levels rose, it penetrated all social strata, with especially large increases in cycling rates among women and children, far exceeding rates in Spanish cities without protected cycling facilities (Huerta and Hernández-Ramírez 2015). Seville's network of safe, low-stress cycling facilities created a cycling culture where cycling is broadly viewed as a normal, everyday way to get around the city. Cyclists do not stand out because of their clothes or accessories—such as Lycra/spandex outfits, reflective vests, helmets, or fancy, expensive bicycles. Cyclists as well as cycling have been normalized and are now a regular feature of Seville's streetscape.

As shown in table 19.2, cycling safety greatly improved over this period. Annual cyclist fatalities and serious injuries per million bike trips fell from an average of 1.71 for the period 2000–2006 to 0.37 for the period 2007–2015, indicating more than a quadrupling in cycling safety. Moreover, the indirect public health benefits of the protected bikeway network (in terms of more physical activity and reduced mortality) have been estimated to be between 20 (Brey et al. 2017) and 24 (Marqués et al. 2015a) deaths avoided per year. This public health benefit corresponds to those described in detail by Garrard et al. (chapter 3, this volume). An economic benefit-cost analysis of the protected bikeway network found that benefits exceeded costs by more than 30% (Brey et al. 2017).

The boom in cycling and great improvement in cycling safety resulting from the pro-cycling policies implemented in Seville led to widespread approval in public opinion polls. Nevertheless, various powerful special interest groups opposed them. For example, owners of private car parking facilities and retailers protested traffic restrictions in the city center (Marqués et al. 2015a). Some neighborhood associations protested the loss of parking spaces on their streets as a consequence of the building of bicycle paths, and some pedestrian associations complained about the presence of cyclists on the sidewalks. Protests against pro-bike policies and especially car-restrictive measures were also politically based. The opposition was particularly strong among political conservatives, who opposed pro-bike policies because they were supported by the opposing coalition of the Social Democrats with the Communists and Greens of Izquierda Unida (Castillo-Manzano and Sánchez-Braza 2013).

After 2011, the number of annual bike trips stagnated, followed by a slow decline (see table 19.2). This reversal in cycling trends coincided with important political changes at the national level in Spain resulting from the economic crisis of 2008. The PSOE was blamed for the recession and lost local elections in the largest Spanish cities, including Seville, where the Conservative Party (Partido Popular) took control of the city government. The new mayor, Juan-Ignacio Zoido, supported by the car lobby and by some city center retailers and neighborhood associations, promised during the political campaign to remove car traffic restrictions in the historical center. Thus, car restrictions were lifted after the elections of 2011, and transport policy in general became more car oriented. The Bicycle Office of the city government was dismantled, and no new specific cycling promotion policies were developed. Nevertheless, the network of protected bicycle paths continued to grow, mainly because of the completion of projects that had already started or already had been planned and financed.

The most important lesson from this interruption in political support for cycling is that urban cycling is highly resilient. With so much protected cycling infrastructure already in place, the strong momentum established in the previous period was enough to keep cycling at fairly high levels, only 20% less than before the Conservatives took power. This difficult period showed that bicycle paths were not just a whim of "the Communists" of Izquierda Unida but rather a valuable and popular infrastructure investment for many citizens, regardless of their political ideology. This point of view was reinforced by the international recognition gained by Seville's cycling infrastructure during this period. For example, in 2013, Seville was ranked fourth in the Copenhagenize index of bicycle-friendly cities (Copenhagenize 2013).

In 2015, the PSOE won the local elections in Seville and returned to power with the support of other leftist parties, including Izquierda Unida, which chose to support the new government but did not officially join it. The new PSOE mayor, Juan Espadas, renewed many of the former pro-bike policies and initiated some new ones. These policies were summarized in a new Bicycle Masterplan (Ayuntamiento de Sevilla 2017) elaborated by the city government and in December 2017 approved unanimously by the city council with the objective of doubling the modal share of cycling.

The new Bicycle Masterplan was created with the participation of local cyclist associations, bicycle retailers and repair shops, bicycle rental and small cycle-logistic companies, and other cyclist representatives. The plan proposes further expansion and improvement of the network of bicycle paths. It goes well beyond that, however, proposing the development of other bicycle facilities that were strongly demanded by the cyclists. For example, it recommends the provision of secure bicycle parking at workplaces, schools, residential buildings, malls, and public transport hubs. It also proposes traffic-calming policies favoring bicycle traffic over car traffic, including contraflow bicycle traffic and special streets with bicycle priority. This master plan, however, has not yet been fully implemented because it was only recently approved. Originally, the plan did not include any kind of traffic restrictions in the city center. In early 2020, however, Mayor Juan Espadas announced the implementation of some car traffic restrictions in the city center as part of a new plan called Sevilla Respira. That is an encouraging development because traffic restrictions in the city center were an important part of the pro-bike policies during 2007–2011, the peak of bicycle use in the city. As can be seen in table 19.2, the change in transport policy since 2015 has already led to a renewed upward trend in cycling levels.

Importantly, all pro-bike policies and increased cycling achievements described in this section were limited to the city of Seville. In the metropolitan ring surrounding the city, pro-bike policies have been absent or minimal. The municipalities of the metropolitan ring have no cycling infrastructure comparable to the bicycle network shown in figure 19.4 for the city of Seville itself. Even pro-bike policies in some municipalities aimed at developing such infrastructure have been difficult to implement. The outer metropolitan ring is composed of many small, independent, uncoordinated, and often competing municipalities. There is no unified mobility authority to integrate cycling facilities across municipal boundaries.

In 2014, the government of the Autonomous Region of Andalusia, which covers most of the south of Spain and of which Seville is the capital, created a master plan for the development of cycling infrastructure in the main metropolitan areas of the region, including the metropolitan area of Seville. The cycling plan has a total budget of €400 million (Junta de

Andalucía 2014). However, because of the financial constraints imposed by yet another economic crisis in Spain, this regional cycling master plan has received less funding than expected. Another key factor missing for the development of utilitarian cycling in the metropolitan ring is the development of an effective policy promoting the integration of cycling with public transport (Marqués et al. 2015b).

## Conclusions

The experiences of Portland and Seville affirm the necessity of a high-quality, connected network of bikeways for effectively increasing bicycling. Their experiences also affirm the benefit of consistent and strong political leadership to reallocate space from the car to the bicycle and to back the funding and implementation of well-designed, connected cycling infrastructure. Both cities also demonstrate the importance of cycling advocacy, as highlighted by Pucher et al. (chapter 20, this volume). Strong, well-organized advocacy was crucial for generating the public and political support for pro-bike policies in general and also for providing feedback on the specific kinds of cycling infrastructure and programs most needed to attract citizens to cycling for daily travel and for utilitarian purposes. The innovative ideas as well as public and political support from cycling advocacy groups were critical in overturning former car-centric policies, practices, and special interests favoring the status quo that places motor vehicle transport above all other means.

The differences between these cities are also noteworthy. In only four years, from 2007 to 2011, Seville's infrastructure program stimulated a doubling in bicycle mode share from 3% to 6% (from 5% to 9% excluding pedestrian trips). Portland's bicycle use grew much more slowly, indeed over more than two decades, from 1% in 1990 to 2% in 2000 and 6% in 2010, reaching a peak of 7% in 2014. The much faster growth in Seville can be attributed to the very rapid construction of Seville's bikeways. In addition, cycling in Seville has benefited from much denser land use than in Portland and a much shorter history of car use in Spain than in the United States.

The restrictions imposed on car use in Seville's city center contributed to the rapid uptake of bicycling. The PSOE-led government that initiated the development of its bikeway network also pedestrianized large areas of Seville's city center and then further restricted car access to the center.

Car trips to the center fell immediately following the imposition of car-restrictive measures, while bicycle use rose rapidly. The Conservative, pro-car government that followed removed the car restrictions, which almost immediately increased car use and decreased cycling. The strong correlation between the renewed ease of car access to the city center and the decrease in bicycle trips is corroborated by many other studies documenting the crucial need to restrict car use, and make it more expensive, in order to encourage modes of transport that are more sustainable, such as cycling, walking, and public transport (Buehler, Pucher, and Altshuler 2017; Buehler et al. 2017).

Seville's experience is somewhat similar to Portland's situation in 2008. During the worldwide economic recession, average gasoline prices in Portland rose from approximately $2.50 per gallon in late 2006 to a peak of $4.29 in July 2008 (AAA Oregon 2018). With this sharp increase in gasoline prices came a strongly correlated increase in bicycle and transit commuting, as reflected in US census data. That is the closest Portland has come in the past 25 years to making driving less attractive. The cycling rate remained steady and even increased during the eight years from 2006 to 2014. Since 2017, however, bicycle use has been slowly declining as a percentage of all trips. In particular, the number of new bicycle riders is not keeping pace with the large increase in work commuters from the rapidly growing suburbs and outer areas of Portland to the city center.

Thus, one important lesson is that building a high-quality bicycle network is a necessary condition for fostering more and safer cycling for a range of trip purposes, but it is not sufficient. It is also crucial to restrict the use and parking of cars and to increase their cost, as is widespread in many northern European cities (Buehler, Pucher, and Altshuler 2017; Buehler et al. 2017). To implement and maintain the necessary restrictions on car use, including higher prices, is a formidable political task, especially in car-oriented countries. It requires coordinated, long-term efforts by political leaders and their parties, as well as professional engineers and planners. Car-restrictive measures require public and political support of a wide range of advocacy groups supporting sustainable transport—including cyclists but also environmental, equity, urban revitalization, and historical preservation groups. Finally, a massive public information campaign would facilitate car-restrictive policies by explaining their benefits for overall livability and sustainability.

## References

AAA Oregon. 2018. *Gas Prices Archive*. Salem, OR: American Automobile Association Oregon.

Ayuntamiento de Sevilla. 2017. *Programa de la Bicicleta Sevilla 2020*. Ayuntamiento de Sevilla. http://www.urbanismosevilla.org/areas/sostenibilidad-innovacion/sevilla -en-bici/programa-bicicleta-2016-2020.

Ayuntamiento de Sevilla. n.d. *Tipología de las Vías Ciclistas de Sevilla*. Ayuntamiento de Sevilla. http://www.sevilla.org/sevillaenbici/Pdf/tipologia_vias_ciclistas.pdf.

Brey, Raúl, José Ignacio Castillo-Manzano, Mercedes Castro-Nuño, Lourdes Lopez-Valpuesta, Manuel Marchena-Gomez, and Antonio Sanchez-Braza. 2017. Is the Widespread Use of Urban Land for Cycling Promotion Policies Cost Effective? A Cost-Benefit Analysis of the Case of Seville. *Land Use Policy* 63: 130–139.

Buehler, Ralph, John Pucher, and Alan Altshuler. 2017. Vienna's Path to Sustainable Transport. *International Journal of Sustainable Transportation* 11(4): 257–271.

Buehler, Ralph, John Pucher, Regine Gerike, and Thomas Götschi. 2017. Reducing Car Dependence in the Heart of Europe: Lessons from Germany, Austria, and Switzerland. *Transport Reviews* 37(1): 4–28.

Castillo-Manzano, Jose Ignacio, Lourdes López-Valpuesta, and Manuel Marchena-Gómez. 2015. Seville: A City with Two Souls. *Cities* 42A: 142–151.

Castillo-Manzano, José Ignacio, and Antonio Sánchez-Braza. 2013. Can Anyone Hate the Bicycle? The Hunt for an Optimal Local Transportation Policy to Encourage Bicycle Usage. *Environmental Politics* 22(6): 1010–1028.

City of Portland. 2018. *2035 Comprehensive Plan*. Portland, OR: City of Portland.

Copenhagenize. 2013. *The 2013 Index*. Copenhagenize. http://www.copenhagenize .com/2013/04/copenhagenize-index-2013-bicycle.html.

Cortright, Joe. 2007. *Portland's Green Dividend: A White Paper*. Chicago: CEOs for Cities.

CROW. 2007. *Design Manual for Bicycle Traffic*. Ede, Netherlands: CROW.

Geller, Roger. 2009. *Four Types of Cyclists*. Portland, OR: Portland Bureau of Transportation. https://www.portlandoregon.gov/transportation/44597?a=237507.

Hernández-Herrador, Vicente, and Ricardo Marqués. 2017. El impacto del "carril-bici" de Sevilla sobre el espacio urbano de la ciudad: Un análisis preliminar. *Hábitat y Sociedad* 10: 181–202.

Huerta, Elena, and Macarena Hernández-Ramírez. 2015. *Etnografía de la bicicleta en Sevilla*. Seville: DAPS Informes y documentos de trabajo. https://rio.upo.es/xmlui /handle/10433/1425.

Junta de Andalucía. 2014. *Plan Andaluz de la Bicicleta.* Seville: Junta de Andalucía. http://www.aopandalucia.es/planandaluzdelabicicleta/.

Marqués, Ricardo, and Vicente Hernández-Herrador. 2017. On the Effect of Networks of Cycle-Tracks on the Risk of Cycling: The Case of Seville. *Accident Analysis and Prevention* 102: 181–190.

Marqués, Ricardo, Vincente Hernández-Herrador, Manuel Calvo-Salazar, and José A. García-Cebrián. 2015a. How Infrastructure Can Promote Cycling in Cities: Lessons from Seville. *Research in Transportation Economics* 43: 31–44.

Marqués, Ricardo, Vincente Hernández-Herrador, Manuel Calvo-Salazar, Javier Herrera-Sánchez, and Manuel López-Peña. 2015b. When Cycle Paths Are Not Enough: Seville's Bicycle-PT Project. *Urban Transport* 21(146): 79–91.

National Organization of City Transportation Officials (NACTO). 2014. *Urban Bikeway Design Guide.* Washington, DC: Island Press.

Portland Bureau of Transportation (PBOT). 2010. *Portland Bicycle Plan for 2030.* Portland, OR: Portland Bureau of Transportation.

# 20 Cycling Advocacy in Europe, North America, and Australia

John Pucher, Bernhard Ensink, Tim Blumenthal, Bill Nesper, Ken McLeod, Andy Clarke, Jean-François Pronovost, Dave Snyder, Robin Stallings, Fiona Campbell, and Peter Bourke

Booming academic research on cycling, as well as increasingly well-trained and dedicated transport planners and engineers, has provided a growing base of expertise on the most effective and safest cycling infrastructure and programs (Pucher and Buehler 2017). Knowledge alone, however, does not lead to implementation. Cycling advocacy at the local level has been crucial for the implementation of expanded and improved cycling infrastructure and the necessary complementary programs and policies that have helped increase cycling levels while making cycling safer. The detailed case studies in this volume all emphasize the important role of advocacy (in this volume, Pardo and Rodriguez, chapter 16; Pucher, Parkin, and de Lanversin, chapter 17; Koglin, te Brömmelstroet, and van Wee, chapter 18; Geller and Marqués, chapter 19).

This chapter examines the crucial role of national and state cycling advocacy in the growth of cycling in cities throughout the world. The chapter is based on firsthand information provided by directors of cycling advocacy organizations in Europe, North America, and Australia, who are included as authors. The chapter starts with a description of bike advocacy in Europe, focusing on the international efforts of the European Cyclists' Federation (ECF) and the three leading national cycling organizations in the European Union (EU): Fietsersbond (FB) in the Netherlands, Cyklistforbundet (CFB) in Denmark, and Allgemeiner Deutscher Fahrrad-Club (ADFC) in Germany. We then turn to cycling advocacy in the United States, which is led at the national level by the League of American Bicyclists (LAB) and PeopleFor-Bikes (PFB). Although almost every American state has a cycling advocacy organization, we focus on BikeTexas and CalBike as examples of especially important and effective state cycling advocacy in the United States. In Canada, most cycling advocacy is at the local level, but there are also provincial organizations and a national organization (Vélo Canada Bikes). We examine in detail Vélo Québec (VQ), which is by far the most important cycling advocacy organization at any level in Canada. Finally, we turn to cycling advocacy in Australia, which is led at the national level by We Ride Australia (WRA) in coordination with state-level organizations such as Bicycle Network (in Victoria) and BicycleNSW (in New South Wales).

## International Advocacy

By far the most important international cycling advocacy organization is the European Cyclists' Federation (ECF), which was founded in 1983 by 12 organizations in six northern European countries, spearheaded by the national cycling organizations in Germany, the Netherlands, the United Kingdom, and Denmark (ECF 2019). In 2019, ECF had 76 member organizations and a professional staff of 20 at its headquarters in Brussels. ECF's growth over the past four decades has been facilitated by increased support from national member organizations, grants from the European Commission, and contributions from the bicycle industry. As part of its growth, ECF initiated its flagship EuroVelo project to help establish, with government financial support, an integrated network of 15 major long-distance bike routes extending over 70,000 km (the world's largest cycling network) and connecting hundreds of cycling routes in 42 countries. In addition, it expanded its Velo-city conference from a biennial, European-focused cycling conference (first held in 1980 in Bremen, Germany) to the world's premier global cycling conference, including non-European cities such as Montreal (Canada), Rio de Janeiro (Brazil), Taipei (Taiwan), Adelaide (Australia), and Vancouver (Canada).

ECF has a history of spearheading international and global efforts to promote cycling (ECF 2019). At the 2010 Velo-city conference in Copenhagen, ECF launched two new global networks to promote information sharing: Cycling Cities, to connect cities around the world that want to promote cycling; and Scientists for Cycling, to connect scientists involved in cycling research and related topics (Gerike and Parkin 2016). In 2011, ECF collaborated with global leaders of the bicycle industry to establish the ECF Cycling Industry Club (currently Cycling Industries Europe), whose members financially support ECF advocacy work. At Velo-city 2014 in Adelaide, ECF led the initiative to form a new World Cycling Alliance (WCA), a network of cycling organizations with a board of directors that includes representatives from all continents. At Velo-city 2018 in Rio de Janeiro, WCA was formally made an independent global association of cycling organizations.

ECF partners closely with international organizations such as the European Union, the United Nations, UN Habitat, the World Health Organization (WHO), and the World Bank to promote cycling as a highly sustainable mode of travel that is healthy, environmentally friendly, and socially equitable, affordable by almost all segments of society (ECF 2019). In 2016, for example, ECF and WCA succeeded in having cycling prominently included in the sustainability policies of the New Urban Agenda adopted at the 2016 UN Habitat III conference in Quito, Ecuador. ECF is a partner in the World Bank's initiative Sustainable Mobility for All. It also successfully campaigned with its global cycling allies to have the UN General Assembly declare an annual World Bicycle Day, which was first celebrated in June 2018. Moreover, ECF helped develop a pan-European master plan for cycling, which is jointly managed by WHO and the United Nations Economic Commission for Europe (UNECE). The master plan provides specific recommendations for national authorities on how best to promote cycling.

In 2017, ECF led a group of scientists, planners, transport experts, government officials, advocacy organizations, and representatives of private industry to draft the *EU Cycling Strategy: Recommendations for Delivering Green Growth and an Effective Mobility System in 2030* (ECF 2017). The comprehensive long-term plan provides valuable guidance for the EU and, most important, for EU countries, states, and cities, on every aspect of cycling. It recognizes the crucial need to coordinate efforts across country boundaries and to establish uniform standards for cycling infrastructure, with recommendations for complementary policies and programs.

Currently, ECF and its member organizations are lobbying the EU to increase its funding of cycling projects to €3 billion for the years 2021–2027 and to €6 billion for the years 2028–2034 (ECF 2019). In addition to raising EU funding for cycling, ECF operates the EU Funds Observatory for Cycling to alert and assist national, regional, and local authorities in finding opportunities to use EU funds for cycling projects (ECF 2019). ECF has also lobbied the EU (responsible for motor vehicle standards) for mandatory intelligent speed assistance in cars and sensor technology on trucks, for example to avoid blind-spot crashes.

## National Advocacy in European Countries

The three most important national cycling advocacy organizations in the EU are the German Cyclists' Association, Allgemeiner Deutscher Fahrrad-Club (ADFC); the Danish Cyclists' Federation, Cyklistforbundet (CFB); and the Dutch Cyclist's Union, Fietsersbond (FB). All three organizations were founding members of ECF. ADFC was founded in 1979 and has about 175,000 individual members, each of whom is also associated with one of the 16 state cycling organizations or with one of about 450 local cycling organizations in Germany's many cities and regions (ADFC 2019). This federal structure of cycling advocacy is typical of many countries throughout the world.

In recent years, ADFC has increasingly advocated for high-quality, separate cycling infrastructure to get people on bikes who are interested in cycling but concerned about safety (ADFC 2019). This echoes the "all ages and abilities" principle of most national and state cycling advocacy organizations, which have been seeking to be as inclusive as possible by making cycling conditions safer, more convenient, and more comfortable (i.e., less stressed by motor vehicle traffic). ADFC has also been a leader in cycle tourism. It publishes an annual market analysis of bicycle tourism in Germany and runs the popular Bett+Bike program, which not only assists with route choices but also helps cycling tourists find appropriate overnight accommodations. ADFC establishes and monitors quality standards for both cycling routes and accommodations, organizes cycling tours, and is the national coordinator of the German portions of ECF's EuroVelo trans-European cycling route network.

The Danish Cyclists' Federation (CFB) was founded in 1905 and currently has 16,000 members (CFB 2019). CFB has lobbied over many decades for improved cycling conditions in Denmark. For example, it convinced the Danish Parliament to establish a €234 million National Cycle Fund for projects across the country from 2009 to 2014. CFB's Bike to Work campaign has been running for over 20 years and served as a best-practice example for ECF's Bike to Work project. CFB is dedicated to teaching children to cycle from an early age. In cooperation with many Danish municipalities, CFB has established permanent "bicycle playgrounds" near schools and in residential neighborhoods to teach safe cycling skills and traffic regulations in a professionally supervised, off-street environment. It also serves as the secretariat of the Cycling Embassy of Denmark (CED 2019), an international information-sharing network not only to promote cycling in Denmark but also to share practical information about Danish best practices (both infra-structure and programs) with cities around the world in order to promote more and safer cycling.

Fietsersbond (FB), the Dutch Cyclists' Union, was founded in 1975 and has 33,000 members (FB 2019). It focuses on promoting the expansion and improvement of cycling infrastructure, cycling-friendly roadway design (especially at intersections), cycling education, and various services for cyclists, such as web-based route planners. Similar to ADFC, FB works at all three levels of government: national, provincial, and local. For cycling advo-cacy at the regional and local levels, FB has more than 1,650 active volunteers in 150 local branches.

FB initiated programs such as the annual Fietstelweek, during which cycling volumes are measured at various locations throughout the Nether-lands. It also initiated the Cycling for All Ages program, which focuses espe-cially on the needs of children and older adults. Over the 45 years since its founding, FB has successfully lobbied government authorities to invest in a fully integrated, comprehensive network of comfortable, safe, and direct cycle routes that are separated and protected from motor vehicle traffic. This might account for the high cycling rates among children, seniors, and women in the Netherlands, which are higher than in almost all other coun-tries (see in this volume Buehler and Pucher, chapter 2; Garrard, chapter 11; McDonald, Kontou, and Handy, chapter 12; Garrard et al., chapter 13).

FB has also pushed governments to provide massive increases in bike parking, especially sheltered and secure bike parking, with a focus on key locations such as train stations, commercial centers, and schools (FB 2019). In addition to better bike parking at train stations, FB helped develop the OV-Fiets bike rental program at train stations, which offers easy access to public transport riders in the form of a combination fare card that can be used throughout the Netherlands. Improved bike parking and OV-Fiets bike rentals at train stations facilitate the use of cycling to access public trans-port, thus greatly increasing the service area of stations at minimal cost.

Similar to CFB's role in establishing the CED, FB played the key role in help-ing to establish the Dutch Cycling Embassy, whose official motto is "cycling for everyone," reflecting the goals of Dutch cycling policies already described (DCE 2019). The Dutch Cycling Embassy is a joint information exchange and cycling promotion network of cycling advocacy organizations, national and local governments, private firms, and research institutions. FB also plays a key role in the Tour de Force 2020, which is a partnership among Dutch national and local governments, the private sector, various nongovernmental organizations (NGOs), and research institutes. The aim of Tour de Force 2020 is to ensure that the Netherlands remains the world's leading cycling country with the goal of increasing the number of kilometers cycled by 20% over the coming decade (see Buehler and Pucher, chapter 2, this volume).

National cycling advocacy organizations are also active in other EU countries, such as Austria, Belgium, and Italy, as well as non-EU countries, such as Norway, Switzerland, and the United Kingdom. Cycling advocacy in the United Kingdom, in particular, has a history of large, well-organized, well-funded, and very active organizations. There are two main advocates: Cycling UK and Sustrans. Cycling UK (formerly the Cyclists' Touring Club), which has 65,000 members, campaigns at both the national and local levels for increased legal protection for cyclists, better cycling conditions, and more training programs (Cycling UK 2019). Sustrans, an organization that promotes both cycling and walking, has over 500 staff members and an annual budget in excess of £45 million. It has worked to expand and improve the 16,575 mi National Cycle Network, which includes UK segments of four EuroVelo routes. Sustrans has been actively involved in promoting safe and low-stress walking and cycling routes to school across the country (Sustrans 2019). Cycling UK and Sustrans work together to advocate for increased funding for walking and cycling at the national government level.

Many European countries, however, have weak or even nonexistent national associations. To help alleviate this problem, ECF initiated a program in 2012 to enhance national cycling advocacy throughout Europe. Initial funding for strengthening cycling advocacy came from the SRAM Cycling Fund and then grew with additional donations from the ECF Cycling Industry Club. By 2018, the ECF-led advocacy-building program had strengthened national advocacy associations in 16 European countries, including new organizational structures in some (ECF 2019). The success of this advocacy-strengthening initiative is reflected in the adoption of national cycling plans in many of those countries and increased government funding for cycling infrastructure and programs.

## Cycling Advocacy in North America

The role of national, state and provincial, and local cycling associations has been perhaps even more important in promoting city cycling in the

United States and Canada than in Europe. Most northern European countries, in particular, have a long history of cycling as a mainstream means of daily travel, both in cities and in rural towns and villages (Oldenziel, de la Bruheze, and Veraart 2016). In sharp contrast, cycling was never a mainstream mode of daily travel in any American or Canadian city (Pucher and Buehler 2008; Pucher and Buehler 2011). Thus, cycling advocacy has been crucial for overcoming the lack of cycling culture. Since 1990, in particular, the bike mode share of daily trips has risen dramatically in many North American cities—up to 4% or more in some cities. Without the massive organizing, networking, research, information exchange, marketing, and government lobbying efforts of advocacy associations, the growth of city cycling in North America would not have been possible.

### National Advocacy in the United States

There are two main cycling advocacy groups at the national level in the United States: the League of American Bicyclists (LAB) and PeopleForBikes (PFB). LAB is primarily an association of state and local advocacy organizations, as well as individual members, whose annual dues are the main source of funding. In contrast, PFB was built around a consortium of firms in the bicycle and related industries, who also provide most of PFB's funding.

**League of American Bicyclists**    The League of American Wheelmen (LAW) was founded in 1880 and renamed the League of American Bicyclists (LAB) in 1994. It is the oldest and largest national cycling association in the United States, whose members include 18,000 individuals, 350 advocacy organizations, and 800 cycling clubs (LAB 2019). Throughout its 140-year history, its main goal has been to represent the interests of bicyclists, although those interests have changed over time (LAB 2019). In the 1890s, during the first bicycling boom in the United States, LAW was a major proponent of improved roads to promote more and safer road cycling.

Even up to the 1980s, LAW mainly served experienced cyclists who rode bikes for recreation, sports, and long-distance touring. Instead of advocating any sort of separate cycling facilities, LAW offered various training programs to teach the skills for safe cycling on roads shared with motor vehicles, the "vehicular cycling" approach criticized by Furth (chapter 5, this volume). The failure to promote separate cycling facilities led to cycling becoming an extremely marginal mode of travel. Urban cycling for daily, utilitarian travel was almost nonexistent. Those who did cycle were mostly well-trained, experienced men, giving cycling an elitist, male-dominated image. The main exception was children riding bikes for recreation, usually on short trips within their residential neighborhoods.

To help enable more people to ride bikes safely on the roads, LAB developed a network of league-certified cycling instructors throughout the country, with courses for varying levels of ability. That program is now called Smart Cycling and has over 6,000 League Cycling Instructors (LCIs). LAB has continued to advocate for safer roads and cyclists' right to ride on the roads. In recent years, however, LAB has increasingly supported separate cycling facilities as essential to low-stress bicycling networks, which also include recommended bike routes on low-speed, low-volume streets. Moreover, LAB's training programs have been broadened to include skills to navigate a wide variety of cycling facilities and not just cycling on roads (LAB 2019).

That shift has been important. Most surveys show that the vast majority of Americans are not willing to risk their lives or their children's lives cycling on heavily traveled roads together with motor vehicles (see in this volume Furth, chapter 5; Garrard, chapter 11; McDonald, Kontou, and Handy, chapter 12; Garrard et al., chapter 13). By rejecting exclusive reliance on "vehicular cycling," LAB has joined cycling advocacy organizations throughout the world supporting cycling for everyone—of all ages and abilities.

LAB's advocacy efforts have focused on lobbying directly with Congress and federal agencies, mainly for increased funding and improved safety standards. Together with a wide range of sustainable transport, environmental, and urban advocates, LAB helped lobby for the 1991 federal transport law, the Intermodal Surface Transportation Efficiency Act, which provided increased funding for public transport and for the first time established special funding for bicycle and pedestrian facilities and programs (USDOT 1991). In 1997, LAB moved its headquarters to Washington, DC, specifically to work more closely with the federal government. Since about 2000, LAB has collaborated with other bicycling organizations and the bicycle industry in the America Bikes campaign. Prior to successive updates to the federal transport law in 1998, 2005, and 2012, they lobbied Congress and federal agencies to increase federal funding for bicycling and walking, which rose from $239 million in 1997 to $1.022 billion in 2019 (USDOT 2019).

LAB has actively supported the establishment and implementation of Safe Routes to School, Vision Zero, and Complete Streets policies at the state and local levels. Vision Zero seeks to reduce all traffic fatalities but especially those of vulnerable road users such as cyclists and pedestrians (see Elvik, chapter 4, this volume). In many cities, Vision Zero has been an important justification for greatly improved cycling infrastructure. The Complete Streets policies adopted by 34 states and hundreds of local jurisdictions have been an important justification for roadway designs that accommodate all road users, including cyclists.

LAB has also worked to strengthen the interlinking network of state and local cycling advocacy groups throughout the country. For example, its Advocacy Advance program and Active Transportation Leadership Institute provide technical assistance to state and local advocacy organizations so that they can encourage state and local governments to increase funding for cycling. Part of this effort is to convince state and local governments to devote more of the federal government's flexible transport funding to cycling initiatives.

LAB has organized the annual National Bike Summit in Washington, DC, since 2000. The conference facilitates information exchange among cycling advocates and professionals from throughout the United States and Canada. As part of the annual summit, groups of attendees from each state visit their senators and representatives in Congress to make the case for more cycling funding and more cycling-friendly policies.

LAB has also promoted nationwide events such as National Bike Month, Bike to Work Day, and Bike to School Day (all three in May). Perhaps LAB's most ambitious and successful promotional effort has been its Bicycle Friendly America program. It started in 1995 with Bicycle Friendly Communities and expanded to include Bicycle Friendly Businesses and Bicycle Friendly States in 2008 and Bicycle Friendly Universities in 2011. Participating communities, businesses, and universities must apply for Bicycle Friendly status, mainly showing how they meet the 5E criteria for effective cycling policy: engineering, education, encouragement, enforcement, and evaluation and planning. By 2019, there were 475 Bicycle Friendly Communities, 1,300 Bicycle Friendly Businesses, and 190 Bicycle Friendly Universities. There are five levels of certification as bicycle friendly: diamond, platinum, gold, silver, and bronze. The rankings are widely published, and there is a friendly competition among communities, businesses, and universities to achieve the highest bicycle-friendly status.

Unlike the three other Bicycle Friendly America programs, states do not formally apply for Bicycle Friendly status, but they are rated by LAB on the same 5E criteria, with information gathered from state departments of transport and other sources. The Bicycle Friendly States program gives states specific feedback on their performance and identifies areas for improvement.

To further enable cities and states to compare their performance on a wide variety of measures, LAB publishes a biannual *Benchmarking Report on Bicycling and Walking in the United States*, funded by the Centers for Disease Control and Prevention (LAB 2018). The report presents comprehensive, comparable statistics for all 50 states and the 50 largest cities, plus 20 smaller cities. Prior to 2018, the report was published by the Alliance for Biking and Walking, which was incorporated into LAB in 2017.

Although ignored for many decades, transport equity has become increasingly important in recent years, both in cycling advocacy and in participatory decision-making at the local level. There is much more concern now with improving cycling infrastructure and extending bikesharing to low-income and minority neighborhoods. Reflecting the importance of bicycling justice examined by Martens, Golub, and Hamre (chapter 14, this volume), LAB has adapted its Smart Cycling program to include more racially and economically diverse instructors and special programs in various languages aimed at marginalized, low-income communities.

**PeopleForBikes**  Formerly known as Bikes Belong, PeopleForBikes (PFB) was founded in 1999 to coordinate the cycling promotion efforts of major firms in the bicycle industry, which generates over $5 billion a year in revenue (PFB 2019). One goal was obviously to sell more bikes, but by achieving the more important goal of increasing cycling levels in the United States. PFB's particular focus was on increasing the percentage of Americans riding a bike at least occasionally by making cycling more attractive to a broader spectrum of the population. There was also a new focus on promoting daily utilitarian cycling in addition to the recreational cycling to which the industry had previously catered (e.g., mountain bikes). During its two decades of existence, PFB's base of support has broadened to include bicycle retailers, firms in related industries, other cycling advocacy organizations, and private foundations. In 2018 and 2019, over 60% of PFB's revenue came from outside the bicycle industry. It has grown from one staff member in 1999, with an annual budget of $400,000, to 24 staff members in 2019, with an annual budget of $8.5 million (PFB 2019). Since its founding, it has spent over $70 million on promoting the expansion and improvement of cycling infrastructure and programs to encourage cycling.

In PFB's early years, its efforts were aimed at providing financial support for other associations promoting cycling, in addition to cycling marketing, informational programs, government relations, and lobbying. For example, PFB was a major funder of the Bicycle Friendly Communities program of

LAB and a sponsor of the annual National Bike Summit. It financed various media campaigns to publicize the benefits of cycling as well as the need for motorists to share the road with cyclists. It also helped fund the America Bikes campaign that lobbied the federal government for more federal funding for pedestrian and bicycling projects, which rose from $204 million in 1999 to $1.022 billion in 2019 (LAB 2018).

Especially over the past 10 years, PFB has been awarding grants to cities for infrastructure projects that can be used to help match increased federal funding for pedestrian and cycling projects. PFB funding has been important to offset insufficient local funding and to finance innovative projects intended to provide lessons about the most effective ways to promote cycling, thus helping to develop guidelines for best practices. PFB has focused especially on infrastructure projects that make cycling safer, less stressful, and more appealing for inexperienced riders.

The Green Lane project was a five-year project (2011–2015) that provided financial assistance and technical advice to cities building protected on-street bike lanes and paths separated from motor vehicle traffic (as Furth describes in chapter 5, this volume). Several major cities (most prominently New York) had been building protected bike lanes since 2007, but the Green Lane project helped boost their expansion. As Furth shows (chapter 5, this volume, figure 5.2), the number of kilometers of protected bike lanes increased twelvefold from 2006 to 2018 (from 55 km to 684 km). Protected bike lanes (also called cycle tracks) have been so successful in attracting more riders and improving safety that in 2015 the US Department of Transportation officially endorsed the concept.

In 2017, PFB initiated its Big Jump project, which provides funding for 10 American cities to help improve the connectivity of their cycling networks by filling in key gaps. The overall aim is to help cities help themselves, focusing on accelerating the pace of bikeway network expansion not only with funding but also with assistance in improving the planning, public engagement, communications, and implementation processes used by these cities. Similar to the Green Lane project, the Big Jump project supports the creation of a safe, low-stress cycling network that will attract not only more cyclists but also a broader range of cyclists, including the vulnerable, risk-averse, and less experienced.

In recent years, PFB has developed the PFB Business Network and the PFB Better Bike Share Partnership. PFB has also taken a leadership role in monitoring federal, state, and local bike legislation in a joint effort with the Bicycle Product Suppliers Association. During 2016, PFB developed the PlacesForBikes program, which focuses on helping cities quickly build complete bike infrastructure networks that serve people of all ages and cycling abilities, financed mainly by a multimillion-dollar grant from Trek, a major bicycle manufacturer.

PFB recently developed a research division that produces bicycle network analyses for cities and towns, which identify both high- and low-stress bicycling locations citywide. PFB's annual City Rating scorecard evaluates cities for their cycling levels, cycling safety, infrastructure improvements, and progress in serving all parts of the community. The organization has also increased its polling and public messaging work in an effort to help elected officials gain strong citizen support for pro-bike initiatives. Adding further to its wide array of cycling promotional programs, PFB launched Ride Spot in 2019, a nationwide program to direct new and inexperienced bike riders to the best rides in their communities and also provide helpful advice on equipment and events.

**Other National Advocacy Groups**   In addition to the two main national cycling organizations, there are several national active travel organizations that have specialized in particular aspects of cycling advocacy. The Rails-to-Trails Conservancy (RTC), founded in 1986, has concentrated on the conversion of abandoned railroad corridors into paved, multiuse paths. Over the past three decades, RTC has built over 24,000 mi (40,000 km) of such rail trails, mostly in rural areas but often extending into urban areas as well. Recently, RTC has increased its efforts in urban areas and has been building trails in public utility corridors and active railroad corridors. In 2018, RTC proposed the creation of the Great American Trail, which would extend over 4,000 mi (6,700 km) across the United States (RTC 2019).

Founded in 1991, the East Coast Greenway Alliance (ECGA 2019) has been promoting and coordinating its planned 2,900 mi (4,843 km) East Coast Greenway from Maine to Florida. Thanks to the investment of over a billion dollars of public funds so far, 975 miles (1,628 km) had been completed by 2018, connecting 15 states and 450 communities along the route and passing through all the major metropolitan areas on the East Coast (ECGA 2019). Completed segments of the greenway host over 38 million bike rides, runs, and walks per year—making the route one of the most popular parks in America. ECGA's 12 full-time staff members and over 300 volunteers have worked with local, state, and federal governments, organizations, and individual supporters to develop this multiuse path (ECGA 2019).

Although it is not an advocacy organization, the Association of Pedestrian and Bicycle Professionals (APBP 2019) has played a crucial role in promoting information exchange to improve the design and engineering of walking and cycling facilities. APBP was established in 1995 following discussions among state and local bicycle and pedestrian coordinators at the 1992 and 1994 Pro Bike conferences. In 1998, APBP hired its first executive director and was a partner in the creation of the Pedestrian and Bicycle Information Center, funded by the US Department of Transportation. By 2018, there were over 1,200 members of APBP, mainly transport planners, engineers, researchers, and government officials, in both the United States and Canada. APBP is important because its members are the ones responsible for developing, implementing, and coordinating cycling and walking infrastructure and programs.

## State Advocacy in the United States

This section focuses on the California Bicycle Coalition (CalBike) and Bike-Texas, two of the largest and most important of the 44 statewide advocacy organizations in the United States. Those 44 state organizations work with over 350 local cycling organizations in cities and towns to promote cycling throughout the country (LAB 2018).

**California Bicycle Coalition (CalBike)**  The California Bicycle Coalition coordinates bicycling advocacy in a state of 40 million people, most of whom are represented by 24 bicycling advocacy organizations at the local level, with a total of over 35,000 members (CalBike 2019). With the fourth-highest bicycle mode share of the 50 states, California is one of the country's leaders in promoting cycling not only overall but especially among low-income groups and communities of color (LAB 2018).

CalBike promotes cycling by lobbying the state government to improve cycling, primarily by increasing funding for infrastructure and by requiring bicycle facilities whenever repaving, widening, or rebuilding a state-owned street. CalBike helped revise state bikeway design guidelines to encourage the construction of protected bike lanes, reversing the previous state guidelines, which had discouraged them. CalBike also worked to change the state vehicle code by requiring motorists to give at least 3 ft of clearance when passing cyclists on the road and led the effort to legalize electric-assist bikes.

CalBike and its associated local cycling organizations have been at the forefront of promoting social justice in bicycling policies, as urged by Martens, Golub, and Hamre (chapter 14, this volume; see also CalBike 2019). It is one of the key priorities of CalBike to give community residents more control over transport policy and spending decisions by changing the composition of decision-making bodies to make them more democratic. It especially advocates for increased transport investments in low-income communities. CalBike has supported improved access to bicycles through the expansion of bikesharing programs, especially in low-income communities, which are usually underserved by private operators. It has also promoted bicycle subsidies and purchase incentives, especially to make e-bikes affordable to low-income Californians. To coordinate these social justice initiatives, CalBike holds mobility justice labs, provides technical training, and organizes leadership retreats.

Like many cycling advocacy organizations around the world, one key priority of CalBike is to make cycling feasible, safe, and convenient for cyclists of all ages and abilities, women as well as men and beginners as well as experienced cyclists. The key to enabling cycling for everyone is the construction of well-designed, well-maintained bikeways that are connected to form an integrated, complete network that provides a safe, low-stress cycling environment, as documented by the analysis elsewhere in this volume (see Buehler and Pucher, chapter 2; Elvik, chapter 4; Furth, chapter 5; Garrard, chapter 11; McDonald, Kontou, and Handy, chapter 12; Garrard et al., chapter 13; Pucher, Parkin, and de Lanversin, chapter 17; Koglin, te Brömmelstroet, and van Wee, chapter 18; Geller and Marqués, chapter 19). On heavily traveled arterial roads, that requires cycle tracks that are protected from motor vehicle traffic with physical barriers, as noted by Furth (chapter 5, this volume). CalBike has pushed to create a Bike Network Grant program that would help finance such cycling networks in selected model communities, which would then expand to cover more parts of those cities and spread to other cities as well.

**BikeTexas** BikeTexas (founded in 1991 as the Texas Bicycle Coalition) promotes bicycle access, safety, and education in the state of Texas, the second-largest state in the country, both in area and population (29 million) (BikeTexas 2019). BikeTexas started off in response to proposed bike bans on rural roads but has since greatly expanded its activities. It now works at the local, state, and national levels to represent individual Texans who ride bikes as well as local cycling advocacy organizations, bike clubs, and bike shops. It also organizes and sponsors a wide range of programs and events.

In 2001, BikeTexas introduced Texas's Safe Routes to School (SRTS) program, one of the country's first statewide programs. BikeTexas was also part of the nationwide coalition that helped pass federal SRTS legislation in 2005. The SRTS program has provided over a billion dollars in funding for infrastructure and programs to facilitate safe walking and cycling for children between their homes and schools. BikeTexas's SafeCyclist education program has trained over 8,000 Texas teachers in bicycle safety instruction. Those teachers, in turn, have taught safe biking and walking skills and distributed brochures with traffic safety information to over two million Texas children. Recently, BikeTexas expanded its educational programming by making its cycling education information and safety training methods available to schools and nonprofits nationwide.

For many years, BikeTexas has been supporting state legislation to improve the safety of cyclists riding on the roads. In 2009, a Safe Passing Law was approved by both chambers of the Texas legislature but was vetoed by the governor. After that setback, BikeTexas shifted its focus to the local level and encouraged cities around the state to enact their own safe-passing ordinances. As of late 2019, 27 Texas municipalities had passed ordinances requiring a 3 ft minimum passing clearance between motor vehicles and vulnerable road users.

Since 2010, BikeTexas has organized the biennial Texas Trails and Active Transportation (TTAT) conference, which attracts hundreds of attendees from around the state to hear talks by national and international cycling, walking, and trail experts. The TTAT conference is a key resource for educating Texas's policy makers and civic leaders about state-of-the-art active transport infrastructure. The success of the TTAT conferences has helped BikeTexas work with government officials and local stakeholders in cities around the state to create, fund, and implement bicycle and trail plans. BikeTexas developed Texas Bicycle Tourism Trails legislation that successfully passed the Texas legislature in 2005 and is the foundation for a future 8,000 mi statewide network that will connect to most Texas cities and towns as well as to the US Bike Route System. BikeTexas continues to advocate for increased and more secure trail funding at the local, state, and federal government levels.

BikeTexas has been a leader in advancing the adoption of electric bicycles via state legislation, utility company rebates, Electric Bike Day at the capitol in Austin, and group rides utilizing a fleet of e-bikes. These guided group bike rides have enabled policy makers and civic leaders to experience firsthand the challenges facing people who ride bikes. BikeTexas organized annual bipartisan bike rides at the National Conference of State Legislators from 2005 to 2016; bipartisan rides during Texas legislative sessions for legislators, staffers, and participants in the Texas Legislative Internship Program since 2003; and bike rides for NAACP state and regional leaders since 2008.

### National Advocacy in Canada

Only since 2012 has Canada had a national cycling advocacy organization: Vélo Canada Bikes (VCB 2019). The Ottawa-based VCB has been encouraging the federal government to establish a national cycling strategy and to invest in cycling infrastructure and programs. In addition, VCB has organized an annual bike summit in Ottawa since 2017. The limited efforts so far at the national level in Canada stem mostly from the Canadian constitution, which relegates urban transport policies, planning, and funding to the provincial levels of government, which in turn can delegate partial responsibility for these issues to the local government level. However, VCB has identified key areas in which the federal government can and should support provincial and local governments: establishing a national forum for communication and research; creating a dedicated infrastructure fund; setting a bicycle mode share target for the country; and requiring Statistics Canada to collect data on cycling levels and cycling safety.

In addition to VCB, the Toronto-based Centre for Active Transportation (TCAT 2019) has been promoting walking and cycling since its founding in 2006. Its current projects include Complete Streets for Canada, an online hub tracking the adoption and development of Complete Streets policies and projects across the country; Scarborough Cycles, which focuses on promoting cycling in suburban communities; and Active Neighborhoods Canada, a pan-Canadian partnership using co-design to create green, active, and healthy neighborhoods.

### Provincial and Local Advocacy in Canada

Vélo Québec (VQ) promotes cycling throughout the province of Quebec (VQ 2019). It is one of the largest, strongest, and most effective cycling advocacy organizations in North America, with 75 full-time staff members (rising to 100 in the busy summer season), over 5,000 dues-paying members, and over 80,000 registered online supporters, dwarfing every other cycling advocacy organization in Canada.

VQ was founded in 1967 mainly to promote physical activity among youths. That aim soon broadened to include Quebec residents of all ages, women as well as men, and all levels of ability, from beginners to experienced cyclists. While other provincial cycling organizations in Canada in the 1960s and 1970s focused on cycling for sports, VQ was the first to focus its efforts on promoting cycling for daily utilitarian travel, recreation, and tourism, seeking to get as many people as possible riding bikes.

Inspired by the cycling infrastructure in the Netherlands, VQ lobbied strongly for the development of a system of protected bike lanes and paths (cycle tracks) in the cities of the province of Quebec, the first such effort in North America (except for the small university town of Davis, California). As a result of VQ's effective lobbying and Dutch-based design advice, Montreal built its first protected bike lane in 1985 and continuously expanded its network of protected on-street bike lanes to 75 km by 2007, the year that the first protected on-street lanes were opened in New York City (VQ 2019). For several decades, Montreal served as the only model of protected on-street cycling facilities for other large North American cities to follow.

The quadrupling of cycling in Montreal since 1990 resulted not only from the expansion of high-quality, low-stress, safe, and connected cycling infrastructure but also from a host of supportive cycling programs (VQ 1995–2015). They include cycling training for various ages and skill levels, special programs for cycling to school and to work, and small as well as large group bike rides, such as the annual Tour de l'Île de Montréal, which had 25,000 participants in 2018. VQ also organizes the annual Grand Tour, a seven-day bike ride with varying routes each year; it had about 1,600 participants in 2018. In 1995, VQ established the Maison des Cyclistes, a combination café/restaurant, bike rental and repair shop, and cycling information center at a key intersection of cycling routes in Montreal. VQ has also conducted extraordinarily effective media campaigns to promote cycling widely among the population of Quebec. It is the crucial liaison connecting 15 local cycling advocacy organizations and 150 cycling clubs throughout the province.

Especially over the past two decades, VQ has worked closely with the Quebec Ministry of Transport on many cycling projects (Quebec Ministry of Transport 2018; VQ 1995–2015). The most famous is the Route Verte, an integrated, high-quality network of on-road facilities, greenways, and trails throughout the province of Quebec, exceeding 5,000 km in total length in 2018. The Route Verte is probably the best long-distance bikeway network in all North America, so good that it has generated a thriving cycling tourism industry for Quebec, drawing thousands of visitors from all over the world. Demonstrating not only activism but also entrepreneurship, VQ also has a special cycling tourism travel agency that organizes travel to cycling destinations throughout the world as well as tours and lodging for cyclists along the Route Verte. VQ also works with the Ministry of Transport and local governments to expand and improve cycling facilities for daily travel in cities, including a regularly updated technical planning guide for building cycling infrastructure.

In cooperation with the Quebec Ministry of Transport and eight of the largest cities in Quebec, VQ has compiled and published *Cycling in Québec* every five years since 1995 (VQ 1995–2015). Similar to the LAB's biannual *Benchmarking Report on Bicycling and Walking in the United States*, it is a comprehensive statistical analysis of cycling levels, cycling safety, cyclist characteristics, infrastructure expansion, programmatic measures, and indicators of the social and economic impacts of cycling in the province of Quebec. The report is available not only in print but also in an interactive online version. The extremely close cooperation between VQ and the Quebec Ministry of Transport is unique in Canada and perhaps in all North America.

VQ has organized high-profile events such as the first World Cycling Conference (Conférence Vélo Mondiale) in 1992, bringing together the Pro-Bike Conference (North America) and the ECF's Velo-city conference (Europe). In celebration of its high cycling rates even during brutally cold Quebec winters, VQ hosted the fifth international Winter Cycling Congress in 2017.

The very strong provincial cycling advocacy in Quebec contrasts sharply with the newer and less developed efforts in other Canadian provinces. There are, however, two other provincial cycling organizations in Canada that have become increasingly influential in recent years: in Ontario and in British Columbia. The Share the Road Cycling Coalition (SRCC) in Ontario developed the first Ontario Cycle Plan, secured C\$80 million in provincial funding for cycling infrastructure at the municipal level, and introduced the Bicycle Friendly Community program to Ontario. SRCC has hosted 10 Ontario Bike Summits, bringing together cycling advocates from around the province (SRCC 2019). Since its inception, SRCC has lobbied for better provincial legislation to protect vulnerable road users. The British Columbia Cycling Coalition (BCCC) works primarily by coordinating the efforts of its 26 member organizations throughout the province. Together with its partners, BCCC encouraged the provincial government to develop its first Active Transportation Strategy as part of its CleanBC climate plan. BCCC hosted the first BC Active Transportation Summit in June 2019 (BCCC 2019).

With the important exception of Quebec, cycling advocacy is mainly at the local level. For example, HUB Cycling in Greater Vancouver is many times larger than BCCC. Indeed, HUB Cycling is by far the largest and most effective local cycling organization in Canada, with over 3,000 dues-paying members, over 40,000 registered supporters, 19 full-time staff members, 18 cycling instructors, and over 500 volunteers (HUB Cycling 2019). HUB Cycling has been the driving force in the massive expansion and improvements in cycling infrastructure and programs in Greater Vancouver, which has the highest cycling mode share of any major Canadian metropolitan area. The Montreal Bike Coalition, Bike Ottawa, Cycle Toronto, and Bike Calgary are examples of other local cycling advocacy organizations in Canada.

### Cycling Advocacy in Australia

Similar to the United States and Canada, cycling advocacy in Australia is present at the national, state, and local levels in Australia, but it is strongest at the state level.

### National Advocacy in Australia

From 1979 to 2010, the Bicycle Federation of Australia (BFA) was the main national cycling advocacy organization in Australia. It was primarily an association of state cycling organizations and was funded by their contributions and the revenues from *Australian Cyclist* magazine, which BFA published. In addition to organizing an annual national bike summit, BFA lobbied for pro-cycling legislation and funding and helped coordinate cycling advocacy in the different states. BFA ceased to exist in 2010 because of insufficient funding. Its role in national bike advocacy has been taken over by We Ride Australia (WRA), formerly known as the Cycling Promotion Fund, which had already existed since 2000 but took on additional advocacy importance at the national level when BFA dissolved.

Similar to PeopleForBikes in the United States, WRA is funded primarily through contributions from the bicycle industry (WRA 2019), supplemented by funds from local governments, not-for-profit organizations, and private firms. For almost two decades, WRA has worked with federal and state ministries as well as foundations, councils, forums, workshops, and task forces to promote the importance of cycling as a healthy, sustainable means of daily urban travel as well as recreation. It works closely with PFB in the United States on programs such as bicycle network analysis and city rating schemes. Like PFB, WRA's advocacy efforts are aimed at making cycling safe, convenient, low stress, and feasible for as many people as possible, including all ages, abilities, and demographic segments of society. Thus, just like PFB, WRA has been a strong proponent of protected bike lanes and paths that are connected to form an integrated network of safe cycling facilities.

Since 2005, the Amy Gillett Foundation (2019) has focused on improving cycling safety in Australia. It has advocated for the cyclist's right to ride on the road, safe passing distances by motorists, and increased punishment of motorists who deliberately endanger cyclists. Over the last decade, various other national advocacy groups, such as the Australian Cycle Alliance, have emerged, with differing areas of interest, such as media or safety.

**State and Local Advocacy in Australia**

Most of the cycling advocacy in Australia occurs at the state level. All states and territories have one or more bicycle advocacy groups, which are financed primarily by individual membership dues and cycling event registration fees and sponsors such as governments, foundations, NGOs, and private firms. In total, state advocacy organizations have 97,000 members. These state organizations include: Bicycle Network (formerly Bicycle Victoria, which recently expanded its cycling advocacy efforts to the Northern Territory and Tasmania and has 45,000 members); Bicycle Queensland (18,000); Bicycle New South Wales (15,000); Bicycle South Australia (6,000) and the Bicycle Institute of South Australia (500); Pedal Power (in the Australian Capital Territory, 7,700); and Bicycling Western Australia (4,500). Most state groups have some paid staff but rely heavily on volunteers.

Local cycling advocacy provides crucial grassroots support for state advocacy. Local cycling organizations are called bicycle user groups, which are often members of the state organization and help with grassroots support and hands-on advocacy. They concentrate on cycling issues in their particular area, providing a neighborhood-level perspective to the cycling movement. They provide the most effective cycling advocacy with the local government councils, the most important government level within metropolitan areas, cities, and towns in Australia. There are over a hundred bicycle user groups throughout Australia: 48 in Victoria, 29 in New South Wales, 22 in Queensland, 19 in Western Australia, and 21 in South Australia.

## Conclusions

The preceding survey of cycling advocacy in Europe, North America, and Australia reveals that there is much overlap in their efforts to promote cycling. Almost all advocacy organizations engage with government officials at various levels to promote increased funding for cycling infrastructure and programs as well as regulations that protect cyclists riding on the road. Many organizations encourage governments to offer traffic safety education in schools, including cycling safety, thus teaching safety skills at a young age. Some advocacy groups directly offer cycling training for various age groups and skill levels. At the national and state levels, advocacy organizations have urged governments to require more comprehensive driver training and testing that emphasizes the responsibility of motorists to avoid endangering cyclists. Especially at the state and local levels, they have prominently publicized instances where cyclists have been killed by motorists, especially when the motorists have been found guilty of deliberately running into cyclists or driving while impaired. That has drawn media attention that has often pressured governments to enforce more strictly motorist driving regulations affecting the safety of cyclists.

Cycling advocacy organizations have been effective at using their media contacts, social networking, informational programs, and government lobbying to emphasize to the public and politicians the benefits of cycling and thus the need for government support. Some of the organizations have a research division that helps disseminate knowledge about best practices to encourage cycling. They have often developed and proposed specific infrastructure improvements or programs, thus providing government officials and planning staff with ideas to consider. In a further attempt to generate more cycling and to raise the profile of cycling, most advocacy groups have organized large public events, such as open streets festivals and large group bike rides that are covered by the media.

The overall impact of cycling advocacy is enhanced through a closely linked, mutually supportive network of local, state, national, and international advocacy organizations. It is through their combined efforts that they have been so effective at promoting government policies to encourage more and safer cycling. Integrating bicycle manufacturers and retail outlets into cycling advocacy has provided not only funding but also a network of thousands of firms that can help make the case that cycling is good for business. Advocacy organizations have given a strong voice to cyclists by creating local, state, national, and international communities of cyclists. Working with each other across city, state, and national boundaries, advocacy organizations have been crucial for generating the public and political support needed to get the right policies implemented. As illustrated in the case studies in this volume, the necessary policies do not get implemented simply because they are good but rather because they have sufficient public and political support (see Pardo and Rodriguez, chapter 16; Pucher, Parkin, and de Lanversin, chapter 17; Koglin, te Brömmelstroet, and van Wee, chapter 18; and Geller and Marqués, chapter 19).

**References**

Allgemeiner Deutscher Fahrrad-Club (ADFC). 2019. Allgemeiner Deutscher Fahrrad-Club. https://www.adfc.de/.

Amy Gillett Foundation. 2019. *Safe Together.* Amy Gillett Foundation. https://www.amygillett.org.au/.

Association of Pedestrian and Bicycle Professionals (APBP). 2019. Association of Pedestrian and Bicycle Professionals. https://www.apbp.org/.

BikeTexas. 2019. BikeTexas. https://www.biketexas.org/.

British Columbia Cycling Coalition (BCCC). 2019. British Columbia Cycling Coalition. https://cyclingbc.net/.

CalBike. 2019. CalBike. https://www.calbike.org/.

Centre for Active Transportation (TCAT). 2019. Centre for Active Transportation. https://www.tcat.ca/.

Cycling Embassy of Denmark (CED). 2019. *Cycling all over the World.* Cycling Embassy of Denmark. http://www.cycling-embassy.dk/.

Cycling UK. 2019. Cycling UK. https://www.cyclinguk.org/.

Cyklistforbundet (CFB). 2019. Cyklistforbundet. https://www.cyklistforbundet.dk.

Dutch Cycling Embassy (DCE). 2019. *Cycling for Everyone.* Dutch Cycling Embassy. https://www.dutchcycling.nl/.

East Coast Greenway Alliance (ECGA). 2019. East Coast Greenway Alliance. https://www.greenway.org/.

European Cyclists' Federation (ECF). 2017. *EU Cycling Strategy: Recommendations for Delivering Green Growth and an Effective Mobility System in 2030.* Brussels: European Cyclists' Federation.

European Cyclists' Federation (ECF). 2019. European Cyclists' Federation. https://ecf .com/.

Fietsersbond (FB). 2019. Fietsersbond. https://www.fietsersbond.nl/.

Gerike, Regine, and John Parkin. 2016. *Cycling Futures: From Research into Practice.* New York: Routledge.

HUB Cycling. 2019. HUB Cycling. https://bikehub.ca/.

League of American Bicyclists (LAB). 2018. *2018 Benchmarking Report on Bicycling and Walking in the United States.* Washington, DC: League of American Bicyclists.

League of American Bicyclists (LAB). 2019. League of American Bicyclists. https://www.bikeleague.org/.

Oldenziel, R., M. Emmanuel, A. A. de la Bruheze, and F. Veraart. 2016. *Cycling Cities: The European Experience.* Einhoven: Foundation for the History of Technology.

PeopleForBikes (PFB). 2019. PeopleForBikes. https://peopleforbikes.org/.

Pucher, John, and Ralph Buehler. 2008. Making Cycling Irresistible: Lessons from the Netherlands, Denmark, and Germany. *Transport Reviews* 28(4): 495–528.

Pucher, John, and Ralph Buehler. 2011. Bicycling Renaissance in North America: An Update and Reappraisal of Cycling Trends and Policies. *Transportation Research Part A: Policy and Practice* 45(6): 451–475.

Pucher, John, and Ralph Buehler. 2017. Cycling towards a More Sustainable Transport Future. *Transport Reviews* 37(6): 689–694.

Quebec Ministry of Transport. 2018. *Sustainable Transport Policy 2030.* Quebec City: Quebec Ministry of Transport.

Rails-to-Trails Conservancy (RTC). 2019. Rails-to-Trails Conservancy. https://www .railstotrails.org/.

Share the Road Cycling Coalition (SRCC). 2019. Share the Road Cycling Coalition. https://www.sharetheroad.ca/.

Sustrans. 2019. Sustrans. https://www.sustrans.org.uk/.

US Department of Transportation (USDOT). 1991. *Intermodal Surface Transportation Efficiency Act*. Washington, DC: US Department of Transportation.

US Department of Transportation (USDOT). 2019. *Federal-Aid Highway Program Funding for Pedestrian and Bicycle Facilities and Programs, FY 1992 to 2018*. Washington, DC: US Department of Transportation. https://www.fhwa.dot.gov/environment /bicycle_pedestrian/funding/bipedfund.cfm.

Vélo Canada Bikes (VCB). 2019. Vélo Canada Bikes. http://www.canadabikes.org/.

Vélo Québec (VQ). 1995–2015. *Cycling in Québec*. Montreal: Vélo Québec/Quebec City: Quebec Ministry of Transport.

Vélo Québec (VQ). 2019. Vélo Québec. http://www.velo.qc.ca.

We Ride Australia (WRA). 2019. We Ride Australia. https://www.weride.org.au/.

# 21   Cycling to a More Sustainable Transport Future

Ralph Buehler and John Pucher

Reinforcing the main findings of our 2012 book *City Cycling*, this expanded and updated book provides additional evidence of the continued strong growth in cycling in many cities around the world (see Pucher and Buehler, chapter 1, this volume, figure 1.1). As shown in several chapters, however, there is much variation in cycling rates among different groups as well as among countries, cities, and different parts of metropolitan areas. Cycling is still male dominated except in countries with high cycling mode shares, such as the Netherlands, Japan, Denmark, and Germany (see in this volume Buehler and Pucher, chapter 2; Garrard, chapter 11). Cycling by children continues its decades-long decline in most countries (McDonald, Kontou, and Handy, chapter 12, this volume), while cycling has been increasing among other age groups, including seniors in some countries (Garrard et al., chapter 13, this volume). The increase in adult cycling has more than offset the decline in children cycling, although that is a serious concern, as McDonald, Kontou, and Handy (chapter 12) examine in detail.

The most easily measurable variation in cycling levels is among countries and cities (see Buehler and Pucher, chapter 2, figures 2.1, 2.2, and 2.3). Differences between countries can be as large as 30 to 1 (e.g., Netherlands vs. United States). Even within countries, differences among cities can be as large as 30 to 1 (Italy), 20 to 1 (United States), or 10 to 1 (India). Differences in cycling rates among cities in the same country tend to be smallest in the countries with the highest overall cycling levels; for example, only about 2 to 1 in the Netherlands (Buehler and Pucher, chapter 2, figure 2.3).

Cycling remains concentrated in urban areas. Cycling rates are highest in central city areas, with rates falling with distance from the city center, being lower in the outer areas of cities and even lower in the suburbs. Thus, there has indeed been a boom in cycling, but in most countries it has been limited mostly to central city areas. There are exceptions to this pattern. In some German, Austrian, Swiss, and Dutch cities, for example, cycling levels in suburbs, small towns, and rural villages are as high as or higher than cycling levels in the centers of large cities (Buehler et al. 2017). India is also an exception; rural areas have a higher cycling mode share than urban areas (22% vs. 18%) (see Pucher et al., chapter 15, this volume). The spatial concentration of cycling in the city center is most pronounced in countries with very low overall levels of cycling, such as the United States, Canada, and Australia, where cycling in outlying suburbs and rural areas is rare. Spatial variation within metropolitan areas is documented in the city case studies in this volume by Pucher, Parkin, and de Lanversin (chapter 17), Koglin, te Brömmelstroet, and van Wee (chapter 18), and Geller and Marqués (chapter 19).

The past decade has produced even more overwhelming scientific evidence of the large and multifaceted health benefits of cycling, which far offset the risks of traffic injury, even for seniors (see in this volume Garrard et al., chapter 3, and Garrard et al., chapter 13). Moreover, per capita cyclist fatality rates in 10 of 11 countries of the Organization for Economic Cooperation and Development (OECD) fell by 50%–68% from 1990 to 2018 (see Buehler and Pucher, chapter 2, this volume, figure 2.6). The least progress was in the United States, with only a 30% decline from 1990 to 2010 and then an alarming 13% increase from 2010 to 2018. The improvement in cycling safety has been mainly in those cities investing in expanded and improved cycling infrastructure, especially protected bike lanes and paths (see in this volume Buehler and Pucher, chapter 2; Furth, chapter 5; Pardo and Rodriguez, chapter 16; Pucher, Parkin, and de Lanversin, chapter 17; Koglin, te Brömmelstroet, and van Wee, chapter 18; Geller and Marqués, chapter 19).

### Crucial Importance of Safe Cycling Infrastructure

Table 21.1 lists 20 cities in Europe, the Americas, and Australia that have greatly expanded their overall network of cycling facilities in recent years, with the time period varying from city to city depending on data availability. Without exception, the large increases in the networks of separate cycling facilities have been accompanied by large increases in cycling levels over the same period (for each city), as well as sharp reductions in rates of cyclist fatalities and serious injuries relative to cycling levels. Although not shown as a separate category in table 21.1, many of these cities have focused their recent

Table 21.1

Better bicycling infrastructure improves cyclist safety and increases cycling

| City | Years | Growth in kilometers of cycling facilities (%) | Growth in cycling (%) | Change in fatalities and severe injuries relative to cycling levels (%) |
|---|---|---|---|---|
| Portland, OR | 2000–2017 | 67 | 375 | −66 |
| Washington, DC | 2000–2017 | 174 | 435 | −54 |
| New York, NY | 2000–2017 | 381 | 207 | −72 |
| Minneapolis, MN | 2000–2017 | 113 | 321 | −79 |
| Chicago, IL | 2005–2017 | 165 | 88 | −42 |
| Seattle, WA | 2000–2017 | 321 | 101 | −60 |
| Los Angeles, CA | 2005–2015 | 130 | 114 | −43 |
| Vancouver, Canada | 2006–2016 | 71 | 65 | −62 |
| Montreal, Canada | 2000–2018 | 125 | 295 | −65 |
| Bogota, Colombia | 2005–2018 | 69 | 213 | −61 |
| Sydney, Australia | 2010–2018 | 30 | 124 | −48 |
| London, United Kingdom | 2000–2018 | 401 | 154 | −65 |
| Paris, France | 2000–2018 | 415 | 385 | −81 |
| Seville, Spain | 2006–2017 | 228 | 465 | −79 |
| Copenhagen, Denmark | 1996–2018 | 55 | 200 | −80 |
| Amsterdam, Netherlands | 1990–2017 | 125 | 71 | −76 |
| Freiburg, Germany | 1982–2018 | 47 | 127 | −82[a] |
| Berlin, Germany | 1998–2017 | 67 | 50 | −45[a] |
| Munich, Germany | 2002–2017 | 90 | 80 | −54[a] |
| Vienna, Austria | 1990–2017 | 133 | 209 | −47 |

[a] Fatalities only.

*Sources*: Based on data collected directly from each city.

infrastructure investments on fully separated on-street bike lanes and off-street paths that are protected from motor vehicle traffic by physical barriers (designated as "cycle tracks" by Furth, chapter 5, this volume). New York, for example, built 244 km of protected bike lanes and off-street paths between 2007 and 2019 (see Pucher, Parkin, and Lanversin, chapter 17, this volume).

It is notable that in some cities the percentage growth in cycling levels has been far larger than the percentage expansion in facilities. That is partly because the percentage increases in kilometers of facilities do not reflect the significant improvement in the quality of cycling facilities and their increasing connectivity, such as in Portland, Oregon (see Geller and Marqués, chapter 19, this volume). Similarly, there are many other complementary programs (such as bikesharing, bike parking, traffic calming, and promotional events) that have contributed to the growth in cycling. Overall, however, the large growth in cycling levels reflects the enormous potential demand for expanded cycling facilities, which once built are heavily used. The other important trend shown in table 21.1 is the dramatic improvement in cycling safety, with fatality and serious injury rates falling by at least 42% in all the cities.

As Elvik (chapter 4, this volume) emphasizes, higher cycling levels do not automatically increase safety, as suggested by the theory of "safety in numbers." Rather, increasing cycling levels mainly result from improved infrastructure and the greater safety and convenience it provides. The most effective way to increase cycling safety, raise overall cycling levels, and encourage more women, children, and seniors to ride a bike is to provide connected networks of low-stress, separate bike lanes and paths that are protected from motor vehicle traffic. Cycling safety, cycling levels, and the diversity of cyclists have increased dramatically in almost every instance where such protected facilities have been installed (see in this volume Buehler and Pucher, chapter 2; Furth, chapter 5; Pardo and Rodriguez, chapter 16; Pucher, Parkin, and de Lanversin, chapter 17; Koglin, te Brömmelstroet, and van Wee, chapter 18; Geller and Marqués, chapter 19).

The vast expansion and connection of protected facilities is one of the most important recommendations made both in this chapter and in almost all the chapters in this book. The number of kilometers of such protected facilities increased more than twelvefold in the United States as a whole from 2006 (55 km) to 2018 (684 km), facilitating impressive cycling growth in cities that built such facilities (see Furth, chapter 5, this volume, figure 5.2). Given the documented increases in cycling levels and improved safety that cycle tracks have facilitated, all major bicycling advocacy organizations in the Americas, Australia, and Europe are increasingly advocating separate, protected cycling facilities with the goal of increasing cycling for all ages and abilities (see Pardo and Rodriguez, chapter 16; Pucher, Parkin, and de Lanversin, chapter 17; Koglin, te Brömmelstroet, and van Wee, chapter 18; Geller and Marqués, chapter 19; Pucher et al., chapter 20).

Cycling facilities protected from motor vehicle traffic have become even more important over the past decade because of the worsening epidemic of distracted driving, which has been shown to be as dangerous as driving under the influence of alcohol or narcotics. Although cell phone use and texting while driving are illegal, they are no longer the exception but have become the norm because they are so difficult for police to detect. In addition, motor vehicle dashboards have become increasingly interactive, complicated, and distracting. GPS navigation screens, multifunctional entertainment consoles, and a wide variety of control features, although not illegal, can obviously be distracting as well. A painted line on the roadway offers cyclists no protection at all from distracted and dangerous driving by motorists. That increases the need for physically separated, traffic-protected bike lanes and paths.

Express bikeways are a promising recent development. Such express bike routes increase the speed and safety of long-distance cycling by providing separate bike paths parallel to major roads, with minimal road crossings and sometimes with a green wave of synchronized traffic signals at intersections timed for faster cycling. Express bikeways are proving to be increasingly popular and necessary to serve the longer-distance bike commutes in metropolitan areas, which are spreading out not only in North America and Australia but in Europe as well. Dozens of European cities already have express bikeways of various kinds, with many more cities planning them.

Another important trend in almost all major American, Canadian, and Australian cities is the rapid expansion of paved off-road, mixed-use (including pedestrians) greenways, trails, and paths throughout metropolitan areas and on their outskirts. Such mixed-use greenways are usually located in parks—often along rivers, creeks, lakes, and seashores—or in abandoned railroad rights-of-way (rail trails). In the United States, there was roughly tenfold growth in the number of kilometers of such facilities from 1990 to 2018. Similarly, in the province of Quebec, with Canada's most extensive cycling network, the number of kilometers of off-road greenways and trails grew twelvefold from 1990 to 2018. Although they are mainly intended for recreational use, greenway and trail networks are increasingly providing useful connections for utilitarian cycling as well, especially when located within urban corridors (such as the Midtown Greenway in Minneapolis and the Hudson River Greenway in Manhattan). Such off-road cycling facilities are especially important for beginning cyclists and for risk-averse cyclists of any kind. They enable beginners to become comfortable riding a bike under stress-free, traffic-free conditions before transitioning to the more challenging cycling required on city streets.

## Need for an Integrated Package of Programs and Policies
## to Complement Bikeways

Building bikeways is not the only important strategy to increase cycling and make it safer. For example, it is crucial to provide ample, secure, convenient, and sheltered bike parking, especially at key destinations such as workplaces, stores, restaurants, public transport stops, universities, and schools (see in this volume Buehler, Heinen, and Nakamura, chapter 6, and McDonald, Kontou, and Handy, chapter 12). In multiunit residential buildings, the lack of secure and conveniently accessible bike parking deters bicycle ownership and use. City zoning codes should encourage cycling by requiring sheltered, secure, convenient, and free bicycle parking for multiunit residential buildings.

A wide range of complementary programs are necessary in any package of policies to promote cycling (see Heinen and Handy, chapter 7, this volume). Land-use policies that promote compact and mixed-use developments are crucial for keeping trip distances short enough for cycling. Traffic calming of residential neighborhood streets and reduced general speed limits in cities make cycling safer and attract more people to cycling. Better police enforcement of existing laws makes cycling safer. This should also include prohibiting texting and cell phone use while driving or riding a bicycle, as they have become serious sources of danger for both cyclists and pedestrians. Improved cyclist and motorist education and training can help increase cycling safety—and should emphasize a respectful traffic culture for all roadway users (see in this volume Elvik, chapter 4; Heinen and Handy, chapter 7; Garrard, chapter 11; McDonald, Kontou, and Handy, chapter 12; Garrard et al., chapter 13; Pucher, Parkin, and de Lanversin, chapter 17). Better integration of bicycling with public transport enables longer trips to be made without a car (see Buehler, Heinen, and Nakamura, chapter 6). Wayfinding systems should not rely solely on GPS and smartphones but should also include better maps, signage, and pavement markings that do not exclude people without modern technological devices—an aspect of the often overlooked importance of equity.

Policy packages should comprise policies that make cycling more attractive as well as measures that make driving a car slower, more expensive, and less convenient. Cities that have implemented mutually reinforcing push-pull policies have the highest levels of cycling. Policies that make driving less attractive almost always promote cycling. For example, traffic calming lowers car traffic volumes and reduces car travel speeds on neighborhood streets. That makes cycling less stressful, safer, and more attractive for a broader spectrum of neighborhood residents—especially for children, who do most of their cycling close to their home (see McDonald, Kontou, and Handy, chapter 12). Most European countries and cities already implement a wide range of policies that make driving less attractive. The most common car-restrictive measures in Europe are high taxes on gasoline and car purchases, parking restrictions (fewer spaces, shorter duration, and higher price), and car-free zones. A more recent, innovative approach is congestion pricing, which charges motorists to access city centers. Cycling increased dramatically in Central London after congestion pricing was introduced in 2003, with a 44% increase by 2005 and a fourfold increase by 2019. Congestion pricing will be implemented in Manhattan in early 2021 and is likely to provide a large boost to cycling in New York.

### Harnessing the Benefits and Limiting the Costs of Technological Developments

The boom in bikesharing has significantly stimulated cycling in most cities where it has been introduced (see Fishman and Shaheen, chapter 10, this volume). Since 2016, so-called floating or dockless bikesharing systems have grown even more rapidly than systems with docking stations, and in many cities they now dominate the bikesharing market. It is crucial, however, for local governments to regulate dockless systems, as their unfettered, uncoordinated, unplanned growth has led to a vast oversupply of such bikes in many cities, causing serious problems of bicycle vandalism, abandonment, and chaotic and dangerous parking. This new generation of more flexible, more convenient bikesharing has unquestionable benefits. However, in order to function well, it requires strict government regulation of the location and supply of bikes, repair and maintenance standards, rebalancing systems, coordination with competing bikesharing systems (both dockless and with docking stations), and parking of bikes in an orderly, safe manner.

Another important new development is the technological evolution and spread of electric-assist bikes (e-bikes) (see Cherry and Fishman, chapter 9, this volume). E-bikes already account for the majority of bicycles in China and up to one-third of bicycles in northern Europe. With increasing speed comes increased risk of serious injuries, especially since e-bikes are allowed to use the same bike paths and lanes as traditional pedal bicycles, with the potential for large differences in cycling speed among cyclists on the same cycling facility. There are a wide range of speeds among the varying types of e-bikes, including some that can travel much too fast to share the same lanes and paths as traditional bicycles. Thus, there is a need for government regulation, with strict enforcement, of the permissible speeds of e-bikes, including requirements that faster e-bikes use the roadway. Similarly, the rapid spread of battery-powered electric stand-up scooters and electric skateboards requires safety regulations as well, since they are usually operated on sidewalks, greenways, or cycling facilities.

## Cycling for Everyone: The Importance of Equity

Several chapters of this book demonstrate how to promote cycling for women, children, and seniors (see Garrard, chapter 11; McDonald, Kontou, and Handy, chapter 12; Garrard et al., chapter 13). The needs of those groups differ from the needs of males aged 16–50 years, who make up the majority of cyclists in low-cycling countries such as the United States, Canada, Australia, and the United Kingdom. Well-designed, connected, and safe cycling infrastructure attracts more groups to cycling, including women, seniors, and children (see in this volume Furth, chapter 5; Garrard, chapter 11; McDonald, Kontou, and Handy, chapter 12; Garrard et al., chapter 13; Pardo and Rodriguez, chapter 16; Pucher, Parkin, and de Lanversin, chapter 17; Koglin, te Brömmelstroet, and van Wee, chapter 18; Geller and Marqués, chapter 19). It is also crucial to implement programs aimed at creating more inclusive cycling cultures because societal expectations, as well as opinions of peers and family members, play an important role in the individual's decision whether to cycle (see in this volume Garrard, chapter 11; McDonald, Kontou, and Handy, chapter 12; Garrard et al., chapter 13). Thus, it is important to create supportive environments where cycling is accepted or even expected. Examples of supportive programs include special cycling training courses for specific groups (e.g., children, seniors, and beginners); bike rodeos, bike playgrounds, and bike-to-school programs for children; bike-to-work events for employees, combined with bike parking, showers, and lockers at the workplace; group rides for older cyclists, women, and all-age family-oriented groups; and open-streets events, where streets are temporarily car-free and thus especially conducive to a low-stress, festive, and sociable cycling environment that invites cycling by truly everyone (see in this volume Buehler, Heinen, and Nakamura, chapter 6; Heinen and Handy, chapter 7; Pardo and Rodriguez, chapter 16; Pucher, Parkin, and de Lanversin, chapter 17; Koglin, te Brömmelstroet, and van Wee, chapter 18; Geller and Marqués, chapter 19).

In chapter 14 of this volume, Martens, Golub, and Hamre expand on the preceding three chapters by highlighting equity considerations for low-income and minority groups who have been marginalized and often disadvantaged in the overall urban transport system, including cycling policy and infrastructure provision. Dealing with such inequities requires active policy responses to the special needs of these groups, both in the type and location of cycling facilities as well as supportive programs. The planning process must prioritize the inclusion of vulnerable, risk-averse, and disadvantaged groups to ensure that their needs are considered fully at every stage. There should be mandatory regulations to ensure that the implemented infrastructure and programs actually meet the needs of all groups and not just young, white males.

## Implementation Strategies

Cycling promotion has been a long-term, multistage process requiring compromises, trial and error, and coalition building among political parties and groups of stakeholders. In successful cycling cities, city governments have used a consensual approach to determine the most appropriate mix of pro-cycling measures and thus have garnered the necessary public and political support. This approach assures citizen input and ongoing feedback from cyclists and other key stakeholders to alter, improve, or fine-tune policies and programs (see in this volume Pardo and Rodriguez, chapter 16; Pucher, Parkin, and de Lanversin, chapter 17; Koglin, te Brömmelstroet, and van Wee, chapter 18; Geller and Marqués, chapter 19).

Several implementation strategies were already mentioned in the previous section, on policy recommendations. For example, many cities combine policies that promote cycling with measures that make driving a car slower, more costly, and less convenient. Moreover, many cities combine their efforts to promote cycling with improvements for pedestrians and public transport users, thus increasing the base of public and political support for cycling improvements. In order to garner the public and political support necessary to implement car-restrictive measures, it is essential to provide more viable and attractive alternatives to the car. In combination, these coordinated multimodal policies can create a favorable environment for a car-free or a car-light lifestyle. By helping to reduce the volume of motorized traffic, policies that promote walking and public transport also help make cycling more attractive because of decreased real and perceived traffic danger from cars, a major deterrent to cycling for much of the population.

As this book documents, there are many good examples of cities that have succeeded at increasing cycling levels. Thus, cities that intend to promote cycling and adopt new bike-friendly policies can learn from the successes of cities that have already implemented the range of bike-friendly policies and programs examined in the previous section and throughout this book. For example, New York City studied cycle tracks and cycling in Copenhagen, Denmark, prior to building its own cycle tracks, and it hired Danish consultants to help with implementation (see Pucher, Parkin, and de Lanversin, chapter 17, this volume). New York adapted design and engineering concepts from Copenhagen to match New York's specific context and regulations. In turn, New York City's experience has provided important lessons for other cities in the United States, stimulating the construction of cycle tracks throughout the country. Similarly, Bogota, Colombia, hired Dutch consultants to build its extensive system of protected bike paths and lanes, which meet the extraordinarily high Dutch standards for bikeway design. The spectacular success in Bogota has made it a leading example for other Latin American cities, indeed for cities throughout the world (see Pardo and Rodriguez, chapter 16, this volume).

Many cities have conducted before-and-after studies to measure the impacts of policy interventions (see, for example, in this volume Pardo and Rodriguez, chapter 16; Pucher, Parkin, and de Lanversin, chapter 17; Koglin, te Brömmelstroet, and van Wee, chapter 18; Geller and Marqués, chapter 19). Cities often install bicycle counters along new bikeways (especially cycle tracks) to measure changes in bicycle traffic volumes. Several cities, such as New York City, also evaluate trends in cyclist fatalities and serious injuries in new bicycle-friendly corridors. Other cities collect data about changes in retail sales along roadways with newly installed cycling infrastructure, usually finding significant growth and thus allaying concerns of local businesses.

Following the example of Copenhagen's Bicycle Account (see Koglin, te Brömmelstroet, and van Wee, chapter 18), several cities undertake periodic surveys of observed cycling levels and safety as well as respondents' opinions about cycling conditions and needed improvements in cycling infrastructure and programs. Such bicycle accounts provide trend data over time to help measure progress in meeting goals established in long-term bike plans and to update and adjust those plans to respond to changing conditions over time. Measurable results can also help other cities decide which measures to implement. In addition, cities adopting new measures can point to empirical evidence for successful policy implementation in other cities when promoting new policies to their city's residents, politicians, traffic engineers, and cyclists.

Many of the case studies examined in this volume started off with uncontroversial bicycle infrastructure projects that almost everyone agreed on and that had a high probability of success (see Furth, chapter 5; Pardo and Rodriguez, chapter 16; Pucher, Parkin, and de Lanversin, chapter 17; Koglin, te Brömmelstroet, and van Wee, chapter 18; Geller and Marqués, chapter 19). One example would be the installation of bike lanes on roadways that are unnecessarily wide overall or have unnecessarily wide traffic lanes for motor vehicles. Similarly, choosing less heavily trafficked routes for bike lanes reduces opposition from motorists while providing safer, less stressful riding environments for cyclists. Reduced speed limits and comprehensive traffic calming were introduced first in neighborhoods where there was widespread support from residents, especially where citizens proactively requested traffic calming to reduce noise, pollution, and traffic dangers (see Furth, chapter 5; Heinen and Handy, chapter 7; Koglin, te Brömmelstroet, and van Wee, chapter 18; Geller and Marqués, chapter 19).

Financing of cycling facilities can also be made less controversial. Many cities, even in the United States, include new cycling facilities as part of roadway improvement projects. In such cases, they are usually financed through the road budget, thus avoiding the need for separate funding for bicycling. In some countries, laws even require the installation or upgrading of cycling facilities when roadways are improved.

Bikesharing also helps increase public support for bicycling, because it reaches more demographically diverse users than current cyclists—especially in low-cycling countries such as the United States (Buck et al. 2013; Fishman and Shaheen, chapter 10, this volume). That greater diversity of cyclists helps garner broader public support for expanding and improving bicycling infrastructure. More cyclists and greater public and political support for cycling subsequently allow cities to implement projects that are more controversial and to develop fully connected bicycle networks.

As Furth (chapter 5) shows, many important bicycle projects are controversial because they require substantial funding and/or road space currently dedicated to automobiles. Strong political leadership by charismatic and powerful individuals has been crucial for implementing pro-bike policies and programs in many cities, especially those that are controversial. There are many examples of politicians whose strong support has enabled the success of cycling in their cities. Examples include Bertrand Delanoë and Anne Hidalgo (Paris); Enrique Peñalosa (Bogota); Ken Livingstone, Boris Johnson, and Sadiq Khan (London); Earl Blumenauer, Sam Adams, and Steve Novick (Portland); Janette Sadik-Khan, Michael Bloomberg, Polly Trottenberg, and Bill de Blasio (New York); and Alfredo Sánchez-Monteseirin, Paula Garvín, and Antonio Rodrigo-Torrijos (Seville). Their political support and inspiring visions of less car-dependent, more equitable, and more livable cities made it possible to commit financial resources to cycling and to implement controversial pro-bicycle policies over a period long enough to actually transform the transport systems in their cities (see in this volume Pardo and Rodriguez, chapter 16; Pucher, Parkin, and de Lanversin, chapter 17; Koglin, te Brömmelstroet, and van Wee, chapter 18; Geller and Marqués, chapter 19).

It has been crucial to form alliances to generate sufficient political support to get the necessary policies adopted and implemented, especially in cities with initially low levels of cycling. City departments of transport often work with bike advocacy groups and other nongovernmental organizations devoted to the environment, livable cities, and sustainable transport. It has taken the coordinated efforts of many individuals and interest groups to offset the minority status of cyclists in car-oriented cities (see in this volume Pardo and Rodriguez, chapter 16; Pucher, Parkin, and de Lanversin, chapter 17; Koglin, te Brömmelstroet, and van Wee, chapter 18; Geller and Marqués, chapter 19; Pucher et al., chapter 20).

Public information campaigns are essential for explaining the wide range of benefits cycling has for individuals and for society at large. Most obviously, cycling provides direct mobility benefits to the cyclist. Cycling is cheaper than driving a car and can also save time. In cycling-oriented cities such as Amsterdam and Copenhagen, cycling often offers the most direct and fastest route. Even in car-oriented American, Canadian, and Australian cities, the generally faster speed of driving is at least partially offset by the time lost in traffic congestion or in finding a parking space. The comparative time savings of cycling are even larger if one includes the hours of employment required to earn enough money to purchase, maintain, drive, and park a car (Tranter 2012; Tranter and Tolley 2020). Moreover, there are significant health benefits to cyclists (see Garrard et al., chapter 3, this volume). Studies show that cycling is so healthy that it can extend one's expected healthy lifetime, which is perhaps an even more important kind of time savings. In recent years, the direct health benefits of cycling have been increasingly highlighted in campaigns to convince more people to cycle on a daily basis. Thus, the first message of public information campaigns is that cycling pays off for the individual.

To generate widespread public support, however, information campaigns must also convey cycling's broader societal benefits, which include reduced noise, air pollution, energy use, and traffic congestion, as well as improved public health and more travel options for everyone. Voters and their elected officials are more likely to support measures necessary to increase cycling if they are convinced that cycling generates population-wide sustainability benefits. Thus, the second message of public information campaigns is that even noncyclists should support cycling because of its many social and environmental benefits.

As emphasized by Martens, Golub, and Hamre (chapter 14), ongoing citizen involvement is important for garnering public support and for developing a package of policies that is most appropriate to the needs and preferences of the local population. The planning process must be transparent and inclusive, sending a clear message that policies are not being imposed from above but rather are generated in close consultation with key stakeholders, neighborhood groups, and all interested citizens, including representatives of marginalized groups. Because the goal is to increase daily cycling among the general population, citizen feedback is crucial for choosing those measures that would most encourage them to cycle more often. In addition to project-specific outreach efforts, many cities have cycling advisory boards composed of cyclists, bike planners, and other stakeholders, who provide frequent feedback to city officials.

As documented by Pucher et al. (chapter 20), most of the countries examined in this book have national cycling organizations that help coordinate the advocacy efforts of state and local cycling organizations. As well-integrated and mutually supportive partners, they jointly lobby national, state, and local governments for funding and pro-bike legislation; organize promotional events and cycling training; help develop design standards for cycling facilities; collect and disseminate information about cycling; and provide advice on national, state, and local bike plans.

## Emerging Challenges and Future Outlook

The coming decades will bring many changes to the transport sector, including increasing automation of vehicles and more sharing of cars, bicycles, and scooters by using mobile apps (Sperling 2018). Automated vehicles rely on technology to take responsibility for some, or eventually all, driving tasks from human drivers. Improved technology has the potential to make roadways safer for motorists and cyclists. However, those safety benefits will not be fully realized for many years. Automation poses significant potential hazards in the short term while the technology is being improved (Botello et al. 2019). For example, it is unclear how robotic vehicles will communicate with cyclists (and pedestrians). This problem will become even more complicated in the coming decades because human-driven cars and motor vehicles with different levels of automation will inevitably have to share the road with cyclists. Neither cyclists nor motorists are prepared for this potentially dangerous interaction among so many different kinds of vehicles. A recent survey of experts concluded that governments need to pass strict regulations ensuring that automated vehicles be built and rigorously tested to ensure that they can operate safely around cyclists and pedestrians (Botello et al. 2019).

Local governments will eventually need to redesign their transport infrastructure, built environment, and land uses to accommodate the needs of increasingly automated vehicles. Possible scenarios for impacts on cycling range widely. Cycling could increase because of predictable robotic vehicles, safer cycling conditions, and more roadway space dedicated to separated bikeways—because automated cars will need less space for operation and parking compared to human-driven cars. On the other hand, automated vehicles could decrease cycling by making driving more attractive and by necessitating infrastructure that bans cyclists from certain roadways, intersections, or entire areas dedicated exclusively to automated vehicles. The alleged fear is that bicyclists may interfere with the traffic flow of automated vehicles designed to yield and not hit cyclists. It will be the task of government, planners, engineers, and cycling advocates to design policies that assure safety and access for cyclists (Botello et al. 2019).

Another important future trend in transport is more sharing of vehicles and integration of sharing services—including bikesharing, commercial ride sharing (such as Uber and Lyft), and carsharing. Fishman and Shaheen (chapter 10) document the boom in bikesharing around the world. An analysis of the 2017 *National Household Travel Survey* found that almost 10% of Americans reported using commercial ridesharing at least once a month (Conway et al. 2018). Recently, commercial ridesharing companies have purchased bikesharing companies to integrate the two services. For example, the commercial ridesharing company Lyft purchased Motivate, an operator of docked bikesharing systems in several large US cities, including New York. The commercial ridesharing company Uber invested in Lime, an electric dockless bikesharing and e-scooter company. Some speculate that, in the future, mobility companies may sell integrated mobility services for trips tailored to the specific needs of customers, allowing them to choose or combine services such as commercial ridesharing, bikesharing, e-scooter sharing, and public transport (MaaS Alliance 2019). The inclusion of bicycling in these mobility services will be crucial to the future of cycling.

Perhaps most promising for the future of cycling is the documented shift in cultural attitudes and preferences toward less reliance on the automobile and increased demand for living in mixed-use, compact developments in or near the city center (Goodwin and van Dender 2013). Many city centers in Europe and North America have experienced a revival and, in particular, the influx of new residents in their twenties and thirties who are more willing than their parents to walk, bike, and ride public transport. That cultural shift in locational and travel preferences is likely to facilitate further growth in cycling.

## References

Botello, Bryan, Ralph Buehler, Steven Hankey, Zhiqui Jiang, and Andrew Mondschein. 2019. Planning for Walking and Cycling in an Autonomous-Vehicle Future. *Transportation Research Interdisciplinary Perspectives* 1. https://doi.org/10.1016/j.trip .2019.100012.

Buck, Darren, Ralph Buehler, Patricia Happ, Bradley Rawls, Payton Chung, and Natalie Borecki. 2013. Are Bikeshare Users Different from Regular Cyclists? First Look at Short-Term Users, Annual Members, and Area Cyclists in the Washington, D.C., Region. *Transportation Research Record* 2387: 112–119.

Buehler, Ralph, John Pucher, Regine Gerike, and Thomas Götschi. 2017. Reducing Car Dependence in the Heart of Europe: Lessons from Germany, Austria, and Switzerland. *Transport Reviews* 37(1): 4–28.

Conway, Matthew, Deborah Salon, and David King. 2018. Trends in Taxi Use and the Advent of Ridehailing, 1995–2017: Evidence from the US National Household Travel Survey. *Urban Science* 2(79): 1–23. https://doi.org/10.3390/urbansci2030079.

Goodwin, Phil, and Kurt van Dender. 2013. Peak Car—Themes and Issues. *Transport Reviews* 33(3): 243–254.

Mobility as a Service (MaaS) Alliance. 2019. *What Is Mobility as a Service?* Brussels: Mobility as a Service (MaaS) Alliance.

Sperling, Daniel. 2018. *Three Revolutions: Steering Automated, Shared, and Electric Vehicles to a Better Future.* Washington, DC: Island Press.

Tranter, Paul. 2012. Effective Speed: Cycling Because It's "Faster." In *City Cycling*, edited by John Pucher and Ralph Buehler, 57–74. Cambridge, MA: MIT Press.

Tranter, Paul, and Rodney Tolley. 2020. *Slow Cities: Conquering Our Speed Addiction for Health and Sustainability.* Oxford: Elsevier.

# Index

Note: Page numbers followed by f and t indicate figures and tables, respectively.

Aarhus, 14f
Abraham, J. E., 105
Active Neighborhoods Canada, 416
Active Transportation Leadership
    Institute, 409
Adams, Sam, 377, 379f, 380–381
Adelaide, 13f
Advocacy, cycling, 10, 401–422. *See also*
    *specific groups and locations*
    growth in, 2–3
    international, 401–403
    national (*see* National advocacy)
    provincial/state, 401–402, 413–416
    vehicular cycling theory in, 85, 87,
        407, 408
Advocacy Advance, 409
African Americans
    disparities in cycling by, 262, 262f,
        263
    injury and fatality rates for, 263,
        268–269
    racial profiling of, 128
Age of cyclists
    and injury risk, 64
    international comparisons of, 16–17,
        17f
    women, 201f, 205–206
Aggregated average risk factors, 139
Ahmedabad, 13f, 284f, 294

Air pollution
    in China, 268, 285, 297–298
    disparities in exposure to, 267–268
    health effects of, 45, 140
    in India, 297–298
    reduced car use and, 45, 48
Alberini, Anna, 143
Alcohol use, 127–128
Aldred, Rachel, 57
Allgemeiner Deutscher Fahrrad-Club
    (ADFC), 401, 404
Alliance for Biking and Walking, 410
America Bikes campaign, 408, 411
American Association of State High-
    way and Transportation Officials
    (AASHTO), 28, 85, 87, 90
*American Community Survey (ACS)*, 11,
    272
American Time Use Survey, 40
Amsterdam, 9, 347–368
    advocacy in, 361–362
    bikesharing in, 173–174, 363, 367
    car use in, 361–363
    cycling rates in, 2, 2f, 14f, 347–348,
        348f, 360, 360t, 427t
    e-bikes in, 367
    future of cycling in, 364–365,
        366–368
    history of cycling in, 358–362

Amsterdam (cont.)
  infrastructure in, 357–358, 358f, 360, 366–367, 427t
  lessons learned from, 365–368
  policies affecting cycling in, 359–363, 365–366
  public transport in, 359, 361, 363–364, 365
  recent trends in, 362–364, 427t
Amy Gillett Foundation, 420
Andalusia, 395–396
Andersen, L. B., 47–48
Antofagasta, 303f, 306f
Appleyard, Donald, 44
Argentina, 304. *See also specific cities*
Arthritis, 238–239
Asian Americans, 261–262, 262f
Association of Pedestrian and Bicycle Professionals (APBP), 413
Australia. *See also specific cities*
  bike parking in, 104
  bikesharing in, 183–185, 188–189, 191
  car ownership rates in, 17–18
  child cyclists in, 220t, 221, 223, 228
  cycling rates in, 11–12, 12f, 13f, 15, 201, 281
  cyclist and driver training in, 26–27, 223
  demographics of cyclists in, 16f
  fatality rates in, 20–21, 20f
  growth in cycling in, 1, 2f
  helmet laws in, 183, 191
  national advocacy in, 402, 419–420
  older cyclists in, 240
  promotional programs in, 123
  speed limits and road design standards in, 28–29
  state and local advocacy in, 402, 420–421
  state policies in, 23
  traffic regulations in, 26, 127
  transportation costs in, 44
  women cyclists in, 16f, 199, 200t, 201f

Austria, 12f, 14f, 16f. *See also specific cities*
Automated vehicles, 438–439
Automobility, 354

Bangalore, 284f, 294
Barajas, Jesus, 263
Barcelona, 178f
Bari, 14f
Barranquilla, 303f, 306f
Basel, 14f
Beale, Sophie, 145
Behavioral Risk Factors Surveillance System, 39
Beijing, 285–287
  air pollution in, 268
  car use in, 285–286
  cycling rates in, 13f, 185, 285, 286f
  public transport in, 287
Belgium
  child cyclists in, 220t, 221, 226
  cycling rates in, 12f
  cyclist training in, 121, 226
  e-bikes in, 248
  incentive programs in, 124
  traffic calming in, 130
Belo Horizonte, 303f, 307f
Benchmarking, 352
*Benchmarking Report on Bicycling and Walking in the United States,* 410, 418
Benefit-cost ratios (BCRs), 143, 150t
Berlin
  bike parking in, 109t, 111, 112f, 114
  cycling rates in, 2f, 14f, 19, 348f, 427t
Better Block PDX, 384–386
Bicirred Mexico, 312
Bicycle Account, 352, 353t, 435
Bicycle Benefits, 125
Bicycle Federation of Australia (BFA), 419
Bicycle Friendly America, 409–411
Bicycle Network, 402
BicycleNSW, 402
Bicycle Product Supplier Association, 160
Bicycle style electric bikes, 159

Bicycle Transportation Alliance (BTA),
    376, 384
Big Box Retail law, 363
Big Jump, 411
Bikeability Program, 25, 226
Bike lanes. *See* Cycle tracks; Lanes,
    advisory bike; Lanes, unprotected/
    conventional bike
Bike Loud PDX, 386
Bike Network Grant program, 414
Bike New York, 339
Bike paths. *See* Paths, dual; Paths,
    stand-alone bike
BikePortland.org, 384
Bike share of daily trips. *See also specific
    locations*
  international comparison of, 11–15,
    12f–14f, 16f, 17f, 201, 281
  for women, 197, 202–204, 260
Bikesharing, 6, 173–192. *See also specific
    locations*
  benefits of, 177–180
  business models for, 175–176
  demographics of users, 180
  design of user-oriented system for,
    173, 187–191, 187f
  of e-bikes, 182
  future of, 192, 431, 435–436, 439
  generations of, 173–174, 174f, 293
  growth of, 173, 174–175, 175f
  lessons learned from failures in,
    183–186
  ridership by city, 177, 178f
  safety of, 179–180, 187, 187f,
    188–189
  technology of, 180–182
BikeTexas (Texas Bicycle Coalition), 401,
    414–416
Bike to Work Days, 121–122
Bjørnskau, Torkel, 65, 74
Blaizot, Stefanie, 60
Blumenauer, Earl, 377–379, 379f
Body mass index (BMI), 36–37

Body weight, impact of cycling on,
    36–37, 38
Bogota
  bicycle theft in, 313
  Ciclovía in, 121, 122f, 311
  cycling rates in, 1, 2f, 13f, 303, 303f,
    307f, 427t
  income and cycling in, 315, 315f
  infrastructure in, 305, 306f, 307, 307f,
    308, 427t, 434
  policies and programs in, 121,
    301–302, 311
Bolivia, 309. *See also specific cities*
Bologna, 14f
Bolzano, 14f
Boston, 2f, 68–69, 178f
Boulder (Colorado), 86
Boulevards, bike
  definition of, 67, 94
  injury risk on, 67
  in Portland, 374, 379f
  speed limits on, 27
Boxes, bike, 67, 131, 380
Brazil, 265–266, 305, 307
Brisbane, 13f, 178f, 183–184, 188–189,
    191
British Columbia, 418–419
British Columbia Cycling Coalition
    (BCCC), 419
Buehler, Ralph, 58, 68, 69
Buenos Aires
  bike parking in, 308
  cycling rates in, 1, 2f, 13f, 303f,
    307f
  infrastructure in, 305, 305f, 306f, 307f
  institutional staff in, 311
Built environment. *See also* Infrastructure
  and older cyclists, 243–244
  policies influencing, 120t, 128–131
  and women cyclists, 205f, 207–209

Calgary, 13f
Cali, 303f, 306f, 307f

California. *See also specific cities*
  bikesharing in, 186
  gentrification in, 270
  state advocacy in, 413–414
California Bicycle Coalition (CalBike),
  273, 401, 413–414
Calvo-Salazar, M., 204
Cambridge, 14f
Camino del Norte (San Diego), 83
Canada. *See also specific cities and
  provinces*
  car ownership rates in, 17
  child cyclists in, 220t, 221, 223, 227
  cycling rates in, 11–12, 12f, 13f, 15,
    16f, 281
  cyclist and driver training in, 26–27,
    223
  demographics of cyclists in, 16, 16f
  fatality rates in, 20–21, 20f
  infrastructure funding in, 99
  injury risk disparities in, 269–270
  national advocacy in, 402, 407
  provincial advocacy in, 401–402,
    416–419
  provincial funding in, 25
  provincial policies in, 23
  speed limits and road design standards
    in, 28–29
  traffic regulations in, 26, 127
  women cyclists in, 16f
Canberra, 13f
CAN-Bike, 27, 223
Cancer, 37
Caracas, 303, 306f
Cardiovascular disease (CVD), 37–38
Car-free zones, 129–130
  of Ciclovía, 121, 122f, 311–312
  effects of, 129
  future of, 434
  goals of, 129
  in New York, London, and Paris, 335,
    336
Cargo bikes, 157, 161

Car-light zones, 129–130
Car ownership and use. *See also specific
  locations*
  adverse health effects of, 40
  automation of cars and, 438–439
  bikesharing as substitute for, 177
  and child cycling, 228–229
  congestion charging and, 336–337,
    357, 431
  costs of, 125
  and cycling rates, 17–19, 201
  distracted driving in, 428–429
  e-bikes as substitute for, 164–165, 168,
    169
  health benefits of reducing, 35, 44–48
  injury risk with, 267–268
  international comparisons of, 17–19
  policies and programs discouraging,
    125, 131–133, 430–431, 433–434
Car parking. *See* Lanes, parking
Carr, Theresa, 125
Castro, Alberto, 58
Cavill, Nick, 146
Census Bureau, US. *See* US Department
  of Commerce
*Census of India*, 282
Centers for Disease Control and
  Prevention, 39, 268–269, 410
Central London Grid, 331
Centre for Active Transportation, 416
Centre for Diet and Activity Research
  (CEDAR), 137
Chengdu, 286, 286f
Chennai, 284f, 294
Chicago, 2f, 13f, 178f, 190, 225, 265,
  427t
Children, cycling by, 7, 219–230
  female, 201f, 205–206, 221
  health benefits of, 36, 219
  helmet laws for, 26, 71, 127, 223
  infrastructure and, 223–225
  parents accompanying, 198–199, 227
  safety of, 222–223

to school, 219–221, 220t, 223–225
social and psychological factors in,
225–230
strategies for increasing, 432–433
training for, 26–27, 120–121, 222–223,
226–227, 415
Child ribbons, 224
Chile, 309. *See also specific cities*
China, 8, 281–298. *See also specific cities*
air pollution in, 268, 285, 297–298
bike parking in, 105
bikesharing in, 292–294
car use in, 285–286, 292
child cyclists in, 220t
cycling rates in, 13f, 281–288, 286f
e-bikes in, 157–159, 248, 287, 288f,
289–290
infrastructure in, 290–292, 294–297
injury and fatality rates in, 288–290
policy recommendations for, 294–298
public transport in, 287
women cyclists in, 261, 287
China Bicycle Association, 158
Chronic disease, 35–38, 46
Ciclovía, 121, 122f, 311–312
Citi Bike, 334
Cities for Cycling Program, 380
City Club of Portland, 381
*City Cycling* (Buehler & Pucher), xv, 1,
3, 4, 425
Climate
of Latin America, 304
of New York, London, and Paris, 323t,
324
of Portland, 372t
of Seville, 372t, 387
Climate change, 48
Cochabamba, 303f, 307f
Cognitive benefits of cycling, 39–40
Colombia, 304, 309–311. *See also specific
cities*
Colorado, 13f, 86
Commercial areas, 93

Community livability, 44
Commuting. *See* Work trips by bike
Complete Streets
in Canada, 416
and gentrification, 270
League of American Bicyclists on,
408–409
in Oregon, 374–376
Concepción, 303f
Congestion charging/pricing, 336–337,
357, 431
Connectivity, of bicycle networks,
97–98, 129, 314
Continuous sidewalks, 92–93, 92f
Contramano, A (Association of Urban
Cyclists), 389, 391
Convenience, of bikesharing, 187, 187f,
189–191
Conventional bike lanes. *See* Lanes,
unprotected/conventional bike
Copenhagen, 9, 347–368
bikesharing in, 174, 367
car use in, 349, 352–357, 354t
current trends in, 352–354, 427t
cycle tracks in, 349–350
cycling rates in, 2, 2f, 14f, 347–348,
348f, 354, 356, 427t
cyclist satisfaction in, 352, 353t, 435
e-bikes in, 367
future of cycling in, 356–357,
366–368
history of cycling in, 349–350
infrastructure in, 349–351, 366–367,
427t
injury and fatality rates in, 350–351,
351f
lessons learned from, 365–368
policies affecting cycling in, 352,
365–366
public transport in, 352–354, 354t,
355–356, 365–366
Cordoba, 303f, 304, 305, 307f
Coronavirus pandemic, xvi

Cost-benefit analysis (CBA), 137–154
    challenges of, 138–141, 152–153
    pros and cons of, 141–144, 153
    real-world examples of, 146–152, 150t,
        151t
Cost-effectiveness analysis (CEA),
        137–154
    challenges of, 138–141, 152–153
    pros and cons of, 141–144, 153
    real-world examples of, 144–145,
        145t
Costs. *See also* Funding
    of bike use, incentives covering,
        124–125
    of car use, 125
    of creating infrastructure, 98–99
    transportation, 44, 264
COVID-19 pandemic, xvi
Crane, Melanie, 40
Creighton, Prudence, 70
Critical Mass rides, 312
Crossings, raised, 67
Crosweller, Stan, 57
Curitiba, 266, 303f, 307f, 312
Cycle Safe, 223
Cycle tracks (protected bike lanes),
        91–93. *See also specific locations*
    advocacy for, 411, 417
    cost of creating, 98
    cycling increases with, 428
    definition of, 81, 81f, 91
    injury risk on, 67–70, 87
    at intersections, 91–93, 92f
    means of separation from cars, 91
    one-way vs. two-way, 91–92
    sidewalk-level vs. street-level, 91
Cycling and Walking Investment
        Strategy (UK), 24
Cycling Cities, 402
Cycling Demonstration Towns, 123
Cycling Embassy of Denmark,
        404–405
Cycling Industry Club, 402

*Cycling in Québec,* 418
Cycling policies. *See* Policies and
        programs
Cycling rates. *See also specific locations*
    as bike share of daily trips, 11–15,
        12f–14f
    car ownership rates and, 17–19, 201
    demographics of, 15–17
    disparities in (*See* Disparities)
    growth since 1990, 1–2, 2f
    infrastructure expansion and,
        426–428, 427t
    injury rates in relation to, 46–48
    international comparisons of, 11–15,
        12f–14f, 201, 281, 425–426
    policy shifts increasing, 19–20, 430–436
    by trip distance and purpose, 15
Cycling UK, 406
Cyclists
    meaning of term in Dutch, 199
    Portland's typology of, 382, 383
Cykelpuljen program, 25
Cyklistforbundet (CFB), 349, 401,
        404–405

Daily trips. *See* Bike share of daily trips
Danish Cyclists' Federation. *See*
        Cyklistforbundet
Data sources
    on bike share of daily trips, 11
    on injuries, 59, 60
    on Latin America, 313–314
Davis (California)
    child cyclists in, 226, 228–229
    infrastructure in, 85–86, 90, 94
Decisio and Transaction Management
        Centre, 139
de Hartog, Jeroen Johan, 149
Delaware, 127
Delhi, 13f, 283, 284f, 291
Demographics, cyclist. *See also specific
        locations*
    of bikesharing users, 180

of e-bike users, 163
and health inequalities, 41–42
and injury risk, 64
international comparisons of, 15–17,
   16f
DeMull, Marisa, 272
Denmark. *See also specific cities*
   advocacy in, 401, 404–405
   car ownership rates in, 18
   child cyclists in, 17f, 227
   cycling rates in, 12, 12f, 14f, 15,
      47–48, 281, 348
   cyclist and driver training in, 26
   demographics of cyclists in, 15–17,
      16f, 17f
   history of policy shifts in, 19–20
   injury and fatality rates in, 20f, 21,
      22f, 47–48
   national funding in, 25
   national policies in, 24
   older cyclists in, 17f, 241–242
   promotional programs in, 123
   traffic regulations in, 26, 127
   women cyclists in, 16f
Denver, 13f
Development Bank of Latin America
   (CAF), 303–304
DiGioia, Jonathan, 67
Dijkstra, Lewis, 128
Dill, Jennifer, 125, 209
Discount rates, 143
Discrimination, against cyclists, 210,
   271
Disease, chronic, 35–38, 46
Disparities, 257–275
   gender, 260–261, 260t (*see also*
      Women cyclists)
   income, 263–265, 264f, 314–315
   in injury risk, 267–270
   meanings and use of term, 258–259
   in pollution exposure, 267–268
   racial and ethnic, 261–263, 262f,
      268–269, 360

Distances to destinations
   for children, 224–225
   cycling rates by, 15
   with e-bikes, 164–165
   policies influencing, 129
Distracted driving, 428–429
Divvy bikes, 190
Dockless bikesharing
   business models for, 176
   in China, 293–294
   design of user-oriented system for,
      187, 189
   emergence of, 173, 174, 175
   future of, 431
   in Latin America, 308
   lessons learned from failures in,
      185–186
   in New York, 334
   technology in, 181–182
Dooring, 83, 93
Driver training, international
   comparisons of, 26–27
Dual paths, 90
Dutch Cycling Embassy, 405
Dutch Cyclists' Union. *See* Fietsersbond
Dutch language, meaning of "cyclists,"
   199

East Coast Greenway Alliance (ECGA),
   412–413
E-bikes, 6, 157–169. *See also specific
   locations*
   advocacy for, 416
   bikesharing of, 182
   challenges of defining, 158–161, 160f,
      161f
   environmental impact of, 168–169
   future of, 168–169, 431–432
   growth in use of, 157–158, 162, 162f
   health effects of, 165
   history of, 157–159
   older adults on, 163, 168, 247–248
   policy frameworks for, 162–163

E-bikes (cont.)
  safety of, 65–66, 165–167, 289–290
  speed of, 65–66, 66f, 160, 166–167
  uses for, 163–165
  women on, 168, 212
E-bikesharing, 182
Economic benefits of cycling, 39
Economic crisis of 2008, 394, 397
Economic evaluation of cycling policies,
    6, 137–154
  challenges of, 138–141, 152–153
  definition of, 138
  need for, 138, 152
  pros and cons of methods for,
    141–144, 153
  real-world examples of, 144–152
Ecuador, 305f. *See also specific cities*
Education. *See* Schools; Training of
    cyclists; Training of drivers
Education levels, and cycling rates, 41
Eesteren, Cornelis Van, 359
Electric bikes. *See* E-bikes
*Electric Bike World Report,* 159, 162
Electric scooters, 161, 314
Elvik, Rune, 47
England. *See* United Kingdom
Equity. *See* Social justice
Espadas, Juan, 394, 395
Ethnic disparities
  in bike parking, 263, 274
  in cycling rates, 261–263, 262f, 360
  in injury risk, 268–269
Ettema, Dick, 154
*EU Cycling Strategy,* 403
Europe. *See also specific cities and
    countries*
  advisory bike lanes in, 96–97
  air pollution in, 45
  car ownership rates in, 17–19
  car restrictions and charges in,
    335–337, 431
  cycle tracks in, 81–82
  cycling rates in, 11, 12f, 14f, 281

demographics of cyclists in, 15–17
  e-bikes in, 157–167, 162f
  fatality rates in, 20–21, 20f, 22f
  growth in cycling in, 1–2, 2f
  history of policy shifts in, 19–20
  international advocacy in, 401,
    402–403
  national advocacy in, 401, 404–406
  stand-alone paths in, 90
European Commission, 222
European Cyclists' Federation (ECF), 10,
    389, 401–404, 406, 418
EuroVelo, 402, 404, 406
Events, promotional, 121–122, 122f
"Everyday cycling" culture, 212–213
Ex ante policy evaluations, 137, 141,
    152, 153
Exercise. *See* Physical activity
Experience, in injury risk, 62, 65
Ex post policy evaluations, 137, 153–154
Express bikeways. *See* Superhighways,
    cycle

Fatality rates. *See also specific locations*
  for children, 222
  on e-bikes, 289–290
  infrastructure and, 22, 69–70, 426,
    427t
  international comparisons of, 20–23,
    20f, 22f, 58, 58f, 426
  racial and ethnic disparities in,
    268–269
Ferenchak, Nicholas, 69
Ferrara, 14f
Fietsersbond (FB), 401, 404–406
Fietstelweek, 405
Filtered permeability, 129
Finland, 12f, 71–72
Florianopolis, 303f, 307f
Flügel, Stefan, 65–66
Fobs, 184
Forester, John, 84–85
Fortaleza, 303

France. *See also specific cities*
  cycling rates in, 11, 12f, 14f
  e-bikes in, 162f
  fatality rates in, 20f
  Freiburg, 14f, 106, 107t, 111, 113, 113f, 427t
  Freudendal-Pedersen, Malene, 356
Funding
  from advocacy groups, 410–411
  for bike parking, 114
  in Complete Streets legislation, 374–376
  for infrastructure, 98–99, 119, 265–267, 411, 435
  international comparisons of, 24–25
Furth, Peter G., 188
Fyhri, Aslak, 65

Garages, bike parking, 104, 111–112, 112f
García-Cebrian, J. A., 204, 390
García-Jaen, Pepa, 390
Garrard, Jan, 62, 165, 219
Garvín, Paula, 389
Gasoline/petrol prices, 18, 125, 397
Gasoline/petrol tax, 381
Geller, Roger, 82–83
Gender. *See also* Women cyclists
  and injury risk, 64
  and speed of cycling, 65–66, 66f
Gender Equity Index, 261
Gender gap in cycling. *See also* Women cyclists
  in India, 283, 287
  infrastructure in, 207–209, 211–212, 261
  in Latin America, 304
  safety in, 206–207, 211–212
  social justice and, 260–261, 260t
  in sports cycling, 199–201
  in transport cycling, 201–202, 212, 260–261
General Expansion Plan, 359
Gentrification, 270

Geofencing, 186
Germany. *See also specific cities*
  bike parking in, 106, 107t
  car ownership rates in, 18
  child cyclists in, 17f, 220t, 222
  cycling rates in, 12, 12f, 14f, 15, 16f, 17f, 281
  cyclist and driver training in, 26–27, 222
  demographics of cyclists in, 15–17, 16f, 17f
  e-bikes in, 162f, 248
  fatality rates in, 20f, 21, 22f
  history of policy shifts in, 19–20
  incentive programs in, 124
  national advocacy in, 401, 404
  national cycling plan of, 23
  national funding in, 24–25
  older cyclists in, 17f, 241–242
  speed limits and road design standards in, 27–28
  traffic regulations in, 26, 128
  women cyclists in, 16f, 198
Ghaziabad, 284f
Ghent, 130
Givoni, Moshe, 139
Glasgow, 14f
Goodman, Anna, 179
Götschi, Thomas, 58
Government policies. *See* Policies and programs
GPS, in bikesharing, 174, 181–182
Graz, 14f
Great American Trail, 412
Green, Judith, 179
Greenhouse gas emissions, 48
Green Lane project, 411
Grenoble, 14f
Groningen, 14f, 106, 107t, 111, 112f, 113
Grootveld, Robert Jasper, 362
Guadalajara, 303f, 306f, 307f
Guangzhou, 286, 286f

*Guide for the Development of Bicycling
    Facilities*, 85
Gulsrut, Natalie Marie, 356

Habits, travel mode, 7, 219, 245
Hales, Charlie, 381
Handerson, Jason, 356
Handy, Susan, 209
Hangzhou, 13f, 286, 286f, 293
Harassment, of women cyclists,
    210–211
Haverman, Roland, 363
Health, definition of, 35
Health behavior, social-ecological model
    of, 204, 205f, 242–246
Health benefits of cycling, 4, 35–49
  for body weight, 36–37
  challenges of measuring, 139–141,
    140f
  in children, 36, 219
  for chronic disease, 35–38
  with e-bikes, 165
  economic, 39
  in economic evaluation of policies,
    137, 139–141, 144
  through increased physical activity,
    42–44, 43f
  mental and cognitive, 39–40
  in older adults, 36, 237–240
  through reduced car use, 35, 44–48
  through reduced health inequalities,
    41–42
  social, 40–41
  tools for measuring, 38–39, 137
  in women, 198
Health Economic Assessment Tool
    (HEAT), 39, 137, 139, 146, 165
Health impact assessments (HIAs), 144
Heesch, Kristiann C., 62
Helmet laws, 70–72, 127
  and bikesharing, 183, 191
  for children, 26, 71, 127, 223
  and injury rates, 70–71

international comparisons of, 26, 127
unintended effects of, 71–72, 127
Helmets, injury rates with, 70–71, 71t
Hernandez-Herrador, V., 69
Hispanic and Latino Americans
  child cycling among, 221
  disparities in cycling by, 262, 262f,
    263
  injury and fatality rates for, 263,
    268–269
  racial profiling of, 128
Hoffman, Melody, 270
Hospital records, 59, 60, 61t
Houston, 13f
Høye, Alena, 67, 70–71, 72
HUB Cycling, 419
Hunt, John Douglas, 105
Hyderabad, 284f

Idaho, 127
Illinois. *See* Chicago
Immigrants, disparities in cycling by,
    262–263, 360
Impacts of Cycling Tool (ICT), 38–39
Incentive programs, 124–125
Income, in cycling rates
  and health inequalities, 41
  in India, 283
  in Latin America, 314–315
  and social justice, 263–265, 264f,
    314–315
India, 8, 281–298. *See also specific cities*
  air pollution in, 297–298
  bike parking in, 106
  bikesharing in, 294
  cycling rates in, 13f, 281–288
  infrastructure in, 290–292, 294–297
  injury and fatality rates in, 288–290
  policy recommendations for, 294–298
  women cyclists in, 261, 283, 287
  work trips by bike in, 282–283, 282t,
    284f
Indian Roads Congress, 291

Indore, 284f
Infrastructure, 5, 81–100. *See also specific types*
  for bikesharing, 188–189
  and children, 223–225
  cost of creating, 98–99
  cycling increases with expansion of, 426–428, 427t
  and fatality rates, 22, 69–70, 426, 427t
  finding space for, 88–89
  funding for, 98–99, 119, 265–267, 411, 435
  and gender gap, 207–209, 211–212, 261
  need for separation from traffic in, 83–88
  and older cyclists, 242, 243–244
  promotional programs combined with, 123
  in racial and ethnic disparities, 263
  recommendations for future of, 428–429
  requirements for networks in, 97–98
  safety influenced by, 66–70, 426
  social justice in, 261, 263, 265–270
  types of routes in, 81–82, 90–97
  vehicular cycling theory in, 84–85, 87–88
Initiative for Pedestrian and Bicycle Innovation (IBPI), 386
Injuries, cyclist, 57–73. *See also* Fatality rates; *specific locations*
  by bicycle type, 65–66
  from car doors, 83, 93
  in children, 222
  classification of, 59–60
  data sources for, 59, 60
  demographics of, 64
  disparities in, 267–270
  by infrastructure type, 66–70
  international comparisons of, 60, 61t
  by kilometers cycled, 60–62, 61t, 62f
  measuring risk of, 57–60, 60t
  in older adults, 240–241, 248
  racial and ethnic disparities in, 268–269
  reduced car use and, 46–48
  right-of-way regulations and, 126
  "safety in numbers" and, 5, 47, 62–64
  in single-bicycle crashes, 63, 240–241, 248
  unreported, 59
Injuries, pedestrian, 46–47
Integrated Transport and Health Impact Modelling Tool (ITHIM), 137, 139, 146
Inter-American Development Bank, 303
Intermodal Surface Transportation Efficiency Act, 378, 408
International advocacy, 401–403
International comparisons, 11–29
  of car ownership, 17–19
  of cycling rates, 11–15, 12f–14f, 201, 281, 425–426
  of cycling safety, 20–23, 426
  of cyclist and driver training, 26–27
  of demographics of cyclists, 15–17, 16f, 17f
  of growth in cycling since 1990, 1–2, 2f
  of history of policy shifts, 19–20
  of national funding, 24–25
  of national policies, 23–24
  of speed limits and road design standards, 27–28
  of traffic regulations, 26
Intersections
  bike boxes at, 67, 131, 380
  in Copenhagen, 350
  cycle tracks at, 91–93, 92f
  cyclist behavior at stop signs at, 127
  dedicated bicycle traffic lights at, 131, 132f, 350
  in New York City, 329
  in Portland, 380

Intoxication, 127–128
Intrapersonal factors
  for older cyclists, 243–244
  for women cyclists, 204–207, 205f
Invisible cyclists, 266, 272
Iran, 220t
Italy, 11, 14f, 162f
Izquierda Unida, 389, 394

Jackets, retroreflective, 65
Jacobsen, Peter L., 76
Jaipur, 284f
Japan. *See also specific cities*
  bike parking in, 111–112
  car ownership rates in, 17–18
  child cyclists in, 220t
  cycling rates in, 11–12, 12f, 13f, 16f,
    17f
  demographics of cyclists in, 15–17,
    16f, 17f
  fatality rates in, 20f, 21, 22f
  older cyclists in, 17f, 241
  women cyclists in, 16f
JC Decaux, 183, 333
Johansson, Ole Jørgen, 65
Johnson, Boris, 68, 330, 339
Joo, Joonwon, 304
JUMP, 176, 186

Kahlmeier, Sonja, 58
Kanpur, 284f
Karlsruhe, 14f
Keseru, Imre, 145–146
Khan, Sadiq, 331
Klein, Gabe, 184
Kobe, 12, 13f
Kolkata, 13f, 283, 284f, 294
Kubesch, Nadine J., 38
Kyoto, 13f, 106, 109t, 111–113

La Habana, 303f
Land-use planning, 129
Lanes, advisory bike, 96–97, 96f

Lanes, parking
  conventional bike lanes beside,
    83, 93
  cycle tracks beside, 87, 91
  width of, 88–89
Lanes, protected bike. *See* Cycle tracks
Lanes, road/travel
  reducing number of, 89–90
  sharrows on, 95–96
  width of, 88–89
Lanes, unprotected/conventional bike,
    93
  in commercial areas, 93
  cost of creating, 98–99
  definition of, 81, 93
  in Europe, 82
  gentrification and, 270
  injury risk on, 67–68
  in New York City, 328
  in Portland, 379f
  traffic stress on, 83, 93
  in US, 82, 93
  width of, 93
  women on, 207, 208
Lanversin, Emmanuel de, 68
La Paz, 303, 303f, 307f
Larsen, Jonas, 103
Latin America, 8–9, 301–317. *See also*
    *specific cities and countries*
  advocacy in, 312
  bike parking in, 307–308, 311
  bikesharing in, 308
  challenges facing, 302, 312–315
  cycling rates in, 1, 2f, 11, 13f,
    302–305, 303f, 307f
  e-bikes in, 310
  future of cycling in, 316–317
  infrastructure in, 305–309, 305f, 306f,
    307f
  injury and fatality rates in, 313
  policies and programs in, 301–302,
    304, 309–312, 310f
  public transport in, 307–308

social justice in, 314–315
women cyclists in, 304
Latino Americans. *See* Hispanic and
    Latino Americans
Laws. *See* Policy-regulatory environment;
    *specific laws*
League Cycling Instructors (LCIs), 408
League of American Bicyclists (LAB), 27,
    85, 380, 401, 407–410
League of American Wheelman (LAW),
    85, 407
Leeds, 14f
Leiden, 14f
Let's Ride, 223
Liability, 126–127, 225
Lights, bicycle traffic, 131, 132f, 350
Lights, on bicycles, 65
Lima
    bike parking in, 308
    cycling rates in, 13f, 303f, 307f
    infrastructure in, 305, 305f, 306f,
        307f
    policies and programs in, 309, 311
Lime, 439
Lintell, Mark, 44
Livable Neighborhoods, 330–331, 335
Livingstone, Ken, 68, 330
Local street bikeways, 94–95, 99
Lockers, bike, 104
Locks, bike, 103
London, 9, 321–343
    advocacy in, 338–339
    bike parking in, 106, 110t, 113f, 114,
        329t, 331–332
    bikesharing in, 178f, 188, 333
    car restrictions and charges in,
        335–337, 431
    challenges facing, 321
    climate of, 323t, 324
    cycling rates in, 2, 2f, 14f, 321,
        324–326, 325f, 427t
    demographics of, 322–324, 323t, 326,
        340

infrastructure in, 68, 329t, 330–331,
    427t
injury and fatality rates in, 68,
    326–328, 327f
policy shifts in, 340–343, 341t
promotional programs in, 122
public transport in, 321–322, 332,
    334–335
traffic laws in, 338
women cyclists in, 203
London Cycling Action Plan, 68, 330,
    331
London Cycling Campaign (LCC),
    339
London Cycling Design Standards, 68,
    330
London Cycling Network Plus, 68,
    330
Los Angeles, 13f, 427t
Love London, Go Dutch, 122, 339
Low-stress connectivity, 97
Lucknow, 284f
Lyft, 334, 439
Lyon, 14f, 178f

Maat, Kees, 105
Madrid, 178f
Manaugh, Kevin, 265–266
*Manual on Uniform Traffic Control
    Devices,* 28
Maps
    bike parking, 114
    bike route, 123
Marketing
    for bikesharing, 191
    social, 122–123
    targeted, 123
Marqués, Ricardo, 69, 204
Marseilles, 14f
Marshall, Wesley E., 69
Massachusetts. *See* Boston
McDonald, Noreen C., 105
Medellín, 303f, 306f, 307f

Melbourne
  bikesharing in, 178f, 183–185, 188
  cycling rates in, 2f, 12, 13f
  social justice in infrastructure of, 265
Mendoza, 303f, 307f
Mental health benefits of cycling, 39–40
Mexico
  infrastructure in, 305, 307–308, 309,
    313
  policies and programs in, 304, 309
Mexico City
  cycling rates in, 13f, 303f, 307f
  infrastructure in, 306f, 307–308, 307f
  traffic calming in, 313
  women cyclists in, 304
Mieux se déplacer à Bicyclette (MDB),
    339–340
Minello, Karla, 239
Minneapolis
  bike parking in, 108t, 111, 112f,
    113–114
  bikesharing in, 178f
  cycling rates in, 2f, 13f, 106, 427t
  infrastructure in, 427t
Mixed-traffic bike routes
  vs. conventional bike lanes, 93
  definition of, 81
  local street bikeways as, 94–95
  traffic stress on, 83
  in US, 82
Mixed-use greenways, 429
Monterrey, 303f
Montevideo, 13f, 303f, 306f, 307f
Montreal
  advocacy in, 417
  bike parking in, 105, 106
  bikesharing in, 178f
  cycle tracks in, 86, 417
  cycling rates in, 2f, 13f, 427t
  infrastructure in, 99, 265, 427t
  injury risk disparities in, 269–270
  social justice in, 265, 269–270
Morgan, Andrei Scott, 139

Motivate (company), 334, 439
Motorcycles, electric, 161
Motorization rates, international
  comparisons of, 17–18
Motor vehicles. See Car ownership and
  use
Multiactor multicriteria analysis
  (MAMCA), 145–146, 147t–148t,
  149t
Multicriteria analysis (MCA), 137–154
  challenges of, 138–141, 152–153
  pros and cons of, 141–144, 153
  real-world examples of, 145–146,
    147t–148t, 149t
Mulvaney, Caroline A., 67
Mumbai, 13f, 283, 284f
Munich, 2f, 14f, 348f, 427t
Munster, 14f

Nagoya, 13f, 109t, 112–113, 112f
Nagpur, 284f
Nanjing, 286f
National advocacy, 404–413
  in Australia, 419–420
  in Canada, 416
  in Europe, 401, 404–406
  in US, 401, 406–413
National Association of City
  Transportation Officials (NACTO),
  28, 87–88, 380
National Bike Summit, 409, 411
National Cycling Investment Strategy of
  2017 (UK), 24
National Health Interview Survey, 237
National Highway Traffic Safety
  Administration (NHTSA), 222
National Household Travel Survey (NHTS),
  11, 223, 261–262, 264, 439
National Institute for Transportation
  and Communities (NITC), 386
National Urban Transport Policy (India),
  291
Neighborhood greenways, 67, 94, 379f

Netherlands. *See also specific cities*
  advisory bike lanes in, 96f, 97
  bicycle networks in, 97–98, 129
  bike parking in, 113–114, 274, 405
  car ownership rates in, 17–18
  child cyclists in, 17f, 220t, 221–222,
    224, 227
  cycle tracks in, 82, 91–92
  cycling rates in, 12, 12f, 14f, 15, 16f,
    17f, 281
  cyclist and driver training in, 26
  demographics of cyclists in, 15–17,
    16f, 17f, 41
  e-bikes in, 162f, 164, 248
  economic evaluation of policies in,
    146–151, 150t, 151t
  history of policy shifts in, 19–20
  incentive programs in, 124–125
  infrastructure guidelines in, 28, 138
  injury and fatality rates in, 20f, 21,
    22f, 63, 222, 268
  local street bikeways in, 94
  national advocacy in, 401, 404–406
  national policies in, 24
  older cyclists in, 17f, 237–238,
    241–242
  racial and ethnic disparities in,
    262–263
  speed limits and road design standards
    in, 27–28
  sports cycling in, 199–201
  traffic regulations in, 26, 125–128
  women cyclists in, 16f, 198, 199–201,
    200t, 260–261
*Netherlands National Travel Survey,* 242
Networks, bicycle
  connectivity of, 97–98, 129, 314
  and injury risk, 69
  requirements for, 97–98
  in suburbs, 207–208
New York City, 9, 321–343
  advocacy in, 338–339
  bike parking in, 329t, 331–332

  bikesharing in, 178f, 184, 334
  car restrictions and charges in,
    335–337, 431
  challenges facing, 321
  climate of, 323t, 324
  cycle tracks in, 82f, 86–87, 328, 331f,
    427, 434
  cycling rates in, 2f, 13f, 321, 324–326,
    325f, 427t
  demographics of, 322–324, 323t, 326,
    340
  infrastructure in, 69, 328–329, 329t,
    331f, 427t
  injury and fatality rates in, 22, 69,
    326–328, 327f
  policy shifts in, 340–343, 341t
  public transport in, 321–322, 332, 334
  road diet in, 89–90
  stand-alone paths in, 90
  traffic laws in, 338
New York City Department of
    Transportation (NYCDOT), 324, 328,
    332, 334, 336, 338
New Zealand
  child cyclists in, 223, 226, 228–230
  injury risk in, 267
Nilsson, Philip, 60
Nixon, Deborah, 239
Noise pollution, 45–46, 267–268
North America. *See also specific cities and
    countries*
  cycling rates in, 11, 12f, 13f
  growth in cycling in, 1, 2f
Norway, 12f, 164
Nottingham, 14f
Novick, Steve, 377, 379f, 381
Nudge theory, 245

Obesity, 36–37, 38, 219
Odense, 14f, 123
Older adults, cycling by, 7, 237–249
  on e-bikes, 163, 168, 247–248
  factors influencing, 242–246

Older adults, cycling by (cont.)
  injuries in, 240–241, 248
  international comparisons of, 16–17
  patterns and trends in, 241–242
  physical health benefits of, 36,
    237–240
  rates of, 17f, 241–242
  social health benefits of, 41, 238–240
  strategies for increasing, 246–248,
    432–433
Olivier, Jake, 70, 71
Ontario, 418–419
Oregon. See also specific cities
  Complete Streets legislation in,
    374–376
  injury risk disparities in, 269
Osaka, 12, 13f
Osteoarthritis, 239
Ottawa, 13f, 99
Out group, cyclists as, 210, 245
Overweight, 36–37
OV-Fiets, 363, 405
Oxford, 14f

Palermo, 14f
Panamá, 306f
Pandemic, COVID-19, xvi
Parents
  cycling with children, 198–199, 227
  influence on child cycling, 225,
    227–229
Paris, 9, 321–343
  advocacy in, 338–340
  bike parking in, 329t, 331–332
  bikesharing in, 178f, 188, 190,
    332–333
  car restrictions and charges in,
    335–337
  challenges facing, 321
  climate of, 323t, 324
  cycle tracks in, 331, 331f
  cycling rates in, 1, 2f, 14f, 321,
    324–326, 325f, 348f, 427t

demographics of, 322–324, 323t, 326,
    340
  infrastructure in, 329t, 331, 427t
  injury and fatality rates in, 326–328,
    327f
  policy shifts in, 340–343, 342t
  public transport in, 321–322, 332
  traffic laws in, 338
Paris en Selle, 339–340
Paris respire program, 335
Parkin, John, 68
Parking, bike, 5, 103–114. See also
    specific locations
  in bikesharing, 185–186, 189–190
  case studies on, 106–114, 107t–110t
  covered, 104, 105, 111–112
  funding for, 114
  at high demand times, 112–113
  informal, 103–104, 113
  locations of, 104–106, 111
  locks used for, 103
  racial and ethnic disparities in, 263, 274
  regulations on, 113–114
  residential, 105, 263, 274, 430
  at schools, 105, 224
  total supply of, 110–111
Parking, car. See Lanes, parking
Parma, 14f
Partido Socialista Obrero Español
    (PSOE), 389, 394, 396
Paths, dual (separate for cyclists and
    pedestrians), 90
Paths, stand-alone bike, 90–91
  children on, 224
  cost of creating, 98
  definition of, 81
  in Europe, 90
  injury risk on, 68
  in New York City, 328
  in Portland, 379f
  in US, 90–91
  width of, 90
Pearson correlation, 306

Pedal-assist bikes, 159–161, 160f
Pedelecs, 159–161
    bikesharing of, 182
    injury risk with, 65–66, 66f
Pedestrian and Bicycle Information
    Center, 413
Pedestrians
    on bike paths vs. dual paths, 90
    injuries to, 46–47
Pedroso, Felipe E., 68–69
Peer influence, on child cycling, 225,
    229–230
PeopleForBikes (PFB), 87, 160, 401, 407,
    410–412, 420
Pereira, 303f, 306f
Perth, 13f
Peru. *See specific cities*
Petrol. *See* Gasoline/petrol prices;
    Gasoline/petrol tax
Physical activity. *See also* Health benefits
    of cycling
    health benefits of, 35–36, 42
    minimum recommended level of, 36,
    42, 198, 237
    by older adults, 237–238
    by women, 198, 205
PlacesForBikes, 412
Playgrounds, bicycle, 404
Police reports, on injuries, 59
Policies and programs, 5–6, 119–133.
    *See also specific locations and policies*
    on built environment, 128–131
    economic evaluation of (*see* Economic
    evaluation of cycling policies)
    in gender gap reductions, 209
    goals of, 119–120
    history of development of, 19–20
    incentive, 124–125
    legal, 125–128
    overview of types of, 120t
    promotional, 120–123, 229, 230
    recommendations for future of,
    430–436

Policy-regulatory environment, 125–128
    bike parking in, 113–114
    components of, 208–209, 243
    international comparisons of, 26–27,
    125–128
    liability in, 126–127
    and older cyclists, 243, 244
    right-of-way in, 126
    and women cyclists, 205f, 208–209
    zoning in, 129
Political leadership, 377–381, 379f,
    436
Pollution. *See* Air pollution; Noise
    pollution
Portland (Oregon), 9–10, 371–397
    advocacy in, 384–386, 396
    car use in, 397
    cycling rates in, 2f, 12, 13f, 371, 396,
    397, 427t
    demographics and climate of, 372t
    design approach in, 383–384, 385f
    future of bicycling in, 386–387
    infrastructure in, 86, 95, 265, 427t
    political leadership in, 377–381,
    379f
    preexisting conditions supporting
    cycling in, 373–376
    professional staff in, 380, 381–383
    public transport in, 376
    research in, 386
    vs. Seville, 371–373, 396–397
    social justice in, 265
    women cyclists in, 203
    work trips by bike in, 374, 375f
Portland Bicycle Master Plan, 378,
    380
*Portland Bicycle Plan for 2030,* 380, 383
Portland Bureau of Transportation
    (PBOT), 376–387
Portland State University (PSU), 386
Poulos, R. G., 64, 240
Prati, Gabriele, 260–261
Precipitation, 324

Programs. *See* Policies and programs
Promotional programs, 120–123, 229, 230
Pronto program, 183
Protected bike lanes. *See* Cycle tracks
Provincial advocacy, in Canada, 401–402, 416–419
Public information campaigns, 436–437
Public transport. *See also specific locations*
  bike parking near/for, 104, 111–113, 113f
  bikesharing combined with, 177, 178, 190
  and car ownership, 18–19
Pucher, John, 58, 68, 69, 128, 161, 209
Puebla, 303f
Pune, 284f, 294

Quality-adjusted life-year (QALY), 144, 145
Quality of life, 40
Quebec
  advocacy in, 402, 416–418
  cycle tracks in, 86
  provincial funding in, 25
  provincial policies in, 23
  speed limits and road design standards in, 27
Quebec Ministry of Transport, 418
Quietways, 330
Quito
  bikesharing in, 308
  cycling rates in, 13f, 303, 303f, 307f
  infrastructure in, 305f, 306f, 307f

Racial disparities
  in bike parking, 263, 274
  in cycling rates, 261–263, 262f
  in injury risk, 268–269
Racial profiling, 128
Radun, Igor, 71

Rails-to-Trails Conservancy (RTC), 412
Raised crossings, 67
Recreational cycling
  by children, 221
  on e-bikes, 163
  international rates of, 15
  by older adults, 238
  sociodemographic patterns in, 41–42
Regulations. *See* Policy-regulatory environment
Research on cycling
  bike parking, 103–106
  in combination with bike advocacy, 401
  increase in, 3
  in Portland, 386
Residential bike parking, 105, 263, 274, 430
Residential segregation, 263
Reverse commuting, 266–267
Reward programs, 124–125
Ride Spot, 412
Right-of-way regulations, 126
Rio de Janeiro, 13f, 266, 303f, 305, 307f, 314
Riots, 361–362
Road diets, 89–90
Roads
  finding space for bike lanes and cycle tracks on, 88–90
  international comparisons of design standards for, 27–28
  lane width of, 88–89
  noise pollution from, 45–46
  sharing with advisory bike lanes vs. sharrows, 95–96
Rodgers, Gregory, B., 62
Rodrigo-Torrijos, Antonio, 389
Rosario, 303f, 304, 305, 306f, 307f
Rotterdam, 14f
Route Verte, 418

Safe Routes to School (SRTS), 25, 222–223, 225, 415

Safety, cycling, 4–5, 57–73. *See also* Fatality rates; Helmets, injury rates with; Injuries, cyclist; *specific locations*
with bikesharing, 179–180, 187, 187f, 188–189
for children, 222–223
cyclist experience and skill in, 62, 65
with e-bikes, 65–66, 165–167, 289–290
and gender gap, 206–207, 211–212
infrastructure type in, 66–70, 426
international comparisons of, 20–23, 426
problems with estimating effects on, 139
racial and ethnic disparities in, 263
as social justice issue, 271–273
training in, 26–27

"Safety in kilometers" effect, 62, 73

"Safety in numbers" effect, 62–64
in bikesharing, 179
in Copenhagen, 351
definition of, 47
in evaluation of cycling policies, 139
in New York, London, and Paris, 327–328
problems with concept, 5, 63–64
research on, 62–64

Sahlqvist, Shannon, 62

Salzburg, 14f

Sanatizadeh, Aida, 21

Sánchez de Madariaga, Inés, 211

Sánchez-Monteseirin, Alfredo, 389

San Diego, 83, 186

San Francisco, 2f, 13f, 178f, 186, 270

San Jose, 303f

Santa Cruz, 306f

Santiago
bike parking in, 308
bikesharing in, 308
cycling rates in, 1, 2f, 13f, 303f, 307f

infrastructure in, 306f, 307f, 314
women cyclists in, 304

Sao Paulo, 303f, 305, 307, 307f, 313

Sapporo, 13f

Scarborough Cycles, 416

Ščasný, Milan, 143

Schepers, Paul, 59, 62, 63, 65

Schneider, Robert, 21

Schools
bike parking at, 105, 224
cycling policies of, 225
cycling rates to, 219–221, 220t, 223–225
cyclist training in, 26–27, 121, 222, 226, 415
location of, 224–225
Safe Routes to School program for, 25, 222–223, 225, 415

Scientists for Cycling, 402

Scooters, electric, 161, 314

Scooter style electric bikes, 159, 287–289

Seattle
bikesharing in, 178f, 183, 186, 189
cycling rates in, 2f, 427t
infrastructure in, 427t
women cyclists in, 261

Segregation, residential, 263

Self-reported injuries, 59, 60, 61t

Self-selection, in evaluation of health benefits, 139–140

Sensory changes, in older adults, 243

Sevilla Respira, 395

Seville, 9–10, 371–397
advocacy in, 389, 391, 396
bike parking in, 106, 108t, 111, 114
bikesharing in, 388t, 389–390
car use in, 373, 387–390, 394–397
climate of, 372t, 387
cycle tracks in, 69, 391, 392f
cycling rates in, 1, 2f, 348f, 371, 387, 388t, 392–394, 396, 427t
demographics of, 372t, 387

Seville (cont.)
  design principles in, 391
  evolution of bike network in,
    389–391, 390f
  future of cycling in, 395–396
  infrastructure in, 388t, 427t
  injury and fatality rates in, 69, 388t,
    393
  opposition to cycling in, 393–394
  vs. Portland, 371–373, 396–397
  public transport in, 390
  women cyclists in, 203–204, 388t,
    392–393
Seville Bicycle Masterplan, 394–395
Shadow travel, 266
Shanghai, 13f, 285–287, 286f, 289
Shared lane treatments, 97
Shared road treatments, 95–97
Share the Road Cycling Coalition
  (SRCC), 418–419
Sharing, bike. See Bikesharing
Sharing, road, with advisory bike lanes
  vs. sharrows, 95–97
Sharrows, 95–96
Shijiazhuang, 286, 286f, 287
Shoulders, road, 89
Side street crossing tables, 92–93, 92f
Sidewalks, continuous, 92–93, 92f
Signage, bicycle route, 131
Singapore, 105
Single-bicycle crashes, 63, 240–241, 248
Smart Cycling, 408, 410
Smart Growth America, 270
SmartTrips, 123
Smovengo, 333
Social-cultural environment
  components of, 209, 243, 245
  and older cyclists, 245–246
  and women cyclists, 205f, 209–211
Social-ecological model of health
  behavior, 204, 205f, 242–246
Social health benefits of cycling, 40–41,
  238–240

Social inclusion, 44–45, 141
Social justice, 7–8, 257–275
  advocacy groups on, 273, 410
  in bikesharing, 180
  gender disparities in, 260–261,
    260t
  income disparities in, 263–265, 264f,
    314–315
  in infrastructure, 261, 263, 265–270
  in Latin America, 314–315
  meanings and use of term "disparities"
    in, 258–259
  participation in decision making in,
    273
  pollution exposure and traffic risk in,
    267–270
  racial and ethnic disparities in,
    261–263
  recommendations for future of,
    274–275, 433
  right to safe cycling in, 271–273
  urban displacement and gentrification
    in, 270
Social marketing, 122–123
Social safety, 208
Social vulnerability, 210
Socioeconomic status
  and disparities in cycling, 263–265,
    264f
  reducing health inequalities based on,
    41–42
SOFIE pedal bikes, 247
Soft interventions, 119, 212
South America. See Latin America
Spain. See also specific cities
  bike parking in, 106
  child cyclists in, 223
  fatality rates in, 20f
Speed bumps and humps, 130, 131f
Speed limits
  international comparisons of, 27–28
  in New York, London, and Paris,
    335–336

Speed of cycling
  with e-bikes, 65–66, 66f, 160,
    166–167
  helmet laws and, 72
  injury risk and, 65–66, 66f
Spontaneity, in bikesharing, 183, 187,
    187f, 191
Sports cycling, gender gap in, 199–201
State advocacy
  in Australia, 402, 420–421
  in US, 401, 413–416
Stehlin, John, 270
Stevenson, Mark, 47
St. Gallen, 14f
Stop lines, advance, 67, 131
Stop signs, cyclist behavior at, 127
Strasbourg, 14f, 106, 107t, 110–112
Street intersections. See Intersections
Street Trust, The, 376, 384
Stress. See Traffic stress
Stuttgart, 14f
Suárez, Liliana, 268
Substance intoxication, 127–128
Suburbs, cycling networks in, 207–208
Sundays, car-free. See Ciclovía
Sundfør, Hanne Beate, 62, 65
Superhighways, cycle, 146, 330, 429
Surat, 284f
Surveys
  cyclist satisfaction, 352, 353t, 435
  national travel, 11, 241–242
Sustainability, xvi, 1
Sustainable Transport Fund (UK), 25
Sustrans, 406
Sweden, 12f, 16f, 20f, 23, 59, 223, 239
Switzerland, 12f, 14f, 121
Sydney (Australia), 13f, 123, 427t

Targeted marketing, 123
Taxes
  car, 18, 336, 354
  gasoline/petrol, 381
Tegucigalpa, 303f

Texas, 13f, 414–416
Texas Bicycle Coalition (BikeTexas), 401,
    414–416
Texas Trails and Active Transportation
    (TTAT) conferences, 415
Theft, 174, 313
Thompson, Samantha Hypatia, 106
Tianjin, 286, 286f
Tokyo, 13f, 106, 110t, 111–114
Topography
  of Latin America, 304–305
  of New York, London, and Paris,
    324
Toronto, 2f, 13f, 221, 227
Tour de Force, 24, 406
Tourism, cycle, 404, 418
Tracks, cycle. See Cycle tracks
Traffic, mixed. See Mixed-traffic bike
    routes
Traffic calming, 130
  in China and India, 295
  injury risk with, 67
  international comparisons of, 27
  in Latin America, 313
  in New York, London, and Paris,
    335–336
  speed bumps in, 130, 131f
Traffic injuries. See Injuries, cyclist;
    Injuries, pedestrian
Traffic laws. See also Policy-regulatory
    environment
  in China and India, 295
  international comparisons of, 26–27,
    125–128
  in New York, London, and Paris, 338
  and older cyclists, 244
Traffic lights, bicycle, 131, 132f, 350.
    See also Intersections
Traffic Monitoring Guide, 267
Traffic stress, 83, 93, 97
Traffic tolerance, 82–83
Trails, bicycle and pedestrian, 144–145,
    145t

Training of cyclists, 120–121
  by advocacy groups, 408
  for children, 26–27, 120–121,
    222–223, 226–227, 415
  international comparisons of, 26–27
  for older adults, 246–247
  at schools, 26–27, 121, 222, 226, 415
Training of drivers, 26–27, 128
Transport, cycling for. *See also* Work
    trips by bike
  on e-bikes, 163–164
  gender gap in, 201–202, 212,
    260–261
  income-based disparities in, 41,
    263–264
  international rates of, 15
  minimum level of physical activity
    through, 198
  by older adults, 238
  sociodemographic patterns in, 41–42
Transport, public. *See* Public transport
Transport Alternatives (TA) (NY
    advocacy group), 339
Transportation Alternatives (TA) (US
    federal program), 24–25
Transportation costs, 44, 264
Transportation Research and Education
    Consortium (TREC), 386
Transportation Research and Injury
    Prevention Program (TRIPP), 291
TravelSmart, 123
Travel surveys, national, 11, 241–242
Treat, Leah, 381, 383–384
Tricycles, adult, 247
TriMet, 376
Trujillo, 303f
Tucker, Bronwen, 265–266

Uber, 176, 439
United Kingdom (UK). *See also specific
    cities*
  advocacy in, 406
  car ownership rates in, 18

child cyclists in, 17f, 220t, 221, 222,
    225, 226
cycling rates in, 11, 12f, 14f, 15, 16f,
    17f, 201
cyclist and driver training in, 26–27
demographics of cyclists in, 15–17,
    16f, 17f
e-bikes in, 162f, 164
fatality rates in, 20–21, 20f, 22f, 222
incentive programs in, 124
national funding in, 24–25
national policies in, 24
older cyclists in, 17f, 242
promotional programs in, 123
traffic regulations in, 26, 125–126, 128
women cyclists in, 16f, 202–203
United Nations, 403
United States. *See also specific cities and
    states*
car ownership rates in, 17
child cyclists in, 17f, 221–225
conventional bike lanes in, 82, 93
cycle tracks in, 82, 86–88, 88f
cycling rates in, 11–12, 12f, 13f, 15,
    16f, 17f, 201, 281
cyclist and driver training in, 26, 223
demographics of cyclists in, 15–17,
    16f, 17f
e-bikes in, 157–167, 162f
fatality rates in, 20–22, 20f, 222
helmet laws in, 223
history of infrastructure in, 83–88, 100
income-based disparities in, 263–265,
    264f
injury risk disparities in, 267–269
national advocacy in, 401, 406–413
national funding in, 24–25
national policies in, 24
older cyclists in, 17f, 242
racial and ethnic disparities in,
    261–263, 262f
sharrows in, 95–96
social justice in infrastructure of, 267

speed limits and road design standards
in, 27–28
stand-alone paths in, 90–91
state advocacy in, 401, 413–416
traffic regulations in, 26, 125–127
women cyclists in, 16f, 260–261, 260t
*Urban Bikeway Design Guide*, 87–88
Urban displacement, 270
Urban fabrics, 357, 362
Urban Transportation Showcase Fund
(Canada), 25
US Department of Commerce (USDOC),
12f, 13f, 16f, 17f, 22f, 325
US Department of Transportation
(USDOT)
on bike parking, 103
on cycle tracks, 411
on demographics of cyclists, 16f, 17f,
260t, 262f, 264f, 322
design guidelines of, 28
on driver training, 26
on fatality rates, 22f, 24
on federal funding, 25, 408
in Pedestrian and Bicycle Information
Center, 413
in Safe Routes to School, 25, 225
on total bike share of daily trips, 12f, 20
on trip distance and purpose, 15
on work trips by bike, 260t, 262f
Utilitarian cycling. *See* Transport,
cycling for
Utrecht, 14f

Valdivia, 303f
Valet bike parking, 113
Valparaiso, 303f
Vancouver (Canada), 2f, 12, 13f, 99,
203, 427t
Vandalism, 174
Van Lierop, Dea, 105, 106
Vargo, Jason, 21
Vehicular cycling (VC) theory, 84–85,
87–88, 407, 408

Vélib' system, 190, 332–333
Vélo Canada Bikes (VCB), 416
Vélo Québec (VQ), 402, 416–418
Vienna, 2, 2f, 14f, 105, 348f, 427t
Vision Zero, 22–23
in Sweden, 23, 59
in US, 384, 408–409

Walker, Peter, 204
Wang, Guijing, 144–145
Washington (state). *See specific cities*
Washington, DC
bike parking in, 106, 108t, 110–111,
113–114, 113f
bikesharing in, 178f, 180, 186
cycling rates in, 2f, 13f, 427t
infrastructure and injury risk in, 68,
427t
Watkins, G., 247
Web of Science, 3
Wee, Bert van, 105
Well-being, 40
We Ride Australia (WRA), 402, 419–420
White Americans
cycling rates of, 262, 262f
injury and fatality rates for, 263,
268–269
Wisconsin, 269
Women cyclists, 6–7, 197–213. *See also
specific locations*
age of, 201f, 205–206, 221
benefits for, 198–199
bike share of transport and, 197,
202–204, 260
on e-bikes, 168, 212
factors influencing, 204–211, 205f
international comparisons of, 15–16,
16f
patterns of, 197, 199, 200t, 260–261
purpose of, 199–202, 206
strategies for increasing, 211–213,
432–433
Woodcock, James, 139, 179

Work trips by bike
  incentive programs for, 124
  in India, 282–283, 282t, 284f
  parking for, 105
  in Portland, 374, 375f
  promotional programs for, 121–122,
    123
  reverse commuting, 266–267
  as share of all work trips, 11
  by women, 203
World Bank, xvi, 403
World Bicycle Day, 403
World Bicycle Forum, 312
World Cycling Alliance (WCA), 402–403
World Cycling Conference, 418
World Health Organization (WHO)
  on air pollution, 285, 297, 298
  and European Cyclists' Federation,
    403
  on fatality rates in India, 289
  Health Economic Assessment Tool of,
    39
  on noise pollution, 45
  and older adults, 245
  physical activity guidelines of, 36, 44
Wuhan, 286f

Xuzhou, 13f, 286f

Years lived with disability (YLD), 59–60

Zander, Alexis, 247
Zimmerman, Sara, 266
Zoido, Juan-Ignacio, 394
Zoning regulations, 129
Zurich, 14f
Zwolle, 14f

**Urban and Industrial Environments**

Series editor: Robert Gottlieb, Henry R. Luce Professor of Urban and Environmental Policy, Occidental College

Maureen Smith, *The U.S. Paper Industry and Sustainable Production: An Argument for Restructuring*

Keith Pezzoli, *Human Settlements and Planning for Ecological Sustainability: The Case of Mexico City*

Sarah Hammond Creighton, *Greening the Ivory Tower: Improving the Environmental Track Record of Universities, Colleges, and Other Institutions*

Jan Mazurek, *Making Microchips: Policy, Globalization, and Economic Restructuring in the Semiconductor Industry*

William A. Shutkin, *The Land That Could Be: Environmentalism and Democracy in the Twenty-First Century*

Richard Hofrichter, ed., *Reclaiming the Environmental Debate: The Politics of Health in a Toxic Culture*

Robert Gottlieb, *Environmentalism Unbound: Exploring New Pathways for Change*

Kenneth Geiser, *Materials Matter: Toward a Sustainable Materials Policy*

Thomas D. Beamish, *Silent Spill: The Organization of an Industrial Crisis*

Matthew Gandy, *Concrete and Clay: Reworking Nature in New York City*

David Naguib Pellow, *Garbage Wars: The Struggle for Environmental Justice in Chicago*

Julian Agyeman, Robert D. Bullard, and Bob Evans, eds., *Just Sustainabilities: Development in an Unequal World*

Barbara L. Allen, *Uneasy Alchemy: Citizens and Experts in Louisiana's Chemical Corridor Disputes*

Dara O'Rourke, *Community-Driven Regulation: Balancing Development and the Environment in Vietnam*

Brian K. Obach, *Labor and the Environmental Movement: The Quest for Common Ground*

Peggy F. Barlett and Geoffrey W. Chase, eds., *Sustainability on Campus: Stories and Strategies for Change*

Steve Lerner, *Diamond: A Struggle for Environmental Justice in Louisiana's Chemical Corridor*

Jason Corburn, *Street Science: Community Knowledge and Environmental Health Justice*

Peggy F. Barlett, ed., *Urban Place: Reconnecting with the Natural World*

David Naguib Pellow and Robert J. Brulle, eds., *Power, Justice, and the Environment: A Critical Appraisal of the Environmental Justice Movement*

Eran Ben-Joseph, *The Code of the City: Standards and the Hidden Language of Place Making*

Nancy J. Myers and Carolyn Raffensperger, eds., *Precautionary Tools for Reshaping Environmental Policy*

Kelly Sims Gallagher, *China Shifts Gears: Automakers, Oil, Pollution, and Development*

Kerry H. Whiteside, *Precautionary Politics: Principle and Practice in Confronting Environmental Risk*

Ronald Sandler and Phaedra C. Pezzullo, eds., *Environmental Justice and Environmentalism: The Social Justice Challenge to the Environmental Movement*

Julie Sze, *Noxious New York: The Racial Politics of Urban Health and Environmental Justice*

Robert D. Bullard, ed., *Growing Smarter: Achieving Livable Communities, Environmental Justice, and Regional Equity*

Ann Rappaport and Sarah Hammond Creighton, *Degrees That Matter: Climate Change and the University*

Michael Egan, *Barry Commoner and the Science of Survival: The Remaking of American Environmentalism*

David J. Hess, *Alternative Pathways in Science and Industry: Activism, Innovation, and the Environment in an Era of Globalization*

Peter F. Cannavò, *The Working Landscape: Founding, Preservation, and the Politics of Place*

Paul Stanton Kibel, ed., *Rivertown: Rethinking Urban Rivers*

Kevin P. Gallagher and Lyuba Zarsky, *The Enclave Economy: Foreign Investment and Sustainable Development in Mexico's Silicon Valley*

David N. Pellow, *Resisting Global Toxics: Transnational Movements for Environmental Justice*

Robert Gottlieb, *Reinventing Los Angeles: Nature and Community in the Global City*

David V. Carruthers, ed., *Environmental Justice in Latin America: Problems, Promise, and Practice*

Tom Angotti, *New York for Sale: Community Planning Confronts Global Real Estate*

Paloma Pavel, ed., *Breakthrough Communities: Sustainability and Justice in the Next American Metropolis*

Anastasia Loukaitou-Sideris and Renia Ehrenfeucht, *Sidewalks: Conflict and Negotiation over Public Space*

David J. Hess, *Localist Movements in a Global Economy: Sustainability, Justice, and Urban Development in the United States*

Julian Agyeman and Yelena Ogneva-Himmelberger, eds., *Environmental Justice and Sustainability in the Former Soviet Union*

Jason Corburn, *Toward the Healthy City: People, Places, and the Politics of Urban Planning*

JoAnn Carmin and Julian Agyeman, eds., *Environmental Inequalities beyond Borders: Local Perspectives on Global Injustices*

Louise Mozingo, *Pastoral Capitalism: A History of Suburban Corporate Landscapes*

Gwen Ottinger and Benjamin Cohen, eds., *Technoscience and Environmental Justice: Expert Cultures in a Grassroots Movement*

Samantha MacBride, *Recycling Reconsidered: The Present Failure and Future Promise of Environmental Action in the United States*

Andrew Karvonen, *Politics of Urban Runoff: Nature, Technology, and the Sustainable City*

Daniel Schneider, *Hybrid Nature: Sewage Treatment and the Contradictions of the Industrial Ecosystem*

Catherine Tumber, *Small, Gritty, and Green: The Promise of America's Smaller Industrial Cities in a Low-Carbon World*

Sam Bass Warner and Andrew H. Whittemore, *American Urban Form: A Representative History*

John Pucher and Ralph Buehler, eds., *City Cycling*

Stephanie Foote and Elizabeth Mazzolini, eds., *Histories of the Dustheap: Waste, Material Cultures, Social Justice*

David J. Hess, *Good Green Jobs in a Global Economy: Making and Keeping New Industries in the United States*

Joseph F. C. DiMento and Clifford Ellis, *Changing Lanes: Visions and Histories of Urban Freeways*

Joanna Robinson, *Contested Water: The Struggle against Water Privatization in the United States and Canada*

William B. Meyer, *The Environmental Advantages of Cities: Countering Commonsense Antiurbanism*

Rebecca L. Henn and Andrew J. Hoffman, eds., *Constructing Green: The Social Structures of Sustainability*

Peggy F. Barlett and Geoffrey W. Chase, eds., *Sustainability in Higher Education: Stories and Strategies for Transformation*

Isabelle Anguelovski, *Neighborhood as Refuge: Community Reconstruction, Place Remaking, and Environmental Justice in the City*

Kelly Sims Gallagher, *The Globalization of Clean Energy Technology: Lessons from China*

Vinit Mukhija and Anastasia Loukaitou-Sideris, eds., *The Informal American City: Beyond Taco Trucks and Day Labor*

Roxanne Warren, *Rail and the City: Shrinking Our Carbon Footprint While Reimagining Urban Space*

Marianne E. Krasny and Keith G. Tidball, *Civic Ecology: Adaptation and Transformation from the Ground Up*

Erik Swyngedouw, *Liquid Power: Contested Hydro-Modernities in Twentieth-Century Spain*

Ken Geiser, *Chemicals without Harm: Policies for a Sustainable World*

Duncan McLaren and Julian Agyeman, *Sharing Cities: A Case for Truly Smart and Sustainable Cities*

Jessica Smartt Gullion, *Fracking the Neighborhood: Reluctant Activists and Natural Gas Drilling*

Nicholas A. Phelps, *Sequel to Suburbia: Glimpses of America's Post-suburban Future*

Shannon Elizabeth Bell, *Fighting King Coal: The Challenges to Micromobilization in Central Appalachia*

Theresa Enright, *The Making of Grand Paris: Metropolitan Urbanism in the Twenty-First Century*

Robert Gottlieb and Simon Ng, *Global Cities: Urban Environments in Los Angeles, Hong Kong, and China*

Anna Lora-Wainwright, *Resigned Activism: Living with Pollution in Rural China*

Scott L. Cummings, *Blue and Green: The Drive for Justice at America's Port*

David Bissell, *Transit Life: Cities, Commuting, and the Politics of Everyday Mobilities*

Javiera Barandiarán, *From Empire to Umpire: Science and Environmental Conflict in Neoliberal Chile*

Benjamin Pauli, *Flint Fights Back: Environmental Justice and Democracy in the Flint Water Crisis*

Karen Chapple and Anastasia Loukaitou-Sideris, *Transit-Oriented Displacement or Community Dividends? Understanding the Effects of Smarter Growth on Communities*

Henrik Ernstson and Sverker Sörlin, eds., *Grounding Urban Natures: Histories and Futures of Urban Ecologies*

Katrina Smith Korfmacher, *Bridging the Silos: Collaborating for Environment, Health, and Justice in Urban Communities*

Jill Lindsey Harrison, *From the Inside Out: The Fight for Environmental Justice within Government Agencies*

Anastasia Loukaitou-Sideris, Dana Cuff, Todd Presner, Maite Zubiaurre, and Jonathan Jae-an Crisman, *Urban Humanities: New Practices for Reimagining the City*

Govind Gopakumar, *Installing Automobility: Emerging Politics of Mobility and Streets in Indian Cities*

Amelia Thorpe, *Everyday Ownership: PARK(ing) Day and the Practice of Property*

Ralph Buehler and John Pucher, eds., *Cycling for Sustainable Cities*